[redacted]

Classification, Parameter Estimation and State Estimation

Classification, Parameter Estimation and State Estimation

An Engineering Approach using MATLAB®

F. van der Heijden
Faculty of Electrical Engineering, Mathematics and Computer Science
University of Twente
The Netherlands

R.P.W. Duin
Faculty of Electrical Engineering, Mathematics and Computer Science
Delft University of Technology
The Netherlands

D. de Ridder
Faculty of Electrical Engineering, Mathematics and Computer Science
Delft University of Technology
The Netherlands

D.M.J. Tax
Faculty of Electrical Engineering, Mathematics and Computer Science
Delft University of Technology
The Netherlands

John Wiley & Sons, Ltd

Telephone (+44) 1243 779777

Email (for orders and customer service enquiries): cs-books@wiley.co.uk
Visit our Home Page on www.wileyeurope.com or www.wiley.com

Reprinted December 2005

Other Wiley Editorial Offices

John Wiley & Sons Inc., 111 River Street, Hoboken, NJ 07030, USA

Jossey-Bass, 989 Market Street, San Francisco, CA 94103-1741, USA

Wiley-VCH Verlag GmbH, Boschstr. 12, D-69469 Weinheim, Germany

John Wiley & Sons Australia Ltd, 33 Park Road, Milton, Queensland 4064, Australia

John Wiley & Sons (Asia) Pte Ltd, 2 Clementi Loop #02-01, Jin Xing Distripark, Singapore 129809

John Wiley & Sons Canada Ltd, 22 Worcester Road, Etobicoke, Ontario, Canada M9W 1L1

Wiley also publishes its books in a variety of electronic formats. Some content that appears in print may not be available in electronic books.

Library of Congress Cataloging in Publication Data

Classification, parameter estimation and state estimation : an engineering approach using MATLAB / F. van der Heijden . . . [et al.].
p. cm.
Includes bibliographical references and index.
ISBN 0-470-09013-8 (cloth : alk. paper)
1. Engineering mathematics—Data processing. 2. MATLAB. 3. Mensuration—Data processing. 4. Estimation theory—Data processing. I. Heijden, Ferdinand van der.
TA331.C53 2004
681′.2—dc22

2004011561

British Library Cataloguing in Publication Data

A catalogue record for this book is available from the British Library

ISBN 13: 978-0-470-09013-8

Typeset in 10.5/13pt Sabon by Integra Software Services Pvt. Ltd, Pondicherry, India

This book is printed on acid-free paper responsibly manufactured from sustainable forestry in which at least two trees are planted for each one used for paper production.

Contents

Preface

Information processing has always been an important factor in the development of human society, and its role is still increasing. The inventions of advanced information devices paved the way for achievements in a diversity of fields like trade, navigation, agriculture, industry, transportation, and communication. The term 'information device' refers here to systems for the sensing, acquisition, processing, and outputting of information from the real world. Usually, they are measurement systems. Sensing and acquisition provide us with signals that bear a direct relation to some of the physical properties of the sensed object or process. Often, the information of interest is hidden in these signals. Further signal processing is needed to reveal the information, and to transform it into an explicit form.

The three topics discussed in this book, classification, parameter estimation, and state estimation, share a common factor in the sense that each topic provides the theory and methodology for the functional design of the signal processing part of an information device. The major distinction between the topics is the type of information that is outputted. In classification problems the output is discrete, i.e. a class, a label, or a category. In estimation problems, it is a real-valued scalar or vector. Since these problems occur either in a static or in a dynamic setting, actually four different topics can be distinguished. The term state estimation refers to the dynamic setting. It covers both discrete and real-valued cases (and sometimes even mixed cases).

The similarity between the topics allows one to use a generic methodology, i.e. Bayesian decision theory. Our aim is to present this material concisely and efficiently, by an integrated treatment of similar topics. We present an overview of the core mathematical constructs and the many resulting techniques. By doing so, we hope that the reader recognizes the connections and the similarities between these constructs, but also becomes aware of the differences. For instance, the phenomenon of overfitting is a threat that ambushes all four cases. In a static classification problem it introduces large classification errors, but in the case of dynamic state estimation it may be the cause of instable behaviour.

Our goal is to emphasize the engineering aspects of the matter. Instead of a purely theoretical and rigorous treatment, we aim at the acquirement of skills to bring theoretical solutions to practice. The models that are needed for the application of the Bayesian framework are often not available in practice. This brings in the paradigm of statistical inference, i.e. learning from examples. MATLAB® is used as a vehicle to implement and to evaluate design concepts.

As alluded to above, the range of application areas is broad. Application fields are found within mechanical engineering, electrical engineering, civil engineering, environmental engineering, process engineering, geo-informatics, bio-informatics, information technology, mechatronics, applied physics, and so on. The book is of interest to a range of users, from the first-year graduate-level student up to the experienced professional. The reader should have some background knowledge with respect to linear algebra, dynamic systems and probability theory. Most educational programmes offer courses on these topics as part of undergraduate education. The appendices contain reviews of the relevant material. Another target group is formed by the experienced engineers working in industrial development laboratories. The numerous examples of MATLAB code allow these engineers to quickly prototype their designs.

The book roughly consists of two parts. The first part, Chapters 2, 3 and 4, covers the theory with respect to classification and estimation problems in the static case, as well as the dynamic case. This part handles problems where it is assumed that accurate models, describing the physical processes, are available. The second part, Chapters 5 up to 8, deals with the more practical situation in which these models are not or only partly available. Either these models must be built using experimental data, or these data must be used directly to train methods for estimation and classification. The final chapter presents three worked out problems. The selected bibliography has been kept short in order not to overwhelm the reader with an enormous list of references.

The material of the book can be covered by two semester courses. A possibility is to use Chapters 2, 3, 5, 6 and 7 for a one-semester course

---

®MATLAB is a registered trademark of The MathWorks, Inc. (http://www.mathworks.com).

on Classification and Estimation. This course deals with the static case. An additional one-semester course handles the dynamic case, i.e. Optimal Dynamic Estimation, and would use Chapters 4 and 8. The prerequisites for Chapters 4 and 8 are mainly concentrated in Chapter 3. Therefore, it is recommended to include a review of Chapter 3 in the second course. Such a review will make the second course independent from the first one.

Each chapter is closed with a number of exercises. The mark at the end of each exercise indicates whether the exercise is considered easy ('0'), moderately difficult ('*') or difficult ('**'). Another possibility to acquire practical skills is offered by the projects that accompany the text. These projects are available at http://www.prtools.org. A project is an extensive task to be undertaken by a group of students. The task is situated within a given theme, for instance, classification using supervised learning, unsupervised learning, parameter estimation, dynamic labelling, and dynamic estimation. Each project consists of a set of instructions together with data which should be used to solve the problem.

The use of MATLAB tools is an integrated part of the book. MATLAB offers a number of standard toolboxes that are useful for parameter estimation, state estimation and data analysis; see also Appendix F. The standard software for classification and unsupervised learning is not complete and not well-structured. This motivated us to develop the PRTools software for all classification tasks and related items. PRTools is a MATLAB toolbox for pattern recognition. It is freely available for non-commercial purposes. The version used in the text is compatible with MATLAB Version 5 and higher. It is available from http://www.prtools.org.

The authors keep an open mind for any suggestions and comments (which should be addressed to ce@prtools.org). A list of errata and any other additional comments will be made available at http://www.prtools.org.

F. van der Heijden
R.P.W. Duin
D. de Ridder
D.M.J. Tax

Foreword

A broad range of contemporary engineering problems requires estimating the class (category) of a sensed object or process, parameters controlling the behavior of a "black box" system, or its internal state. The goal of many of these systems is to interact in an intelligent manner with their environment. While the technological advances in sensor design and processors have enabled development of low-cost and real-time systems, algorithms for classification and parameter estimation still need continued development in order to have a more accurate object classification and robust parameter estimation. A variety of disciplines – automatic control, signal processing, statistics, pattern recognition, machine learning – offer a spectrum of solutions to these problems, yet exhibit a convergence to several key approaches. A comprehensive treatment of these approaches is the main objective of this book.

This book emphasizes a unified mathematical treatment of model-based classification and estimation problems across different engineering applications. It provides a practical guide for implementing a wide range of algorithms for supervised and unsupervised classification, feature selection, system identification, and state estimation. The text covers both classical and state-of-the-art algorithms by utilizing MATLAB software that is routinely used in engineering design. One of the main contributions of this book, that distinguishes it from other pattern recognition books, is that it follows a top-down approach to designing a pattern recognition system. Mathematical concepts such as state estimation and parameter estimation are nicely introduced to help a practitioner. Examples in Chapter 9 clearly present, in a step-by-step fashion, various stages in classification and estimation. The software package PRTools, available as a part of this book, is an excellent vehicle for readers to evaluate different competing approaches on their datasets.

This book encompasses all the major aspects of designing a pattern recognition system and is an excellent addition to the collection of pattern recognition books that are available in the market.

Anil K. Jain
Michigan State University

1

Introduction

Engineering disciplines are those fields of research and development that attempt to create products and systems operating in, and dealing with, the real world. The number of disciplines is large, as is the range of scales that they typically operate in: from the very small scale of nanotechnology up to very large scales that span whole regions, e.g. water management systems, electric power distribution systems, or even global systems (e.g. the global positioning system, GPS). The level of advancement in the fields also varies wildly, from emerging techniques (again, nanotechnology) to trusted techniques that have been applied for centuries (architecture, hydraulic works). Nonetheless, the disciplines share one important aspect: engineering aims at designing and manufacturing systems that interface with the world around them.

Systems designed by engineers are often meant to influence their environment: to manipulate it, to move it, to stabilize it, to please it, and so on. To enable such actuation, these systems need information, e.g. values of physical quantities describing their environments and possibly also describing themselves. Two types of information sources are available: *prior* knowledge and *empirical* knowledge. The latter is knowledge obtained by sensorial observation. Prior knowledge is the knowledge that was already there before a given observation became available (this does not imply that prior knowledge is obtained without any observation). The combination of prior knowledge and empirical knowledge leads to *posterior* knowledge.

Classification, Parameter Estimation and State Estimation: An Engineering Approach using MATLAB
F. van der Heijden, R.P.W. Duin, D. de Ridder and D.M.J. Tax
 ISBN: 0-470-09013-8

The sensory subsystem of a system produces measurement signals. These signals carry the empirical knowledge. Often, the direct usage of these signals is not possible, or inefficient. This can have several causes:

- The information in the signals is not represented in an explicit way. It is often hidden and only available in an indirect, encoded form.
- Measurement signals always come with noise and other hard-to-predict disturbances.
- The information brought forth by posterior knowledge is more accurate and more complete than information brought forth by empirical knowledge alone. Hence, measurement signals should be used in combination with prior knowledge.

Measurement signals need processing in order to suppress the noise and to disclose the information required for the task at hand.

1.1 THE SCOPE OF THE BOOK

In a sense, classification and estimation deal with the same problem: given the measurement signals from the environment, how can the information that is needed for a system to operate in the real world be inferred? In other words, how should the measurements from a sensory system be processed in order to bring maximal information in an explicit and usable form? This is the main topic of this book.

Good processing of the measurement signals is possible only if some knowledge and understanding of the environment and the sensory system is present. Modelling certain aspects of that environment – like objects, physical processes or events – is a necessary task for the engineer. However, straightforward modelling is not always possible. Although the physical sciences provide ever deeper insight into nature, some systems are still only partially understood; just think of the weather. But even if systems are well understood, modelling them exhaustively may be beyond our current capabilities (i.e. computer power) or beyond the scope of the application. In such cases, approximate general models, but adapted to the system at hand, can be applied. The development of such models is also a topic of this book.

1.1.1 Classification

The title of the book already indicates the three main subtopics it will cover: classification, parameter estimation and state estimation. In classification, one tries to assign a class label to an object, a physical process, or an event. Figure 1.1 illustrates the concept. In a speeding detector, the sensors are a radar speed detector and a high-resolution camera, placed in a box beside a road. When the radar detects a car approaching at too high a velocity (a parameter estimation problem), the camera is signalled to acquire an image of the car. The system should then recognize the license plate, so that the driver of the car can be fined for the speeding violation. The system should be robust to differences in car model, illumination, weather circumstances etc., so some pre-processing is necessary: locating the license plate in the image, segmenting the individual characters and converting it into a binary image. The problem then breaks down to a number of individual classification problems. For each of the locations on the license plate, the input consists of a binary image of a character, normalized for size, skew/rotation and intensity. The desired output is the label of the true character, i.e. one of 'A', 'B',..., 'Z', '0',..., '9'.

Detection is a special case of classification. Here, only two class labels are available, e.g. 'yes' and 'no'. An example is a quality control system that approves the products of a manufacturer, or refuses them. A second problem closely related to classification is identification: the act of proving that an object-under-test and a second object that is previously seen, are the same. Usually, there is a large database of previously seen objects to choose from. An example is biometric identification, e.g.

Figure 1.1 License plate recognition: a classification problem with noisy measurements

fingerprint recognition or face recognition. A third problem that can be solved by classification-like techniques is retrieval from a database, e.g. finding an image in an image database by specifying image features.

1.1.2 Parameter estimation

In parameter estimation, one tries to derive a parametric description for an object, a physical process, or an event. For example, in a beacon-based position measurement system (Figure 1.2), the goal is to find the position of an object, e.g. a ship or a mobile robot. In the two-dimensional case, two beacons with known reference positions suffice. The sensory system provides two measurements: the distances from the beacons to the object, r_1 and r_2. Since the position of the object involves two parameters, the estimation seems to boil down to solving two equations with two unknowns. However, the situation is more complex because measurements always come with uncertainties. Usually, the application not only requires an estimate of the parameters, but also an assessment of the uncertainty of that estimate. The situation is even more complicated because some prior knowledge about the position must be used to resolve the ambiguity of the solution. The prior knowledge can also be used to reduce the uncertainty of the final estimate.

In order to improve the accuracy of the estimate the engineer can increase the number of (independent) measurements to obtain an overdetermined system of equations. In order to reduce the cost of the sensory system, the engineer can also decrease the number of measurements leaving us with fewer measurements than parameters. The system

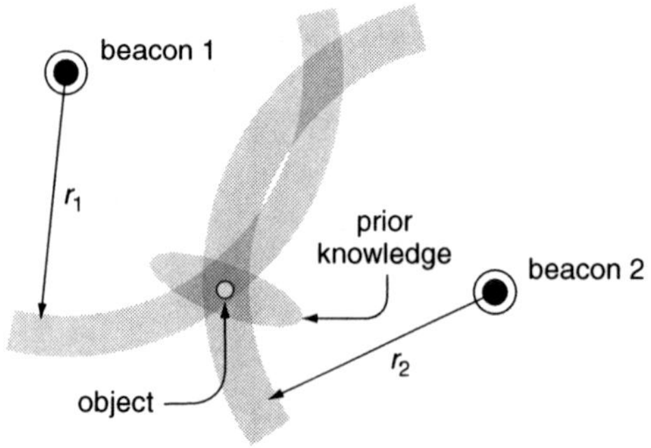

Figure 1.2 Position measurement: a parameter estimation problem handling uncertainties

of equations is underdetermined then, but estimation is still possible if enough prior knowledge exists, or if the parameters are related to each other (possibly in a statistical sense). In either case, the engineer is interested in the uncertainty of the estimate.

1.1.3 State estimation

In state estimation, one tries to do either of the following – either assigning a class label, or deriving a parametric (real-valued) description – but for processes which vary in time or space. There is a fundamental difference between the problems of classification and parameter estimation on the one hand, and state estimation on the other hand. This is the ordering in time (or space) in state estimation, which is absent from classification and parameter estimation. When no ordering in the data is assumed, the data can be processed in any order. In time series, ordering in time is essential for the process. This results in a fundamental difference in the treatment of the data.

In the discrete case, the states have discrete values (classes or labels) that are usually drawn from a finite set. An example of such a set is the alarm stages in a safety system (e.g. 'safe', 'pre-alarm', 'red alert', etc.). Other examples of discrete state estimation are speech recognition, printed or handwritten text recognition and the recognition of the operating modes of a machine.

An example of real-valued state estimation is the water management system of a region. Using a few level sensors, and an adequate dynamical model of the water system, a state estimator is able to assess the water levels even at locations without level sensors. Short-term prediction of the levels is also possible. Figure 1.3 gives a view of a simple water management system of a single canal consisting of three linearly connected compartments. The compartments are filled by the precipitation in the surroundings of the canal. This occurs randomly but with a seasonal influence. The canal drains its water into a river. The measurement of the level in one compartment enables the estimation of the levels in all three compartments. For that, a dynamic model is used that describes the relations between flows and levels. Figure 1.3 shows an estimate of the level of the third compartment using measurements of the level in the first compartment. Prediction of the level in the third compartment is possible due to the causality of the process and the delay between the levels in the compartments.

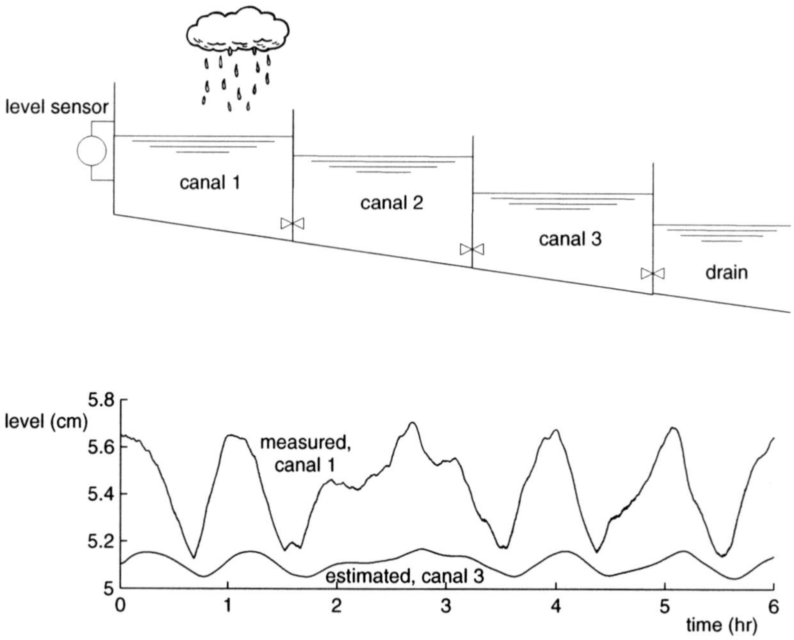

Figure 1.3 Assessment of water levels in a water management system: a state estimation problem (the data is obtained from a scale model)

1.1.4 Relations between the subjects

The reader who is familiar with one or more of the three subjects might wonder why they are treated in one book. The three subjects share the following factors:

- In all cases, the engineer designs an instrument, i.e. a system whose task is to extract information about a real-world object, a physical process or an event.
- For that purpose, the instrument will be provided with a sensory subsystem that produces measurement signals. In all cases, these signals are represented by vectors (with fixed dimension) or sequences of vectors.
- The measurement vectors must be processed to reveal the information that is required for the task at hand.
- All three subjects rely on the availability of models describing the object/physical process/event, and of models describing the sensory system.
- Modelling is an important part of the design stage. The suitability of the applied model is directly related to the performance of the resulting classifier/estimator.

Since the nature of the questions raised in the three subjects is similar, the analysis of all three cases can be done using the same framework. This allows an economical treatment of the subjects. The framework that will be used is a probabilistic one. In all three cases, the strategy will be to formulate the posterior knowledge in terms of a conditional probability (density) function:

$$P(\text{quantities of interest}|\text{measurements available})$$

This so-called posterior probability combines the prior knowledge with the empirical knowledge by using Bayes' theorem for conditional probabilities. As discussed above, the framework is generic for all three cases. Of course, the elaboration of this principle for the three cases leads to different solutions, because the natures of the 'quantities of interest' differ.

The second similarity between the topics is their reliance on models. It is assumed that the constitution of the object/physical process/event (including the sensory system) can be captured by a mathematical model. Unfortunately, the physical structures responsible for generating the objects/process/events are often unknown, or at least partly unknown. Consequently, the model is also, at least partly, unknown. Sometimes, some functional form of the model is assumed, but the free parameters still have to be determined. In any case, empirical data is needed in order to establish the model, to tune the classifier/estimator-under-development, and also to evaluate the design. Obviously, the training/evaluation data should be obtained from the process we are interested in.

In fact, all three subjects share the same key issue related to modelling, namely the selection of the appropriate generalization level. The empirical data is only an example of a set of possible measurements. If too much weight is given to the data at hand, the risk of overfitting occurs. The resulting model will depend too much on the accidental peculiarities (or noise) of the data. On the other hand, if too little weight is given, nothing will be learned and the model completely relies on the prior knowledge. The right balance between these opposite sides depends on the statistical significance of the data. Obviously, the size of the data is an important factor. However, the statistical significance also holds a relation with dimensionality.

Many of the mathematical techniques for modelling, tuning, training and evaluation can be shared between the three subjects. Estimation procedures used in classification can also be used in parameter estimation or state estimation with just minor modifications. For instance, probability density estimation can be used for classification purposes, and also for estimation. Data-fitting techniques are applied in both classification and estimation problems. Techniques for statistical inference can also be shared. Of course, there are also differences between the three subjects. For instance, the modelling of dynamic systems, usually called *system identification*, involves aspects that are typical for dynamic systems (i.e. determination of the order of the system, finding an appropriate functional structure of the model). However, when it finally comes to finding the right parameters of the dynamic model, the techniques from parameter estimation apply again.

Figure 1.4 shows an overview of the relations between the topics. Classification and parameter estimation share a common foundation indicated by 'Bayes'. In combination with models for dynamic systems (with random inputs), the techniques for classification and parameter estimation find their application in processes that proceed in time, i.e. state estimation. All this is built on a mathematical basis with selected topics from mathematical analysis (dealing with abstract vector spaces, metric spaces and operators), linear algebra and probability theory. As such, classification and estimation are not tied to a specific application. The engineer, who is involved in a specific application, should add the individual characteristics of that application by means of the models and prior knowledge. Thus, apart from the ability to handle empirical data, the engineer must also have some knowledge of the physical background related to the application at hand and to the sensor technology being used.

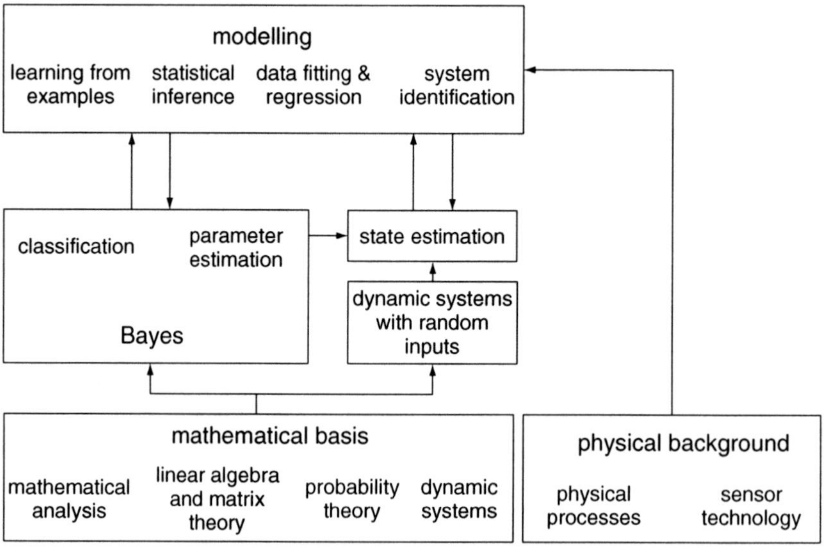

Figure 1.4 Relations between the subjects

All three subjects are mature research areas, and many overview books have been written. Naturally, by combining the three subjects into one book, it cannot be avoided that some details are left out. However, the discussion above shows that the three subjects are close enough to justify one integrated book, covering these areas.

The combination of the three topics into one book also introduces some additional challenges if only because of the differences in terminology used in the three fields. This is, for instance, reflected in the difference in the term used for 'measurements'. In classification theory, the term 'features' is frequently used as a replacement for 'measurements'. The number of measurements is called the 'dimension', but in classification theory the term 'dimensionality' is often used.[1] The same remark holds true for notations. For instance, in classification theory the measurements are often denoted by **x**. In state estimation, two notations are in vogue: either **y** or **z** (MATLAB uses **y**, but we chose **z**). In all cases we tried to be as consistent as possible.

1.2 ENGINEERING

The top-down design of an instrument always starts with some primary need. Before starting with the design, the engineer has only a global view of the system of interest. The actual need is known only at a high and abstract level. The design process then proceeds through a number of stages during which progressively more detailed knowledge becomes available, and the system parts of the instrument are described at lower and more concrete levels. At each stage, the engineer has to make design decisions. Such decisions must be based on explicitly defined evaluation criteria. The procedure, the elementary design step, is shown in Figure 1.5. It is used iteratively at the different levels and for the different system parts.

An elementary design step typically consists of collecting and organizing knowledge about the design issue of that stage, followed by an explicit formulation of the involved task. The next step is to associate

---

[1] Our definition complies with the mathematical definition of 'dimension', i.e. the maximal number of independent vectors in a vector space. In MATLAB the term 'dimension' refers to an index of a multidimensional array as in phrases like: 'the first dimension of a matrix is the row index', and 'the number of dimensions of a matrix is two'. The number of elements along a row is the 'row dimension' or 'row length'. In MATLAB the term 'dimensionality' is the same as the 'number of dimensions'.

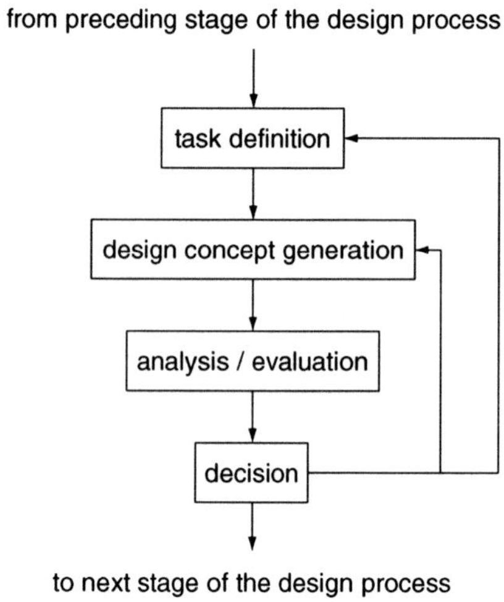

Figure 1.5 An elementary step in the design process (Finkelstein and Finkelstein, 1994)

the design issue with an evaluation criterion. The criterion expresses the suitability of a design concept related to the given task, but also other aspects can be involved, such as cost of manufacturing, computational cost or throughput. Usually, there is a number of possible design concepts to select from. Each concept is subjected to an analysis and an evaluation, possibly based on some experimentation. Next, the engineer decides which design concept is most appropriate. If none of the possible concepts are acceptable, the designer steps back to an earlier stage to alter the selections that have been made there.

One of the first tasks of the engineer is to identify the actual need that the instrument must fulfil. The outcome of this design step is a description of the functionality, e.g. a list of preliminary specifications, operating characteristics, environmental conditions, wishes with respect to user interface and exterior design. The next steps deal with the principles and methods that are appropriate to fulfil the needs, i.e. the *internal functional structure* of the instrument. At this level, the system under design is broken down into a number of functional components. Each component is considered as a subsystem whose input/output relations are mathematically defined. Questions related to the actual construction, realization of the functions, housing, etc., are later concerns.

The functional structure of an instrument can be divided roughly into sensing, processing and outputting (displaying, recording). This book focuses entirely on the design steps related to *processing*. It provides:

- Knowledge about various methods to fulfil the processing tasks of the instrument. This is needed in order to generate a number of different design concepts.
- Knowledge about how to evaluate the various methods. This is needed in order to select the best design concept.
- A tool for the experimental evaluation of the design concepts.

The book does not address the topic 'sensor technology'. For this, many good textbooks already exist, for instance see Regtien *et al.* (2004) and Brignell and White (1996). Nevertheless, the sensory system does have a large impact on the required processing. For our purpose, it suffices to consider the sensory subsystem at an abstract functional level such that it can be described by a mathematical model.

1.3 THE ORGANIZATION OF THE BOOK

The first part of the book, containing Chapters 2, 3 and 4, considers each of the three topics – classification, parameter estimation and state estimation – at a theoretical level. Assuming that appropriate models of the objects, physical process or events, and of the sensory system are available, these three tasks are well defined and can be discussed rigorously. This facilitates the development of a mathematical theory for these topics.

The second part of the book, Chapters 5 to 8, discusses all kinds of issues related to the deployment of the theory. As mentioned in Section 1.1, a key issue is modelling. Empirical data should be combined with prior knowledge about the physical process underlying the problem at hand, and about the sensory system used. For classification problems, the empirical data is often represented by labelled training and evaluation sets, i.e. sets consisting of measurement vectors of objects together with the true classes to which these objects belong. Chapters 5 and 6 discuss several methods to deal with these sets. Some of these techniques – probability density estimation, statistical inference, data fitting – are also applicable to modelling in parameter estimation. Chapter 7 is devoted to unlabelled training sets. The purpose is to find structures underlying these sets that explain the data in a statistical sense. This is useful for both classification and parameter estimation problems. The practical aspects related to state estimation are considered in Chapter 8. In the last chapter all the topics are applied in some fully worked out examples. Four appendices are added in order to refresh the required mathematical background knowledge.

The subtitle of the book, 'An Engineering Approach using MATLAB', indicates that its focus is not just on the formal description of classification, parameter estimation and state estimation methods. It also aims to provide practical implementations of the given algorithms. These implementations are given in MATLAB. MATLAB is a commercial software package for matrix manipulation. Over the past decade it has become the *de facto* standard for development and research in data-processing applications. MATLAB combines an easy-to-learn user interface with a simple, yet powerful language syntax, and a wealth of functions organized in toolboxes. We use MATLAB as a vehicle for experimentation, the purpose of which is to find out which method is the most appropriate for a given task. The final construction of the instrument can also be implemented by means of MATLAB, but this is not strictly necessary. In the end, when it comes to realization, the engineer may decide to transform his design of the functional structure from MATLAB to other platforms using, for instance, dedicated hardware, software in embedded systems or virtual instrumentation such as LabView.

For classification we will make use of PRTools (described in Appendix E), a pattern recognition toolbox for MATLAB freely available for non-commercial use. MATLAB itself has many standard functions that are useful for parameter estimation and state estimation problems. These functions are scattered over a number of toolboxes. Appendix F gives a short overview of these toolboxes. The toolboxes are accompanied with a clear and crisp documentation, and for details of the functions we refer to that.

Each chapter is followed by a few exercises on the theory provided. However, we believe that only working with the actual algorithms will provide the reader with the necessary insight to fully understand the matter. Therefore, a large number of small code examples are provided throughout the text. Furthermore, a number of data sets to experiment with are made available through the accompanying website.

1.4 REFERENCES

Brignell, J. and White, N., *Intelligent Sensor Systems*, Revised edition, IOP Publishing, London, UK, 1996.

Finkelstein, L. and Finkelstein A.C.W., *Design Principles for Instrument Systems in Measurement and Instrumentation* (eds. L. Finkelstein and K.T.V. Grattan), Pergamon Press, Oxford, UK, 1994.

Regtien, P.P.L., van der Heijden, F., Korsten, M.J. and Olthuis, W., *Measurement Science for Engineers*, Kogan Page Science, London, UK, 2004.

2

Detection and Classification

Pattern classification is the act of assigning a class label to an object, a physical process or an event. The assignment is always based on measurements that are obtained from that object (or process, or event). The measurements are made available by a sensory system. See Figure 2.1. Table 2.1 provides some examples of application fields in which classification is the essential task.

The definition of the set of relevant *classes* in a given application is in some cases given by the nature of the application, but in other cases the definition is not trivial. In the application 'character reading for license plate recognition', the choice of the classes does not need much discussion. However, in the application 'sorting tomatoes into "class A", "class B", and "class C"' the definition of the classes is open for discussion. In such cases, the classes are defined by a generally agreed convention that

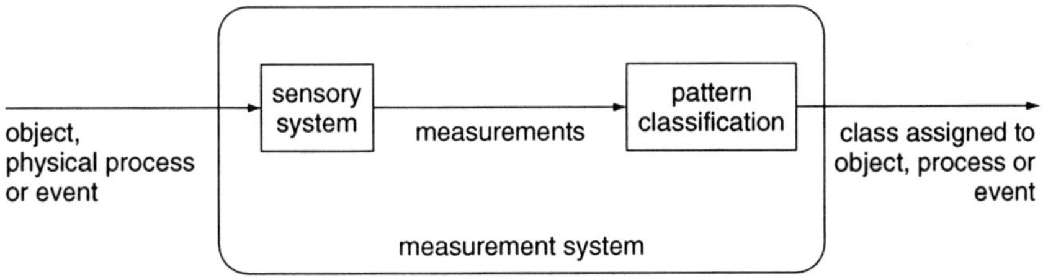

Figure 2.1 Pattern classification

Classification, Parameter Estimation and State Estimation: An Engineering Approach using MATLAB
F. van der Heijden, R.P.W. Duin, D. de Ridder and D.M.J. Tax
 ISBN: 0-470-09013-8

Table 2.1 Some application fields of pattern classification

Application field	Possible measurements	Possible classes
Object classification		
Sorting electronic parts	Shape, colour	'resistor', 'capacitor', 'transistor', 'IC'
Sorting mechanical parts	Shape	'ring', 'nut', 'bolt'
Reading characters	Shape	'A', 'B', 'C',
Mode estimation in a physical process		
Classifying manoeuvres of a vehicle	Tracked point features in an image sequence	'straight on', 'turning'
Fault diagnosis in a combustion engine	Cylinder pressures, temperature, vibrations, acoustic emissions, crank angle resolver,	'normal operation', 'defect fuel injector', 'defect air inlet valve', 'leaking exhaust valve',
Event detection		
Burglar alarm	Infrared	'alarm', 'no alarm'
Food inspection	Shape, colour, temperature, mass, volume	'OK', 'NOT OK'

the object is qualified according to the values of some attributes of the object, e.g. its size, shape and colour.

The sensory system measures some physical properties of the object that, hopefully, are relevant for classification. This chapter is confined to the simple case where the measurements are static, i.e. time independent. Furthermore, we assume that for each object the number of measurements is fixed. Hence, per object the outcomes of the measurements can be stacked to form a single vector, the so-called measurement vector. The dimension of the vector equals the number of measurements. The union of all possible values of the measurement vector is the measurement space. For some authors the word 'feature' is very close to 'measurement', but we will reserve that word for later use in Chapter 6.

The sensory system must be designed so that the measurement vector conveys the information needed to classify all objects correctly. If this is the case, the measurement vectors from all objects behave according to some pattern. Ideally, the physical properties are chosen such that all objects from one class form a cluster in the measurement space without overlapping the clusters formed by other classes.

Example 2.1 Classification of small mechanical parts
Many workrooms have a spare part box where small, obsolete mechanical parts such as bolts, rings, nuts and screws are kept. Often, it is difficult to find a particular part. We would like to have the parts sorted out. For automated sorting we have to classify the objects by measuring some properties of each individual object. Then, based on the measurements we decide to what class that object belongs.

As an example, Figure 2.2(a) shows an image with rings, nuts, bolts and remaining parts, called scrap. These four types of objects will be classified by means of two types of shape measurements. The first type expresses to what extent the object is six-fold rotational symmetric. The second type of measurement is the eccentricity of the object. The image-processing technique that is needed to obtain these measurements is a topic that is outside the scope of this book.

The 2D measurement vector of an object can be depicted as a point in the 2D measurement space. Figure 2.2(b) shows the graph of the points of all objects. Since the objects in Figure 2.2(a) are already sorted manually, it is easy here to mark each point with a symbol that indicates the true class of the corresponding object. Such a graph is called a scatter diagram of the data set.

The measure for six-fold rotational symmetry is suitable to discriminate between rings and nuts since rings and nuts have a similar shape except for the six-fold rotational symmetry of a nut. The measure for eccentricity is suitable to discriminate bolts from the nuts and the rings.

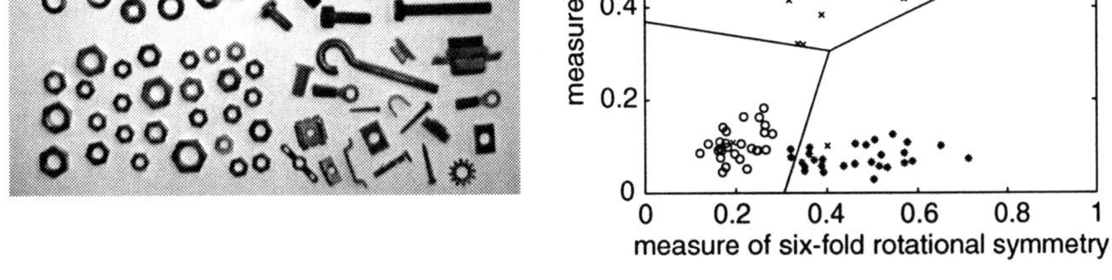

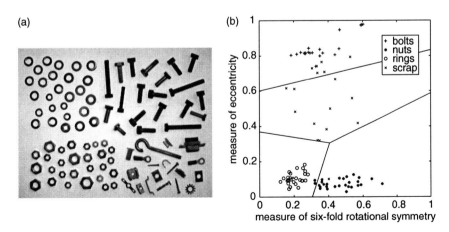

Figure 2.2 Classification of mechanical parts. (a) Image of various objects, (b) Scatter diagram

The shapes of scrap objects are difficult to predict. Therefore, their measurements are scattered all over the space.

In this example the measurements are more or less clustered according to their true class. Therefore, a new object is likely to have measurements that are close to the cluster of the class to which the object belongs. Hence, the assignment of a class boils down to deciding to which cluster the measurements of the object belong. This can be done by dividing the 2D measurement space into four different partitions; one for each class. A new object is classified according to the partitioning to which its measurement vector points.

Unfortunately, some clusters are in each other's vicinity, or even overlapping. In these regions the choice of the partitioning is critical.

This chapter addresses the problem of how to design a pattern classifier. This is done within a Bayesian-theoretic framework. Section 2.1 discusses the general case. In Sections 2.1.1 and 2.1.2 two particular cases are dealt with. The so-called 'reject option' is introduced in Section 2.2. Finally, the two-class case, often called 'detection', is covered by Section 2.3.

2.1 BAYESIAN CLASSIFICATION

Probability theory is a solid base for pattern classification design. In this approach the pattern-generating mechanism is represented within a probabilistic framework. Figure 2.3 shows such a framework. The starting point is a stochastic experiment (Appendix C.1) defined by a set $\Omega = \{\omega_1, \ldots, \omega_K\}$ of K classes. We assume that the classes are mutually exclusive. The probability $P(\omega_k)$ of having a class ω_k is called the *prior*

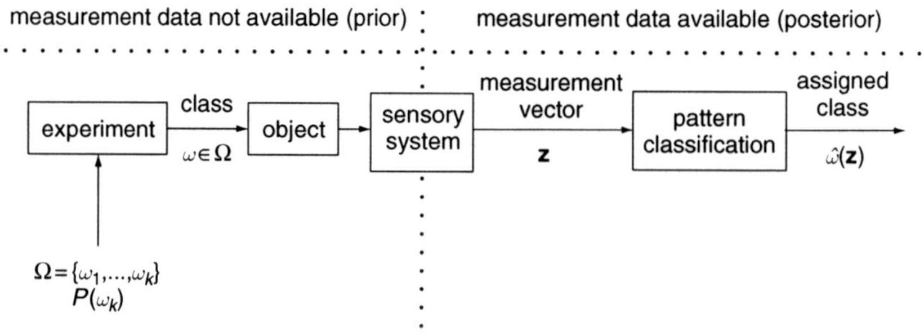

Figure 2.3 Statistical pattern classification

probability. It represents the knowledge that we have about the class of an object before the measurements of that object are available. Since the number of possible classes is K, we have:

$$\sum_{k=1}^{K} P(\omega_k) = 1 \tag{2.1}$$

The sensory system produces a *measurement vector* $\mathbf{z}$ with dimension N. Objects from different classes should have different measurement vectors. Unfortunately, the measurement vectors from objects within the same class also vary. For instance, the eccentricities of bolts in Figure 2.2 are not fixed since the shape of bolts is not fixed. In addition, all measurements are subject to some degree of randomness due to all kinds of unpredictable phenomena in the sensory system, e.g. quantum noise, thermal noise, quantization noise. The variations and randomness are taken into account by the probability density function of $\mathbf{z}$.

The *conditional probability density function* of the measurement vector $\mathbf{z}$ is denoted by $p(\mathbf{z}|\omega_k)$. It is the density of $\mathbf{z}$ coming from an object with known class ω_k. If $\mathbf{z}$ comes from an object with unknown class, its density is indicated by $p(\mathbf{z})$. This density is the unconditional density of $\mathbf{z}$. Since classes are supposed to be mutually exclusive, the unconditional density can be derived from the conditional densities by weighting these densities by the prior probabilities:

$$p(\mathbf{z}) = \sum_{k=1}^{K} p(\mathbf{z}|\omega_k)P(\omega_k) \tag{2.2}$$

The pattern classifier casts the measurement vector in the class that will be assigned to the object. This is accomplished by the so-called decision function $\hat{\omega}(.)$ that maps the measurement space onto the set of possible classes. Since $\mathbf{z}$ is an N-dimensional vector, the function maps $\mathbb{R}^N$ onto Ω. That is: $\hat{\omega}(.)$: $\mathbb{R}^N \to \Omega$.

Example 2.2 Probability densities of the 'mechanical parts' data
Figure 2.4 is a graphical representation of the probability densities of the measurement data from Example 2.1. The unconditional density $p(\mathbf{z})$ is derived from (2.2) by assuming that the prior probabilities $P(\omega_k)$ are reflected in the frequencies of occurrence of each type of object in Figure 2.2. In that figure, there are 94 objects with frequencies bolt:nut:ring:scrap = 20:28:27:19. Hence the corresponding prior

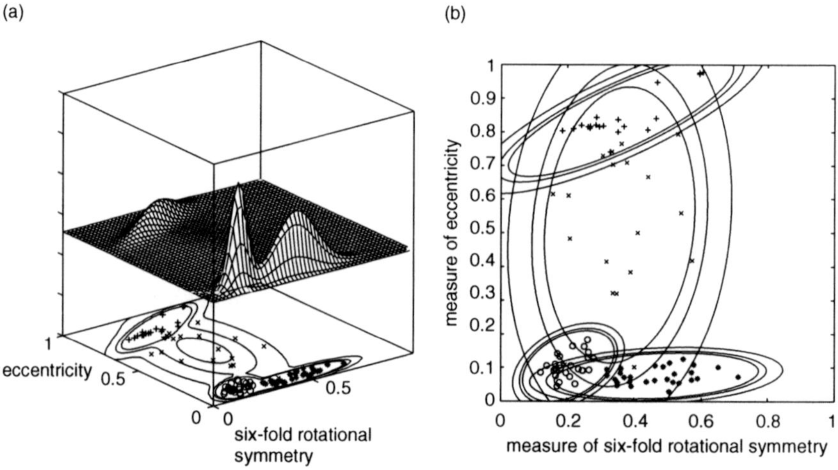

Figure 2.4 Probability densities of the measurements shown in Figure 2.2. (a) The 3D plot of the unconditional density together with a 2D contour plot of this density on the ground plane. (b) 2D contour plots of the conditional probability densities

probabilities are assumed to be 20/94, 28/94, 27/94 and 19/94, respectively.

The probabilities densities shown in Figure 2.4 are in fact not the real densities, but they are estimates obtained from the samples. The topic of density estimation will be dealt with in Chapter 5. PRTools code to plot 2D-contours and 3D-meshes of a density is given in Listing 2.1.

Listing 2.1
PRTools code for creating density plots.

```
load nutsbolts;                      % Load the dataset; see listing 5.1
w=gaussm(z,1);                       % Estimate a mixture of Gaussians
figure(1); scatterd (z); hold on;
plotm(w,6,[0.1 0.5 1.0]);            % Plot in 3D
figure(2); scatterd (z); hold on;
for c = 1:4
  w=gaussm(seldat(z,c),1);           % Estimate a Gaussian per class
  plotm(w,2,[0.1 0.5 1.0]);          % Plot in 2D
end;
```

In some cases, the measurement vectors coming from objects with different classes show some overlap in the measurement space. Therefore, it cannot always be guaranteed that the classification is free from mistakes.

An erroneous assignment of a class to an object causes some damage, or some loss of value of the misclassified object, or an impairment of its usefulness. All this depends on the application at hand. For instance, in the application 'sorting tomatoes into classes A, B or C', having a class B tomato being misclassified as 'class C' causes a loss of value because a 'class B' tomato yields more profit than a 'class C' tomato. On the other hand, if a class C tomato is misclassified as a 'class B' tomato, the damage is much more since such a situation may lead to malcontent customers.

A *Bayes classifier* is a pattern classifier that is based on the following two prerequisites:

- The damage, or loss of value, involved when an object is erroneously classified can be quantified as a cost.
- The expectation of the cost is acceptable as an optimization criterion.

If the application at hand meets these two conditions, then the development of an optimal pattern classification is theoretically straightforward. However, the Bayes classifier needs good estimates of the densities of the classes. These estimates can be problematic to obtain in practice.

The damage, or loss of value, is quantified by a *cost function* (or loss function) $C(\hat{\omega}|\omega_k)$. The function $C(.|.): \Omega \times \Omega \rightarrow \mathbb{R}$ expresses the cost that is involved when the class assigned to an object is $\hat{\omega}$, while the true class of that object is ω_k. Since there are K classes, the function $C(\hat{\omega}|\omega_k)$ is fully specified by a $K \times K$ matrix. Therefore, sometimes the cost function is called a cost matrix. In some applications, the cost function might be negative, expressing the fact that the assignment of that class pays off (negative cost = profit).

Example 2.3 Cost function of the mechanical parts application
In fact, automated sorting of the parts in a 'bolts-and-nuts' box is an example of a recycling application. If we are not collecting the mechanical parts for reuse, these parts would be disposed of. Therefore, a correct classification of a part saves the cost of a new part, and thus the cost of such a classification is negative. However, we have to take into account that:

- The effort of classifying and sorting a part also has to be paid. This cost is the same for all parts regardless of its class and whether it has been classified correctly or not.
- A bolt that has been erroneously classified as a nut or a ring causes more trouble than a bolt that has been erroneously misclassified as scrap. Likewise arguments hold for a nut and a ring.

Table 2.2 is an example of a cost function that might be appropriate for this application.

The concepts introduced above, i.e. prior probabilities, conditional densities and cost function, are sufficient to design optimal classifiers. However, first another probability has to be derived: the *posterior probability* $P(\omega_k|\mathbf{z})$. It is the probability that an object belongs to class ω_k given that the measurement vector associated with that object is $\mathbf{z}$. According to Bayes' theorem for conditional probabilities (Appendix C.2) we have:

$$P(\omega_k|\mathbf{z}) = \frac{p(\mathbf{z}|\omega_k)P(\omega_k)}{p(\mathbf{z})} \tag{2.3}$$

If an arbitrary classifier assigns a class $\hat{\omega}_i$ to a measurement vector $\mathbf{z}$ coming from an object with true class ω_k, then a cost $C(\hat{\omega}_i|\omega_k)$ is involved. The posterior probability of having such an object is $P(\omega_k|\mathbf{z})$. Therefore, the expectation of the cost is:

$$R(\hat{\omega}_i|\mathbf{z}) = \mathrm{E}[C(\hat{\omega}_i|\omega_k)|\mathbf{z}] = \sum_{k=1}^{K} C(\hat{\omega}_i|\omega_k)P(\omega_k|\mathbf{z}) \tag{2.4}$$

This quantity is called the *conditional risk*. It expresses the expected cost of the assignment $\hat{\omega}_i$ to an object whose measurement vector is $\mathbf{z}$.

From (2.4) it follows that the conditional risk of a decision function $\hat{\omega}(\mathbf{z})$ is $R(\hat{\omega}(\mathbf{z})|\mathbf{z})$. The *overall risk* can be found by averaging the conditional risk over all possible measurement vectors:

$$R = \mathrm{E}[R(\hat{\omega}(\mathbf{z})|\mathbf{z})] = \int_{\mathbf{z}} R(\hat{\omega}(\mathbf{z})|\mathbf{z})p(\mathbf{z})d\mathbf{z} \tag{2.5}$$

Table 2.2 Cost function of the 'sorting mechanical part' application

$C(\hat{\omega}_i|\omega_k)$ in \$		**True class**			
		ω_1 = bolt	ω_2 = nut	ω_3 = ring	ω_4 = scrap
Assigned class	$\hat{\omega}$ = bolt	−0.20	0.07	0.07	0.07
	$\hat{\omega}$ = nut	0.07	−0.15	0.07	0.07
	$\hat{\omega}$ = ring	0.07	0.07	−0.05	0.07
	$\hat{\omega}$ = scrap	0.03	0.03	0.03	0.03

The integral extends over the entire measurement space. The quantity R is the overall risk (average risk, or briefly, risk) associated with the decision function $\hat{\omega}(\mathbf{z})$. The overall risk is important for cost price calculations of a product.

The second prerequisite mentioned above states that the optimal classifier is the one with minimal risk R. The decision function that minimizes the (overall) risk is the same as the one that minimizes the conditional risk. Therefore, the Bayes classifier takes the form:

$$\hat{\omega}_{BAYES}(\mathbf{z}) = \hat{\omega}_i \quad \text{such that:} \quad R(\hat{\omega}_i|\mathbf{z}) \leq R(\hat{\omega}_j|\mathbf{z}) \quad i,j = 1,\ldots,K \qquad (2.6)$$

This can be expressed more briefly by:

$$\hat{\omega}_{BAYES}(\mathbf{x}) = \underset{\omega \in \Omega}{\operatorname{argmin}}\{R(\omega|\mathbf{z})\} \qquad (2.7)$$

The expression argmin{} gives the element from Ω that minimizes $R(\omega|\mathbf{z})$. Substitution of (2.3) and (2.4) yields:

$$\begin{aligned}
\hat{\omega}_{BAYES}(\mathbf{z}) &= \underset{\omega \in \Omega}{\operatorname{argmin}}\left\{\sum_{k=1}^{K} C(\omega|\omega_k)P(\omega_k|\mathbf{z})\right\} \\
&= \underset{\omega \in \Omega}{\operatorname{argmin}}\left\{\sum_{k=1}^{K} C(\omega|\omega_k)\frac{p(\mathbf{z}|\omega_k)P(\omega_k)}{p(\mathbf{z})}\right\} \\
&= \underset{\omega \in \Omega}{\operatorname{argmin}}\left\{\sum_{k=1}^{K} C(\omega|\omega_k)p(\mathbf{z}|\omega_k)P(\omega_k)\right\}
\end{aligned} \qquad (2.8)$$

Pattern classification according to (2.8) is called Bayesian classification or minimum risk classification.

Example 2.4 Bayes classifier for the mechanical parts application
Figure 2.5(a) shows the decision boundary of the Bayes classifier for the application discussed in the previous examples. Figure 2.5(b) shows the decision boundary that is obtained if the prior probability of scrap is increased to 0.50 with an evenly decrease of the prior probabilities of the other classes. Comparing the results it can be seen that such an increase introduces an enlargement of the compartment for the scrap at the expense of the other compartments.

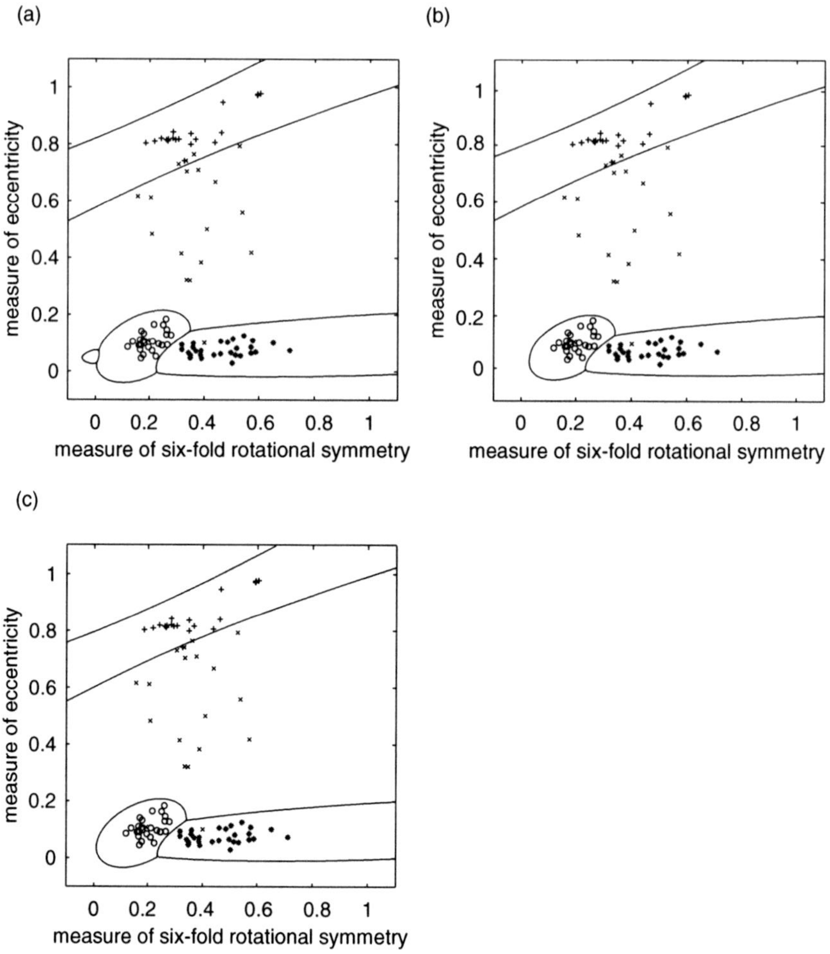

Figure 2.5 Bayes classification. (a) With prior probabilities: $P(bolt) = 0.21$, $P(nut) = 0.30$, $P(ring) = 0.29$, and $P(scrap) = 0.20$. (b) With increased prior probability for scrap: $P(scrap) = 0.50$. (c) With uniform cost function

The overall risk associated with the decision function in Figure 2.5(a) appears to be −\$0.092; the one in Figure 2.5(b) is −\$0.036. The increase of cost (= decrease of profit) is due to the fact that scrap is unprofitable. Hence, if the majority of a bunch of objects consists of worthless scrap, recycling pays off less.

The total cost of all classified objects as given in Figure 2.5(a) appears to be −\$8.98. Since the figure shows 94 objects, the average cost is −\$8.98/94 = −\$0.096. As expected, this comes close to the overall risk.

Listing 2.2
PRTools code for estimating decision boundaries taking account of the cost.

```
load nutsbolts;
cost = [ -0.20    0.07    0.07    0.07 ; ...
          0.07   -0.15    0.07    0.07 ; ...
          0.07    0.07   -0.05    0.07 ; ...
          0.03    0.03    0.03    0.03];
w1 = qdc(z);             % Estimate a single Gaussian per class
                         % Change output according to cost
w2 = w1*classc*costm([],cost);
scatterd(z);
plotc(w1);               % Plot without using cost
plotc(w2);               % Plot using cost
```

2.1.1 Uniform cost function and minimum error rate

A *uniform cost function* is obtained if a unit cost is assumed when an object is misclassified, and zero cost when the classification is correct. This can be written as:

$$C(\hat{\omega}_i|\omega_k) = 1 - \delta(i,k) \quad \text{with:} \; \delta(i,k) = \begin{cases} 1 & \text{if } i = k \\ 0 & \text{elsewhere} \end{cases} \tag{2.9}$$

$\delta(i,k)$ is the Kronecker delta function. With this cost function the conditional risk given in (2.4) simplifies to:

$$R(\hat{\omega}_i|\mathbf{z}) = \sum_{k=1,k\neq i}^{K} P(\omega_k|\mathbf{z}) = 1 - P(\hat{\omega}_i|\mathbf{z}) \tag{2.10}$$

Minimization of this risk is equivalent to maximization of the posterior probability $P(\hat{\omega}_i|\mathbf{z})$. Therefore, with a uniform cost function, the Bayes decision function (2.8) becomes the maximum a posteriori probability classifier (MAP classifier):

$$\hat{\omega}_{MAP}(\mathbf{z}) = \underset{\omega\in\Omega}{\operatorname{argmax}}\{P(\omega|\mathbf{z})\} \tag{2.11}$$

Application of Bayes' theorem for conditional probabilities and cancellation of irrelevant terms yield a classification, equivalent to a MAP classification, but fully in terms of the prior probabilities and the conditional probability densities:

$$\hat{\omega}_{MAP}(\mathbf{z}) = \underset{\omega \in \Omega}{\operatorname{argmax}}\{p(\mathbf{z}|\omega)P(\omega)\} \tag{2.12}$$

The functional structure of this decision function is given in Figure 2.6.

Suppose that a class $\hat{\omega}_i$ is assigned to an object with measurement vector $\mathbf{z}$. The probability of having a correct classification is $P(\hat{\omega}_i|\mathbf{z})$. Consequently, the probability of having a classification error is $1 - P(\hat{\omega}_i|\mathbf{z})$. For an arbitrary decision function $\hat{\omega}(\mathbf{z})$, the *conditional error probability* is:

$$e(\mathbf{z}) = 1 - P(\hat{\omega}(\mathbf{z})|\mathbf{z}) \tag{2.13}$$

It is the probability of an erroneous classification of an object whose measurement is $\mathbf{z}$. The error probability averaged over all objects can be found by averaging $e(\mathbf{z})$ over all the possible measurement vectors:

$$E = \mathrm{E}[\mathrm{e}(\mathbf{z})] = \int_{\mathbf{z}} e(\mathbf{z})p(\mathbf{z})d\mathbf{z} \tag{2.14}$$

The integral extends over the entire measurement space. E is called the *error rate*, and is often used as a performance measure of a classifier.

The classifier that yields the minimum error rate among all other classifiers is called the *minimum error rate classifier*. With a uniform cost function, the risk and the error rate are equal. Therefore, the minimum error rate classifier is a Bayes classifier with uniform cost

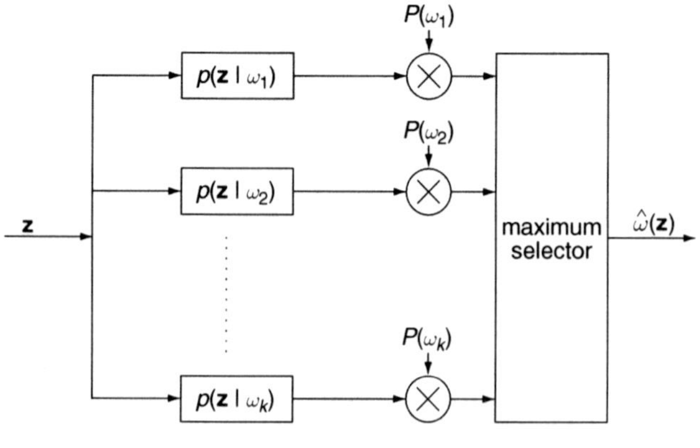

Figure 2.6 Bayes decision function with uniform cost function (MAP classification)

function. With our earlier definition of MAP classification we come to the following conclusion:

$$\begin{array}{c}\text{Minimum error rate}\\ \text{classification}\end{array} \equiv \begin{array}{c}\text{Bayes classification}\\ \text{with unit cost function}\end{array} \equiv \begin{array}{c}\text{MAP}\\ \text{classification}\end{array}$$

The conditional error probability of a MAP classifier is found by substitution of (2.11) in (2.13):

$$e_{\min}(\mathbf{z}) = 1 - \max_{\omega \in \Omega}\{P(\omega|\mathbf{z})\} \tag{2.15}$$

The minimum error rate $E_{\min}$ follows from (2.14):

$$E_{\min} = \int_{\mathbf{z}} e_{\min}(\mathbf{z})p(\mathbf{z})d\mathbf{z} \tag{2.16}$$

Of course, phrases like 'minimum' and 'optimal' are strictly tied to the given sensory system. The performance of an optimal classification with a given sensory system may be less than the performance of a non-optimal classification with another sensory system.

Example 2.5 MAP classifier for the mechanical parts application
Figure 2.5(c) shows the decision function of the MAP classifier. The error rate for this classifier is 4.8%, whereas the one of the Bayes classifier in Figure 2.5(a) is 5.3%. In Figure 2.5(c) four objects are misclassified. In Figure 2.5(a) that number is five. Thus, with respect to error rate, the MAP classifier is more effective compared with the Bayes classifier of Figure 2.5(a). On the other hand, the overall risk of the classifier shown in Figure 2.5(c) and with the cost function given in Table 2.2 is −\$0.084 which is a slight impairment compared with the −\$0.092 of Figure 2.5(a).

2.1.2 Normal distributed measurements; linear and quadratic classifiers

A further development of Bayes classification with uniform cost function requires the specification of the conditional probability densities. This section discusses the case in which these densities are modelled as normal. Suppose that the measurement vectors coming from an object with class ω_k are normally distributed with expectation vector $\boldsymbol{\mu}_k$ and covariance matrix $\mathbf{C}_k$ (see Appendix C.3):

$$p(\mathbf{z}|\omega_k) = \frac{1}{\sqrt{(2\pi)^N|\mathbf{C}_k|}}\exp\left(\frac{-(\mathbf{z}-\boldsymbol{\mu}_k)^T\mathbf{C}_k^{-1}(\mathbf{z}-\boldsymbol{\mu}_k)}{2}\right) \tag{2.17}$$

where N is the dimension of the measurement vector.

Substitution of (2.17) in (2.12) gives the following minimum error rate classification:

$$\hat{\omega}(\mathbf{z}) = \omega_i \quad \text{with}$$
$$i = \underset{k=1,\ldots,K}{\operatorname{argmax}}\left\{\frac{1}{\sqrt{(2\pi)^N|\mathbf{C}_k|}}\exp\left(\frac{-(\mathbf{z}-\boldsymbol{\mu}_k)^T\mathbf{C}_k^{-1}(\mathbf{z}-\boldsymbol{\mu}_k)}{2}\right)P(\omega_k)\right\} \tag{2.18}$$

We can take the logarithm of the function between braces without changing the result of the argmax{ } function. Furthermore, all terms not containing k are irrelevant. Therefore (2.18) is equivalent to

$$\hat{\omega}(\mathbf{z}) = \omega_i \quad \text{with}$$
$$\begin{aligned} i &= \underset{k=1,\ldots,K}{\operatorname{argmax}}\left\{-\frac{1}{2}\ln|\mathbf{C}_k| + \ln P(\omega_k) - \frac{1}{2}(\mathbf{z}-\boldsymbol{\mu}_k)^T\mathbf{C}_k^{-1}(\mathbf{z}-\boldsymbol{\mu}_k)\right\} \\ &= \underset{k=1,\ldots,K}{\operatorname{argmax}}\left\{-\ln|\mathbf{C}_k| + 2\ln P(\omega_k) - \boldsymbol{\mu}_k^T\mathbf{C}_k^{-1}\boldsymbol{\mu}_k + 2\mathbf{z}^T\mathbf{C}_k^{-1}\boldsymbol{\mu}_k - \mathbf{z}^T\mathbf{C}_k^{-1}\mathbf{z}\right\} \end{aligned} \tag{2.19}$$

Hence, the expression of a minimum error rate classification with normally distributed measurement vectors takes the form of:

$$\hat{\omega}(\mathbf{z}) = \omega_i \quad \text{with} \quad i = \underset{k=1,\ldots,K}{\operatorname{argmax}}\{w_k + \mathbf{z}^T\mathbf{w}_k + \mathbf{z}^T\mathbf{W}_k\mathbf{z}\} \tag{2.20}$$

with:

$$w_k = -\ln|\mathbf{C}_k| + 2\ln P(\omega_k) - \boldsymbol{\mu}_k^T \mathbf{C}_k^{-1} \boldsymbol{\mu}_k$$
$$\mathbf{w}_k = 2\mathbf{C}_k^{-1}\boldsymbol{\mu}_k \tag{2.21}$$
$$\mathbf{W}_k = -\mathbf{C}_k^{-1}$$

A classifier according to (2.20) is called a *quadratic classifier* and the decision function is a quadratic decision function. The boundaries between the compartments of such a decision function are pieces of quadratic hypersurfaces in the N-dimensional space. To see this, it suffices to examine the boundary between the compartments of two different classes, e.g. ω_i and ω_j. According to (2.20) the boundary between the compartments of these two classes must satisfy the following equation:

$$w_i + \mathbf{z}^T\mathbf{w}_i + \mathbf{z}^T\mathbf{W}_i\mathbf{z} = w_j + \mathbf{z}^T\mathbf{w}_j + \mathbf{z}^T\mathbf{W}_j\mathbf{z} \tag{2.22}$$

or:

$$w_i - w_j + \mathbf{z}^T(\mathbf{w}_i - \mathbf{w}_j) + \mathbf{z}^T(\mathbf{W}_i - \mathbf{W}_j)\mathbf{z} = 0 \tag{2.23}$$

Equation (2.23) is quadratic in $\mathbf{z}$. In the case that the sensory system has only two sensors, i.e. $N = 2$, then the solution of (2.23) is a quadratic curve in the measurement space (an ellipse, a parabola, an hyperbola, or a degenerated case: a circle, a straight line, or a pair of lines). Examples will follow in subsequent sections. If we have three sensors, $N = 3$, then the solution of (2.23) is a quadratic surface (ellipsoid, paraboloid, hyperboloid, etc.). If $N > 3$, the solutions are hyperquadrics (hyperellipsoids, etc.).

If the number of classes is more than two, $K > 2$, then (2.23) is a necessary condition for the boundaries between compartments, but not a sufficient one. This is because the boundary between two classes may be intersected by a compartment of a third class. Thus, only pieces of the surfaces found by (2.23) are part of the boundary. The pieces of the surface that are part of the boundary are called decision boundaries. The assignment of a class to a vector exactly on the decision boundary is ambiguous. The class assigned to such a vector can be arbitrarily selected from the classes involved.

As an example we consider the classifications shown in Figure 2.5. In fact, the probability densities shown in Figure 2.4(b) are normal. Therefore, the decision boundaries shown in Figure 2.5 must be quadratic curves.

Class-independent covariance matrices

In this subsection, we discuss the case in which the covariance matrices do not depend on the classes, i.e. $\mathbf{C}_k = \mathbf{C}$ for all $\omega_k \in \Omega$. This situation occurs when the measurement vector of an object equals the (class-dependent) expectation vector corrupted by sensor noise, that is $\mathbf{z} = \boldsymbol{\mu}_k + \mathbf{n}$. The noise $\mathbf{n}$ is assumed to be class-independent with covariance matrix $\mathbf{C}$. Hence, the class information is brought forth by the expectation vectors only.

The quadratic decision function of (2.19) degenerates into:

$$
\begin{aligned}
&\hat{\omega}(\mathbf{x}) = \omega_i \quad \text{with} \\
i &= \underset{k=1,\ldots,K}{\operatorname{argmax}}\{2\ln P(\omega_k) - (\mathbf{z}-\boldsymbol{\mu}_k)^T\mathbf{C}^{-1}(\mathbf{z}-\boldsymbol{\mu}_k)\} \\
&= \underset{k=1,\ldots,K}{\operatorname{argmin}}\{-2\ln P(\omega_k) + (\mathbf{z}-\boldsymbol{\mu}_k)^T\mathbf{C}^{-1}(\mathbf{z}-\boldsymbol{\mu}_k)\}
\end{aligned}
\tag{2.24}
$$

Since the covariance matrix $\mathbf{C}$ is self-adjoint and positive definite (Appendix B.5) the quantity $(\mathbf{z}-\boldsymbol{\mu}_k)^T\mathbf{C}^{-1}(\mathbf{z}-\boldsymbol{\mu}_k)$ can be regarded as a distance measure between the vector $\mathbf{z}$ and the expectation vector $\boldsymbol{\mu}_k$. The measure is called the squared Mahalanobis distance. The function of (2.24) decides for the class whose expectation vector is nearest to the observed measurement vector (with a correction factor $-2\ln P(\omega_k)$ to account for prior knowledge). Hence, the name *minimum Mahalonobis distance classifier*.

The decision boundaries between compartments in the measurement space are linear (hyper)planes. This follows from (2.20) and (2.21):

$$
\hat{\omega}(\mathbf{z}) = \omega_i \quad \text{with} \quad i = \underset{k=1,\ldots,K}{\operatorname{argmax}}\{w_k + \mathbf{z}^T\mathbf{w}_k\} \tag{2.25}
$$

where:

$$\begin{aligned} w_k &= 2\ln P(\omega_k) - \boldsymbol{\mu}_k^T \mathbf{C}^{-1}\boldsymbol{\mu}_k \\ \mathbf{w}_k &= 2\mathbf{C}^{-1}\boldsymbol{\mu}_k \end{aligned} \tag{2.26}$$

A decision function which has the form of (2.25) is linear. The corresponding classifier is called a *linear classifier*. The equations of the decision boundaries are $w_i - w_j + \mathbf{z}^T(\mathbf{w}_i - \mathbf{w}_j) = 0$.

Figure 2.7 gives an example of a four-class problem ($K = 4$) in a two-dimensional measurement space ($N = 2$). A scatter diagram with the contour plots of the conditional probability densities are given (Figure 2.7(a)), together with the compartments of the minimum Mahalanobis distance classifier (Figure 2.7(b)). These figures were generated by the code in Listing 2.3.

Listing 2.3
PRTools code for minimum Mahalanobis distance classification

```
mus = [0.2 0.3; 0.35 0.75; 0.65 0.55; 0.8 0.25];
C = [0.018 0.007; 0.007 0.011]; z = gauss(200,mus,C);
w = ldc(z);     % Normal densities, identical covariances
figure(1); scatterd(z); hold on; plotm(w);
figure(2); scatterd(z); hold on; plotc(w);
```

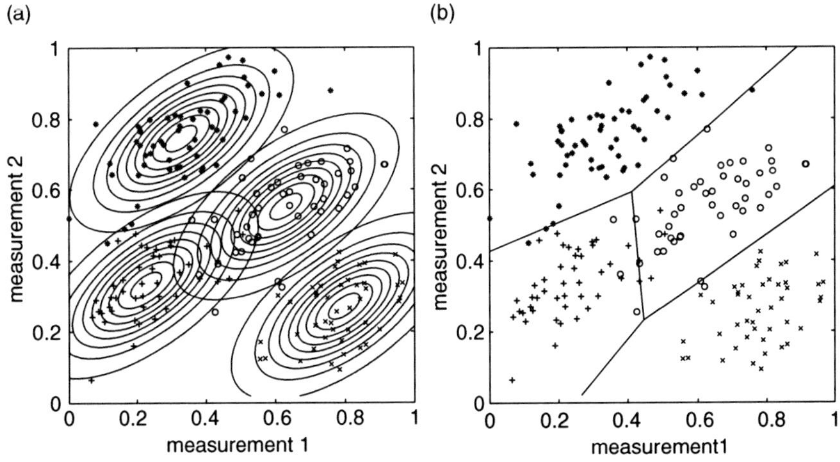

Figure 2.7 Minimum Mahalanobis distance classification. (a) Scatter diagram with contour plot of the conditional probability densities. (b) Decision boundaries

Minimum distance classification

A further simplification is possible when the measurement vector equals the class-dependent vector $\boldsymbol{\mu}_k$ corrupted by class-independent white noise with covariance matrix $\mathbf{C} = \sigma^2\mathbf{I}$.

$$\hat{\omega}(\mathbf{z}) = \omega_i \quad \text{with} \quad i = \operatorname*{argmin}_{k=1,\ldots,K} \left\{ -2\ln P(\omega_k) + \frac{\|\mathbf{z} - \boldsymbol{\mu}_k\|^2}{\sigma^2} \right\} \tag{2.27}$$

The quantity $\|(\mathbf{z} - \boldsymbol{\mu}_k)\|$ is the normal (Euclidean) distance between $\mathbf{z}$ and $\boldsymbol{\mu}_k$. The classifier corresponding to (2.27) decides for the class whose expectation vector is nearest to the observed measurement vector (with a correction factor $-2\sigma^2 \log P(\omega_k)$ to account for the prior knowledge). Hence, the name *minimum distance classifier*. As with the minimum Mahalanobis distance classifier, the decision boundaries between compartments are linear (hyper)planes. The plane separating the compartments of two classes ω_i and ω_j is given by:

$$\sigma^2 \log\frac{P(\omega_i)}{P(\omega_j)} + \frac{1}{2}(\|\boldsymbol{\mu}_j\|^2 - \|\boldsymbol{\mu}_i\|^2) + \mathbf{z}^T(\boldsymbol{\mu}_i - \boldsymbol{\mu}_j) = 0 \tag{2.28}$$

The solution of this equation is a plane perpendicular to the line segment connecting $\boldsymbol{\mu}_i$ and $\boldsymbol{\mu}_j$. The location of the hyperplane depends on the factor $\sigma^2 \log(P(\omega_i)/P(\omega_j))$. If $P(\omega_i) = P(\omega_j)$, the hyperplane is the perpendicular bisector of the line segment (see Figure 2.8).

Figure 2.9 gives an example of the decision function of the minimum distance classification. PRTools code to generate these figures is given in Listing 2.4.

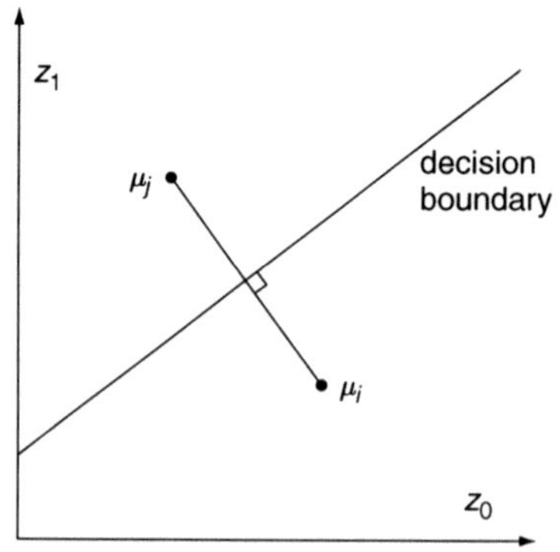

Figure 2.8 Decision boundary of a minimum distance classifier

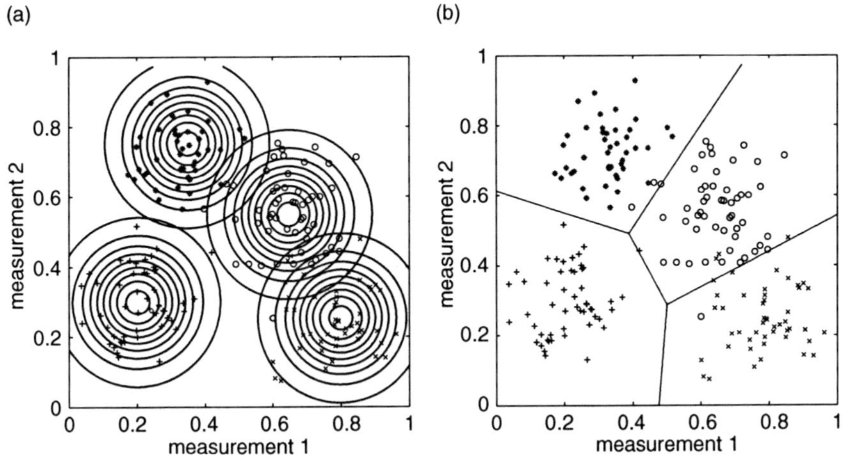

Figure 2.9 Minimum distance classification. (a) Scatter diagram with contour plot of the conditional probability densities. (b) Decision boundaries

Listing 2.4
PRTools code for minimum distance classification

```
mus=[0.2 0.3; 0.35 0.75; 0.65 0.55; 0.8 0.25];
C=0.01*eye(2); z=gauss(200,mus,C);
% Normal densities, uncorrelated noise with equal variances
w=nmsc(z);
figure (1); scatterd (z); hold on; plotm (w);
figure (2); scatterd (z); hold on; plotc (w);
```

Class-independent expectation vectors

Another interesting situation is when the class information is solely brought forth by the differences between covariance matrices. In that case, the expectation vectors do not depend on the class: $\boldsymbol{\mu}_k = \boldsymbol{\mu}$ for all k. Hence, the central parts of the conditional probability densities overlap. In the vicinity of the expectation vector, the probability of making a wrong decision is always largest. The decision function takes the form of:

$$\hat{\omega}(\mathbf{x}) = \omega_i \quad \text{with}$$
$$i = \underset{k=1,\ldots,K}{\operatorname{argmax}} \left\{ -\ln|\mathbf{C}_k| + 2\ln P(\omega_k) - (\mathbf{z}-\boldsymbol{\mu})^T \mathbf{C}_k^{-1} (\mathbf{z}-\boldsymbol{\mu}) \right\} \tag{2.29}$$

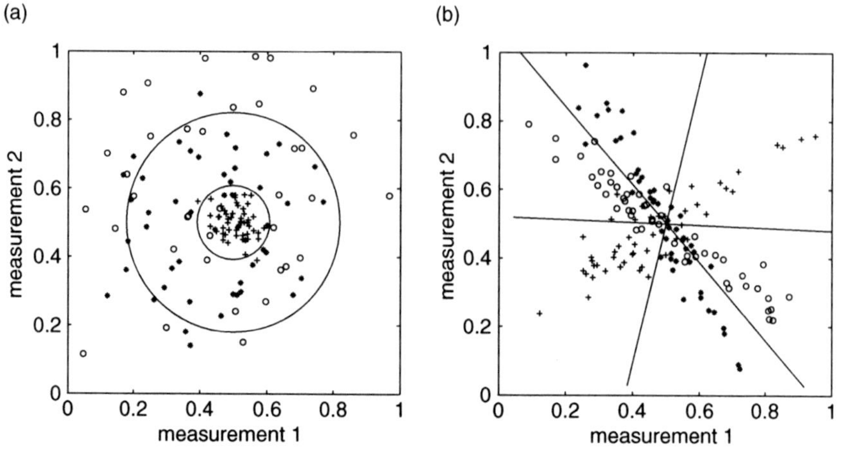

Figure 2.10 Classification of objects with equal expectation vectors. (a) Rotational symmetric conditional probability densities. (b) Conditional probability densities with different orientations; see text

If the covariance matrices are of the type $\sigma_k^2\mathbf{I}$, the decision boundaries are concentric circles or (hyper)spheres. Figure 2.10(a) gives an example of such a situation. If the covariance matrices are rotated versions of one prototype, the decision boundaries are hyperbolae. If the prior probabilities are equal, these hyperbolae degenerate into a number of linear planes (or, if $N = 2$, linear lines). An example is given in Figure 2.10(b).

2.2 REJECTION

Sometimes, it is advantageous to provide the classification with a so-called reject option. In some applications, an erroneous decision may lead to very high cost, or even to a hazardous situation. Suppose that the measurement vector of a given object is in the vicinity of the decision boundary. The measurement vector does not provide much class information then. Therefore, it might be beneficial to postpone the classification of that particular object. Instead of classifying the object, we reject the classification. On rejection of a classification we have to classify the object either manually, or by means of more involved techniques (for instance, by bringing in more sensors or more advanced classifiers).

We may take the reject option into account by extending the range of the decision function by a new element: the rejection class ω_0. The range of the decision function becomes: $\Omega^+ = \{\omega_0, \omega_1, \ldots, \omega_K\}$. The decision function itself is a mapping $\hat{\omega}(\mathbf{z})$: $\mathbb{R}^N \to \Omega^+$. In order to develop a Bayes classifier, the definition of the cost function must also be extended: $C(\hat{\omega}|\omega)$: $\Omega^+ \times \Omega \to \mathbb{R}$. The definition is extended so as to express the cost of rejection. $C(\omega_0|\omega_k)$ is the cost of rejection while the true class of the object is ω_k.

With these extensions, the decision function of a Bayes classifier becomes (2.8):

$$\hat{\omega}_{BAYES}(\mathbf{z}) = \underset{\omega\in\Omega^+}{\text{argmin}}\left\{\sum_{k=1}^{K} C(\omega|\omega_k)P(\omega_k|\mathbf{z})\right\} \tag{2.30}$$

The further development of the classifier follows the same course as in (2.8).

2.2.1 Minimum error rate classification with reject option

The minimum error rate classifier can also be extended with a reject option. Suppose that the cost of a rejection is C_{rej} regardless of the true class of the object. All other costs are uniform and defined by (2.9).

We first note that if the reject option is chosen, the risk is C_{rej}. If it is not, the minimal conditional risk is the $e_{\min}(\mathbf{z})$ given by (2.15). Minimization of C_{rej} and $e_{\min}(\mathbf{z})$ yields the following optimal decision function:

$$\hat{\omega}(\mathbf{z}) = \begin{cases} \omega_0 & \text{if } \mathrm{C_{rej}} < e_{\min}(\mathbf{z}) \\ \hat{\omega}(\mathbf{z}) = \underset{\omega\in\Omega}{\text{argmax}}\{P(\omega|\mathbf{z})\} & \text{otherwise} \end{cases} \tag{2.31}$$

The maximum posterior probability $\max\{P(\omega|\mathbf{z})\}$ is always greater than or equal to $1/K$. Therefore, the *minimal conditional error probability* is bounded by $(1 - 1/K)$. Consequently, in (2.31) the reject option never wins if $C_{rej} \geq 1 - 1/K$.

The overall probability of having a rejection is called the reject rate. It is found by calculating the fraction of measurements that fall inside the reject region:

$$\text{Rej-Rate} = \int_{\{\mathbf{z}|C_{rej}<e_{\min}(\mathbf{z})\}} p(\mathbf{z})d\mathbf{z} \tag{2.32}$$

The integral extends over those regions in the measurement space for which $C_{rej} < e(\mathbf{z})$. The error rate is found by averaging the conditional error over all measurements except those that fall inside the reject region:

$$E_{\min} = \int_{\{\mathbf{z}|C_{rej} \geq e_{\min}(\mathbf{z})\}} e_{\min}(\mathbf{z})p(\mathbf{z})d\mathbf{z} \tag{2.33}$$

Comparison of (2.33) with (2.16) shows that the error rate of a classification with reject option is bounded by the error rate of a classification without reject option.

Example 2.6 The reject option in the mechanical parts application In the classification of bolts, nuts, rings and so on, discussed in the previous examples, it might be advantageous to manually inspect those parts whose automatic classification is likely to fail. We assume that the cost of manual inspection is about \$0.04. Table 2.3 tabulates the cost function with the reject option included (compare with Table 2.2).

The corresponding classification map is shown in Figure 2.11. In this example, the reject option is advantageous only between the regions of the rings and the nuts. The overall risk decreases from −\$0.092 per classification to −\$0.093 per classification. The benefit of the reject option is only marginal because the scrap is an expensive item when offered to manual inspection. In fact, the assignment of an object to the scrap class is a good alternative for the reject option.

Listing 2.5 shows the actual implementation in MATLAB. Clearly it is very similar to the implementation for the classification including the costs. To incorporate the reject option, not only the cost matrix has to be extended, but `clabels` has to be redefined as well. When these labels are not supplied explicitly, they are copied from the data set. In the reject case, an extra class is introduced, so the definition of the labels cannot be avoided.

Table 2.3 Cost function of the mechanical part application with the reject option included

$C(\hat{\omega}_i\|\omega_k)$ in \$		True class		
Assigned class	ω_1 = bolt	ω_2 = nut	ω_3 = ring	ω_4 = scrap
$\hat{\omega}$ = bolt	−0.20	0.07	0.07	0.07
$\hat{\omega}$ = ring	0.07	−0.15	0.07	0.07
$\hat{\omega}$ = nut	0.07	0.07	−0.05	0.07
$\hat{\omega}$ = scrap	0.03	0.03	0.03	0.03
$\hat{\omega} = \omega_0$ = rejection	−0.16	−0.11	0.01	0.07

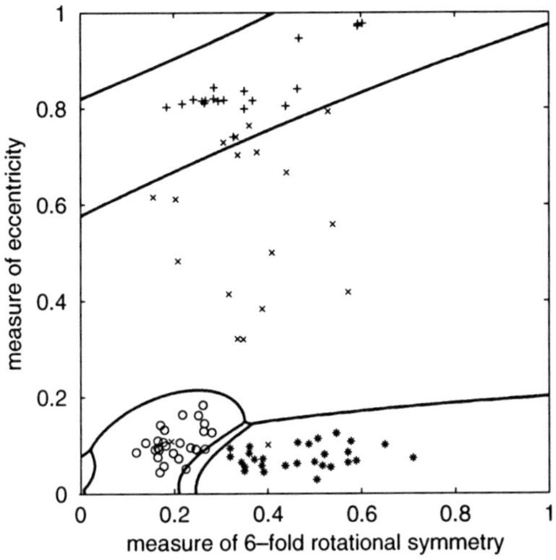

Figure 2.11 Bayes classification with the reject option included

Listing 2.5
PRTools code for minimum risk classification including a reject option

```
load nutsbolts;
cost = [ -0.20    0.07    0.07    0.07 ; ...
          0.07   -0.15    0.07    0.07 ; ...
          0.07    0.07   -0.05    0.07 ; ...
          0.03    0.03    0.03    0.03 ; ...
         -0.16   -0.11    0.01    0.07 ];
clabels = str2mat(getlablist(z),'reject');
w1 = qdc(z);            % Estimate a single Gaussian per class
scatterd(z);
                        % Change output according to cost
w2 = w1*classc*costm([],cost',clabels);
plotc(w1);              % Plot without using cost
plotc(w2);              % Plot using cost
```

2.3 DETECTION: THE TWO-CLASS CASE

The *detection* problem is a classification problem with two possible classes: $K = 2$. In this special case, the Bayes decision rule can be moulded into a simple form. Assuming a uniform cost function the MAP classifier, expressed in (2.12), reduces to the following test:

$$p(\mathbf{z}|\omega_1)P(\omega_1) > p(\mathbf{z}|\omega_2)P(\omega_2) \tag{2.34}$$

If the test fails, it is decided for ω_2, otherwise for ω_1. We write symbolically:

$$p(\mathbf{z}|\omega_1)P(\omega_1) \underset{\omega_2}{\overset{\omega_1}{\gtrless}} p(\mathbf{z}|\omega_2)P(\omega_2) \tag{2.35}$$

Rearrangement gives:

$$\frac{p(\mathbf{z}|\omega_1)}{p(\mathbf{z}|\omega_2)} \underset{\omega_2}{\overset{\omega_1}{\gtrless}} \frac{P(\omega_2)}{P(\omega_1)} \tag{2.36}$$

Regarded as a function of ω_k the conditional probability density $p(\mathbf{z}|\omega_k)$ is called the likelihood function of ω_k. Therefore, the ratio:

$$L(\mathbf{z}) = \frac{p(\mathbf{z}|\omega_1)}{p(\mathbf{z}|\omega_2)} \tag{2.37}$$

is called the likelihood ratio. With this definition the classification becomes a simple likelihood ratio test:

$$L(\mathbf{z}) \underset{\omega_2}{\overset{\omega_1}{\gtrless}} \frac{P(\omega_2)}{P(\omega_1)} \tag{2.38}$$

The test is equivalent to a threshold operation applied to $L(\mathbf{z})$ with threshold $P(\omega_2)/P(\omega_1)$.

Even if the cost function is not uniform, the Bayes detector retains the structure of (2.38), only the threshold should be adapted so as to reflect the change of cost. The proof of this is left as an exercise for the reader.

In case of measurement vectors with normal distributions, it is convenient to replace the likelihood ratio test with a so-called log-likelihood ratio test:

$$\Lambda(\mathbf{z}) \underset{\omega_2}{\overset{\omega_1}{\gtrless}} T \quad \text{with } \Lambda(\mathbf{z}) = \ln L(\mathbf{z}) \text{ and } T = \ln\left(\frac{P(\omega_2)}{P(\omega_1)}\right) \tag{2.39}$$

For vectors drawn from normal distributions, the log-likelihood ratio is:

$$\Lambda(\mathbf{z}) = -\frac{1}{2}\Big(\ln|\mathbf{C}_1| - \ln|\mathbf{C}_2| + (\mathbf{z}-\boldsymbol{\mu}_1)^T\mathbf{C}_1^{-1}(\mathbf{z}-\boldsymbol{\mu}_1) \\ -(\mathbf{z}-\boldsymbol{\mu}_2)^T\mathbf{C}_2^{-1}(\mathbf{z}-\boldsymbol{\mu}_2)\Big) \tag{2.40}$$

which is much easier than the likelihood ratio. When the covariance matrices of both classes are equal ($\mathbf{C}_1 = \mathbf{C}_2 = \mathbf{C}$) the log-likelihood ratio simplifies to:

$$\Lambda(\mathbf{z}) = \left(\mathbf{z} - \frac{1}{2}(\boldsymbol{\mu}_1 + \boldsymbol{\mu}_2)\right)^T \mathbf{C}^{-1}(\boldsymbol{\mu}_1 - \boldsymbol{\mu}_2) \tag{2.41}$$

Two types of errors are involved in a detection system. Suppose that $\hat{\omega}(\mathbf{z})$ is the result of a decision based on the measurement $\mathbf{z}$. The true (but unknown) class ω of an object is either ω_1 or ω_2. Then the following four states may occur:

	$\omega = \omega_1$	$\omega = \omega_2$
$\hat{\omega}(z) = \omega_1$	correct decision I	type II error
$\hat{\omega}(z) = \omega_2$	type I error	correct decision II

Often, a detector is associated with a device that decides whether an object is present ($\omega = \omega_2$) or not ($\omega = \omega_1$), or that an event occurs or not. These types of problems arise, for instance, in radar systems, medical diagnostic systems, burglar alarms, etc. Usually, the nomenclature for the four states is as follows then:

	$\omega = \omega_1$	$\omega = \omega_2$
$\hat{\omega}(z) = \omega_1$	true negative	missed event or false negative
$\hat{\omega}(z) = \omega_2$	false alarm or false positive	detection (or hit) or true positive

Sometimes, the true negative is called 'rejection'. However, we have reserved this term for Section 2.2, where it has a different denotation.

The probabilities of the two types of errors, i.e. the false alarm and the missed event, are performance measures of the detector. Usually these probabilities are given conditionally with respect to the true classes, i.e. $P_{miss} \stackrel{def}{=} P(\hat{\omega}_1|\omega_2)$ and $P_{fa} \stackrel{def}{=} P(\hat{\omega}_2|\omega_1)$. In addition, we may define the probability of a detection $P_{det} \stackrel{def}{=} P(\hat{\omega}_2|\omega_2)$.

The overall probability of a false alarm can be derived from the prior probability using Bayes' theorem, e.g. $P(\hat{\omega}_2, \omega_1) = P(\hat{\omega}_2|\omega_1)P(\omega_1) = P_{fa}P(\omega_1)$. The probabilities P_{miss} and P_{fa}, as a function of the threshold T, follow from (2.39):

$$\begin{aligned} P_{fa}(T) &= P(\Lambda(\mathbf{z}) < T|\omega_1) = \int_{-\infty}^{T} p(\Lambda|\omega_1)d\Lambda \\ P_{miss}(T) &= P(\Lambda(\mathbf{z}) > T|\omega_2) = \int_{T}^{\infty} p(\Lambda|\omega_2)d\Lambda \\ P_{det}(T) &= 1 - P_{miss}(T) \end{aligned} \tag{2.42}$$

In general, it is difficult to find analytical expressions for $P_{miss}(T)$ and $P_{fa}(T)$. In the case of Gaussian distributed measurement vectors, with $\mathbf{C}_1 = \mathbf{C}_2 = \mathbf{C}$, expression (2.42) can be further developed. Equation (2.41) shows that $\Lambda(z)$ is linear in $\mathbf{z}$. Since $\mathbf{z}$ has a normal distribution, so has $\Lambda(\mathbf{z})$; see Appendix C.3.1. The posterior distribution of $\Lambda(\mathbf{z})$ is fully specified by its conditional expectation and its variance. As $\Lambda(\mathbf{z})$ is linear in $\mathbf{z}$, these parameters are obtained as:

$$\begin{aligned} \mathrm{E}[\Lambda(\mathbf{z})|\omega_1] &= \left(\mathrm{E}[\mathbf{z}|\omega_1] - \frac{1}{2}(\boldsymbol{\mu}_1 + \boldsymbol{\mu}_2)\right)^T \mathbf{C}^{-1}(\boldsymbol{\mu}_1 - \boldsymbol{\mu}_2) \\ &= \left(\boldsymbol{\mu}_1 - \frac{1}{2}(\boldsymbol{\mu}_1 + \boldsymbol{\mu}_2)\right)^T \mathbf{C}^{-1}(\boldsymbol{\mu}_1 - \boldsymbol{\mu}_2) \\ &= \frac{1}{2}(\boldsymbol{\mu}_1 - \boldsymbol{\mu}_2)^T \mathbf{C}^{-1}(\boldsymbol{\mu}_1 - \boldsymbol{\mu}_2) \end{aligned} \tag{2.43}$$

Likewise:

$$\mathrm{E}[\Lambda(\mathbf{z})|\omega_2] = -\frac{1}{2}(\boldsymbol{\mu}_1 - \boldsymbol{\mu}_2)^T \mathbf{C}^{-1}(\boldsymbol{\mu}_1 - \boldsymbol{\mu}_2) \tag{2.44}$$

and:

$$\mathrm{Var}[\Lambda(\mathbf{z})|\omega_1] = (\boldsymbol{\mu}_1 - \boldsymbol{\mu}_2)^T \mathbf{C}^{-1}(\boldsymbol{\mu}_1 - \boldsymbol{\mu}_2) = \mathrm{Var}[\Lambda(\mathbf{z})|\omega_2] \tag{2.45}$$

With that, the signal-to-noise ratio is:

$$SNR = \frac{(\mathrm{E}[\Lambda|\omega_2] - \mathrm{E}[\Lambda|\omega_1])^2}{\mathrm{Var}[\Lambda|\omega_2]} = (\boldsymbol{\mu}_1 - \boldsymbol{\mu}_2)^T \mathbf{C}^{-1}(\boldsymbol{\mu}_1 - \boldsymbol{\mu}_2) \tag{2.46}$$

The quantity $(\mu_1 - \mu_2)^T C^{-1}(\mu_1 - \mu_2)$ is the squared Mahalanobis distance between μ_1 and μ_2 with respect to C. The square root, $d \stackrel{def}{=} \sqrt{SNR}$ is called the discriminability of the detector. It is the signal-to-noise ratio expressed as an amplitude ratio.

The conditional probability densities of Λ are shown in Figure 2.12. The two overlapping areas in this figure are the probabilities of false alarm and missed event. Clearly, these areas decrease as d increases. Therefore, d is a good indicator of the performance of the detector.

Knowing that the conditional probabilities are Gaussian, it is possible to evaluate the expressions for $P_{miss}(T)$ and $P_{fa}(T)$ in (2.42) analytically. The distribution function of a Gaussian random variable is given in terms of the error function erf():

$$P_{fa}(T) = \frac{1}{2} + \frac{1}{2}\text{erf}\left(\frac{T - \frac{1}{2}d^2}{d\sqrt{2}}\right)$$
$$P_{miss}(T) = \frac{1}{2} - \frac{1}{2}\text{erf}\left(\frac{T + \frac{1}{2}d^2}{d\sqrt{2}}\right) \tag{2.47}$$

Figure 2.13(a) shows a graph of P_{miss}, P_{fa}, and $P_{det} = 1 - P_{miss}$ when the threshold T varies. It can be seen that the requirements for T are contradictory. The probability of a false alarm (type I error) is small if the

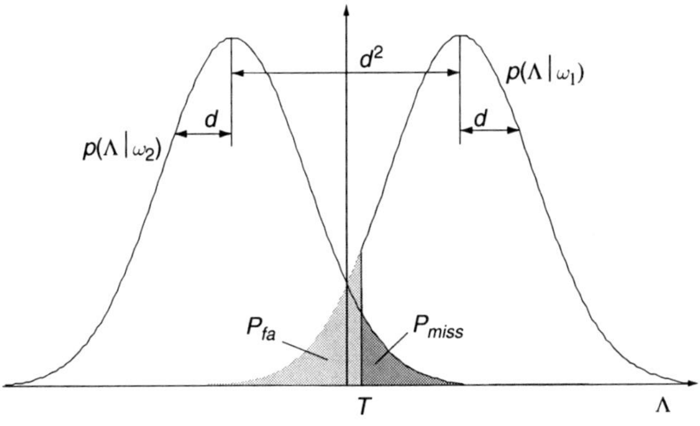

Figure 2.12 The conditional probability densities of the log-likelihood ratio in the Gaussian case with $C_1 = C_2 = C$

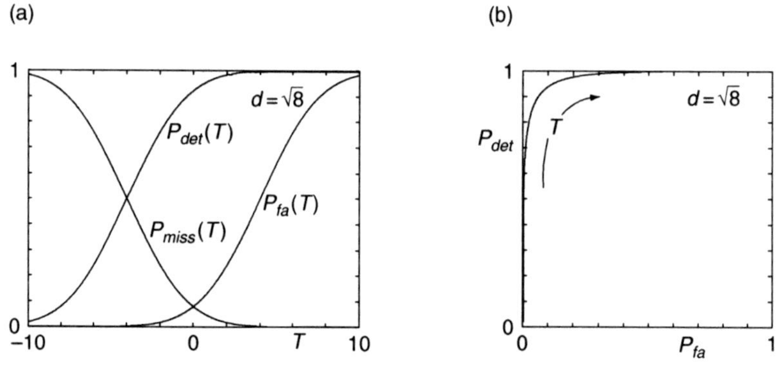

Figure 2.13 Performance of a detector in the Gaussian case with equal covariance matrices. (a) P_{miss}, P_{det} and P_{fa} versus the threshold T. (b) P_{det} versus P_{fa} as a parametric plot of T

threshold is chosen small. However, the probability of a missed event (type II error) is small if the threshold is chosen large. A trade-off must be found between both types of errors.

The trade-off can be made explicitly visible by means of a parametric plot of P_{det} versus P_{fa} with varying T. Such a curve is called a receiver operating characteristic curve (ROC curve). Ideally, $P_{fa} = 0$ and $P_{det} = 1$, but the figure shows that no threshold exists for which this occurs. Figure 2.13(b) shows the ROC curve for a Gaussian case with equal covariance matrices. Here, the ROC curve can be obtained analytically, but in most other cases the ROC curve of a given detector must be obtained either empirically or by numerical integration of (2.42).

In Listing 2.6 the MATLAB implementation for the computation of the ROC curve is shown. To avoid confusion about the roles of the different classes (which class should be considered positive and which negative) in PRTools the ROC curve shows the fraction false positive and false negative. This means that the resulting curve is a vertically mirrored version of Figure 2.13(b). Note also that in the listing the training set is used to both train a classifier and to generate the curve. To have a reliable estimate, an independent data set should be used for the estimation of the ROC curve.

Listing 2.6
PRTools code for estimation of a ROC curve

```
z=gendats(100,1,2);            % Generate a 1D dataset
w=qdc(z);                      % Train a classifier
```

```
r=roc(z*w);                          % Compute the ROC curve
plotr(r);                            % Plot it
```

The merit of a ROC curve is that it specifies the intrinsic ability of the detector to discriminate between the two classes. In other words, the ROC curve of a detector depends neither on the cost function of the application nor on the prior probabilities.

Since a false alarm and a missed event are mutually exclusive, the error rate of the classification is the sum of both probabilities:

$$\begin{aligned} E &= P(\hat{\omega}_2, \omega_1) + P(\hat{\omega}_1, \omega_2) \\ &= P(\hat{\omega}_2|\omega_1)P(\omega_1) + P(\hat{\omega}_1|\omega_2)P(\omega_2) \\ &= P_{fa}P(\omega_1) + P_{miss}P(\omega_2) \end{aligned} \tag{2.48}$$

In the example of Figure 2.13, the discriminability d equals $\sqrt{8}$. If this indicator becomes larger, P_{miss} and P_{fa} become smaller. Hence, the error rate E is a monotonically decreasing function of d.

Example 2.7 Quality inspection of empty bottles

In the bottling industry, the detection of defects of bottles (to be recycled) is relevant in order to assure the quality of the product. A variety of flaws can occur: cracks, dirty bottoms, fragments of glass, labels, etc. In this example, the problem of detecting defects of the mouth of an empty bottle is addressed. This is important, especially in the case of bottles with crown caps. Small damages of the mouth may cause a non-airtight enclosure of the product which subsequently causes an untimely decay.

The detection of defects at the mouth is a difficult task. Some irregularities at the mouth seem to be perturbing, but in fact are harmless. Other irregularities (e.g. small intrusions at the surface of the mouth) are quite harmful. The inspection system (Figure 2.14) that performs the task consists of a stroboscopic, specular 'light field' illuminator, a digital camera, a detector, an actuator and a sorting mechanism. The illumination is such that in the absence of irregularities at the mouth, the bottle is seen as a bright ring (with fixed size and position) on a dark background. Irregularities at the mouth give rise to disturbances of the ring. See Figure 2.15.

The decision of the inspection system is based on a measurement vector that is extracted from the acquired image. For this purpose the area of the ring is divided into 256 equally sized sectors. Within each

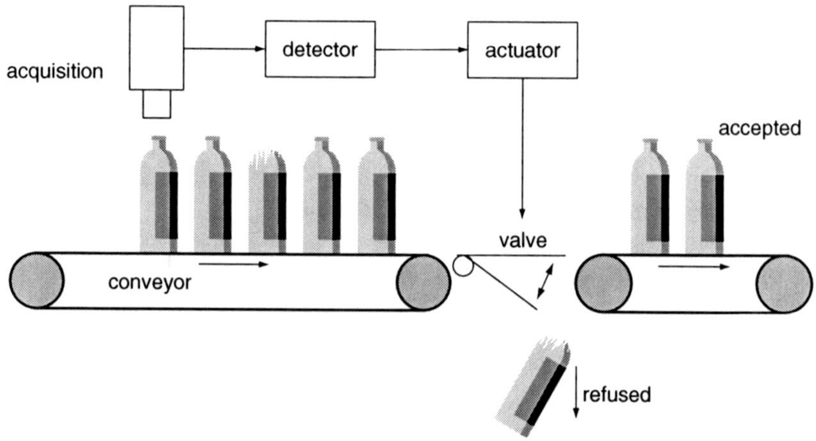

Figure 2.14 Quality inspection system for the recycling of bottles

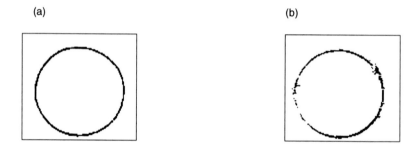

Figure 2.15 Acquired images of two different bottles. (a) Image of the mouth of a new bottle. (b) Image of the mouth of an older bottle with clearly visible intrusions

sector the average of the grey levels of the observed ring is estimated. These averages (as a function of running arc length along the ring) are made rotational invariant by a translation invariant transformation, e.g. the amplitude spectrum of the discrete Fourier transform.

The transformed averages form the measurement vector **z**. The next step is the construction of a log-likelihood ratio $\Lambda(\mathbf{z})$ according to (2.40). Comparing the likelihood ratio against a suitable threshold value gives the final decision.

Such a detector should be trained with a set of bottles that are manually inspected so as to determine the parameters μ_1, μ_2 etc. (see Chapter 5). Another set of manually inspected bottles is used for evaluation. The result for a particular application is shown in Figure 2.16.

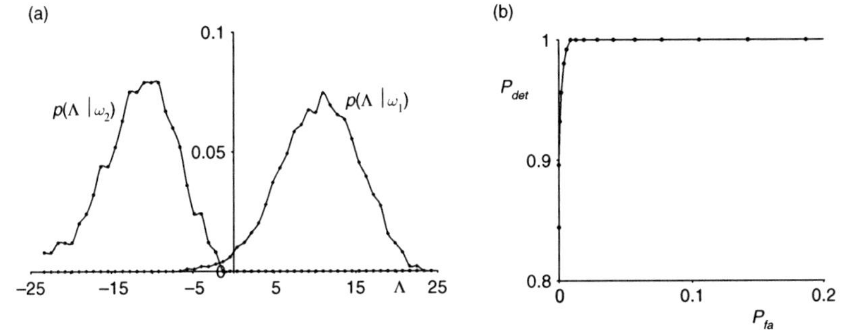

Figure 2.16 Estimated performance of the bottle inspector. (a) The conditional probability densities of the log-likelihood ratio. (b) The ROC curve

It seems that the Gaussian assumption with equal covariance matrices is appropriate here. The discriminability appears to be $d = 4.8$.

2.4 SELECTED BIBLIOGRAPHY

Many good textbooks on pattern classification have been written. These books go into more detail than is possible here and approach the subject from different points of view. The following list is a selection.

Duda, R.O., Hart, P.E. and Stork, D.G., *Pattern Classification*, Wiley, London, UK, 2001.

Fukanaga, K., *Statistical Pattern Recognition*, Academic Press, New York, NY, 1990.

Ripley, B.D., *Pattern Recognition and Neural Networks*, Cambridge University Press, Cambridge, UK, 1996.

Webb, A.R., *Statistical Pattern Recognition*, 2nd edition, Wiley, London, UK, 2002.

2.5 EXERCISES

1. Give at least two more examples of classification systems. Also define possible measurements and the relevant classes. (0)
2. Give a classification problem where the class definitions are subjective. (0)
3. Assume we have three classes of tomato with decreasing quality, class 'A', class 'B' and class 'C'. Assume further that the cost of misclassifying a tomato to a higher quality is twice as expensive as vice versa. Give the cost matrix. What extra information do you need in order to fully determine the matrix? (0)
4. Assume that the number of scrap objects in Figure 2.2 is actually twice as large. How should the cost matrix, given in Table 2.2, be changed, such that the decision function remains the same? (0)

5. What quantities do you need to compute the Bayes classifier? How would you obtain these quantities? (0)
6. Derive a decision function assuming that objects come from normally distributed classes as in Section 2.1.2, but now with an arbitrary cost function (∗).
7. Can you think of a physical measurement system in which it can be expected that the class distributions are Gaussian and where the covariance matrices are independent of the class? (0)
8. Construct the ROC curve for the case that the classes have no overlap, and classes which are completely overlapping. (0)
9. Derive how the ROC curve changes when the class prior probabilities are changed. (0)
10. Reconstruct the class conditional probabilities for the case that the ROC curve is not symmetric around the axis which runs from (1,0) to (0,1). (0)

3

Parameter Estimation

Parameter estimation is the process of attributing a parametric description to an object, a physical process or an event based on measurements that are obtained from that object (or process, or event). The measurements are made available by a sensory system. Figure 3.1 gives an overview. Parameter estimation and pattern classification are similar processes because they both aim to describe an object using measurements. However, in parameter estimation the description is in terms of a *real-valued* scalar or vector, whereas in classification the description is in terms of just one class selected from a finite number of classes.

Example 3.1 Estimation of the backscattering coefficient from SAR images

In earth observation based on airborne *SAR* (synthetic aperture radar) imaging, the physical parameter of interest is the backscattering coefficient. This parameter provides information about the condition of the surface of the earth, e.g. soil type, moisture content, crop type, growth of the crop.

The mean backscattered energy of a radar signal in a direction is proportional to this backscattering coefficient. In order to reduce so-called speckle noise the given direction is probed a number of times. The results are averaged to yield the final measurement. Figure 3.2 shows a large number of realizations of the true backscattering

Classification, Parameter Estimation and State Estimation: An Engineering Approach using MATLAB
F. van der Heijden, R.P.W. Duin, D. de Ridder and D.M.J. Tax
 ISBN: 0-470-09013-8

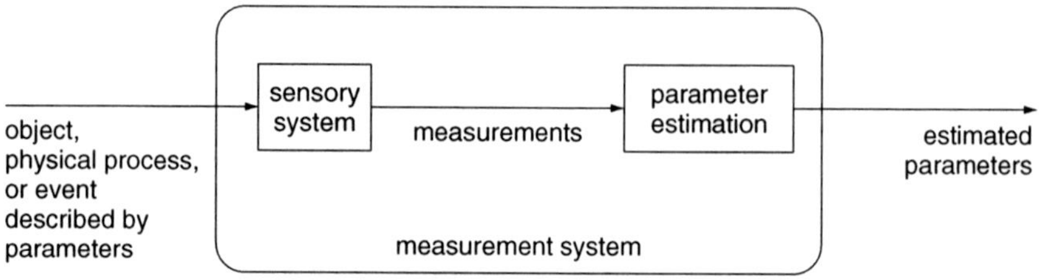

Figure 3.1 Parameter estimation

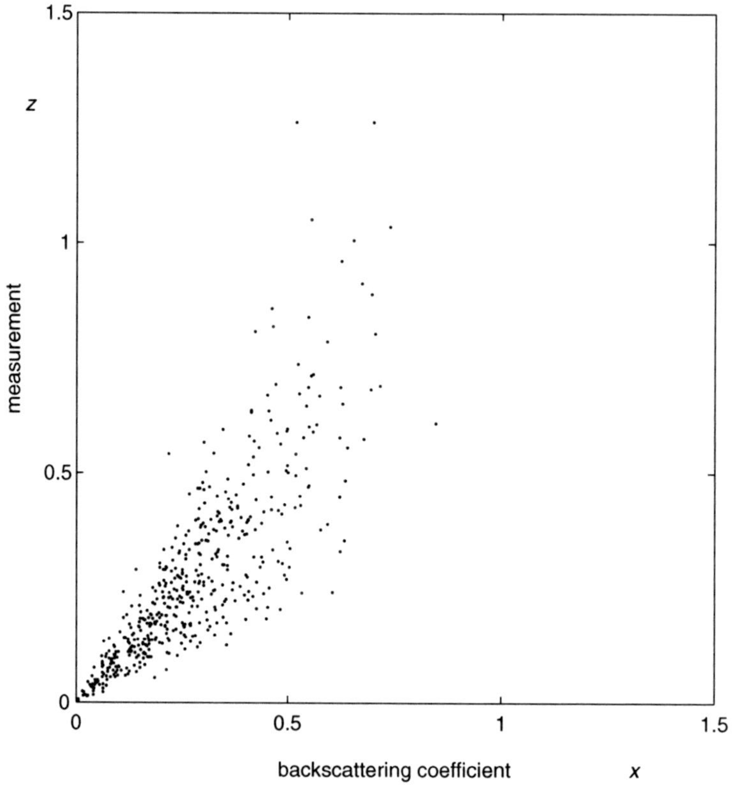

Figure 3.2 Different realizations of the backscattering coefficient and its corresponding measurement

coefficient and its corresponding measurement.[1] In this example, the number of probes per measurement is eight. It can be seen that, even after averaging, the measurement is still inaccurate. Moreover, although the true backscattering coefficient is always between 0 and 1, the measurements can easily violate this constraint (some measurements are greater than 1).

The task of a parameter estimator here is to map each measurement to an estimate of the corresponding backscattering coefficient.

This chapter addresses the problem of how to design a parameter estimator. For that, two approaches exist: Bayesian estimation (Section 3.1) and data-fitting techniques (Section 3.3). The Bayesian-theoretic framework for parameter estimation follows the same line of reasoning as the one for classification (as discussed in Chapter 2). It is a probabilistic approach. The second approach, data fitting, does not have such a probabilistic context. The various criteria for the evaluation of an estimator are discussed in Section 3.2.

3.1 BAYESIAN ESTIMATION

Figure 3.3 gives a framework in which parameter estimation can be defined. The starting point is a probabilistic experiment where the outcome is a random vector $\mathbf{x}$ defined in $\mathbb{R}^M$, and with probability density

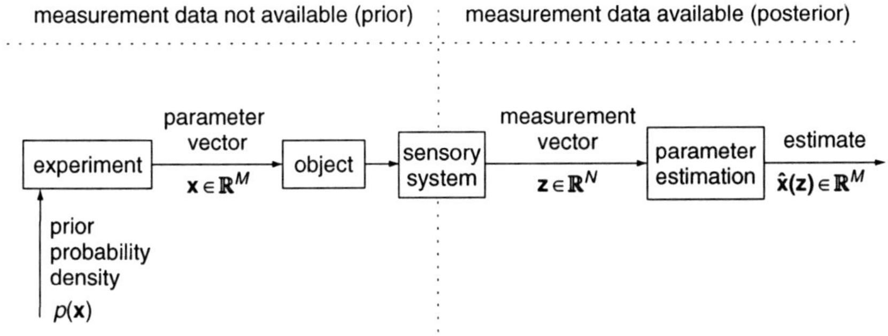

Figure 3.3 Parameter estimation

[1] The data shown in Figure 3.2 is the result of a simulation. Therefore, in this case, the true backscattering coefficients are known. Of course, in practice, the true parameter of interest is always unknown. Only the measurements are available.

$p(\mathbf{x})$. Associated with $\mathbf{x}$ is a physical object, process or event (in short: 'physical object'), of which $\mathbf{x}$ is a property. $\mathbf{x}$ is called a *parameter vector*, and its density $p(\mathbf{x})$ is called the *prior probability density*.

The object is sensed by a sensory system which produces an N-dimensional measurement vector $\mathbf{z}$. The task of the parameter estimator is to recover the original parameter vector $\mathbf{x}$ given the measurement vector $\mathbf{z}$. This is done by means of the *estimation function* $\hat{\mathbf{x}}(\mathbf{z})$: $\mathbb{R}^N \to \mathbb{R}^M$. The *conditional probability density* $p(\mathbf{z}|\mathbf{x})$ gives the connection between the parameter vector and measurements. With fixed $\mathbf{x}$, the randomness of the measurement vector $\mathbf{z}$ is due to physical noise sources in the sensor system and other unpredictable phenomena. The randomness is characterized by $p(\mathbf{z}|\mathbf{x})$. The *overall probability density* of $\mathbf{z}$ is found by averaging the conditional density over the complete parameter space:

$$p(\mathbf{z}) = \int_{\mathbf{x}} p(\mathbf{z}|\mathbf{x})p(\mathbf{x})d\mathbf{x} \tag{3.1}$$

The integral extends over the entire M-dimensional space $\mathbb{R}^M$.

Finally, Bayes' theorem for conditional probabilities gives us the *posterior probability density* $p(\mathbf{x}|\mathbf{z})$:

$$p(\mathbf{x}|\mathbf{z}) = \frac{p(\mathbf{z}|\mathbf{x})p(\mathbf{x})}{p(\mathbf{z})} \tag{3.2}$$

This density is most useful since $\mathbf{z}$, being the output of the sensory system, is at our disposal and thus fully known. Thus, $p(\mathbf{x}|\mathbf{z})$ represents exactly the knowledge that we have on $\mathbf{x}$ after having observed $\mathbf{z}$.

Example 3.2 Estimation of the backscattering coefficient

The backscattering coefficient x from Example 3.1 is within the interval [0,1]. In most applications, however, lower values of the coefficient occur more frequently than higher ones. Such a preference can be taken into account by means of the prior probability density $p(x)$. We will assume that for a certain application x has a beta distribution:

$$p(x) = \frac{(a+b+1)!}{a!b!}x^a(1-x)^b \quad \text{for} \quad 0 \le x \le 1 \tag{3.3}$$

The parameters a and b are the shape parameters of the distribution. In Figure 3.4(a) these parameters are set to $a = 1$ and $b = 4$. These

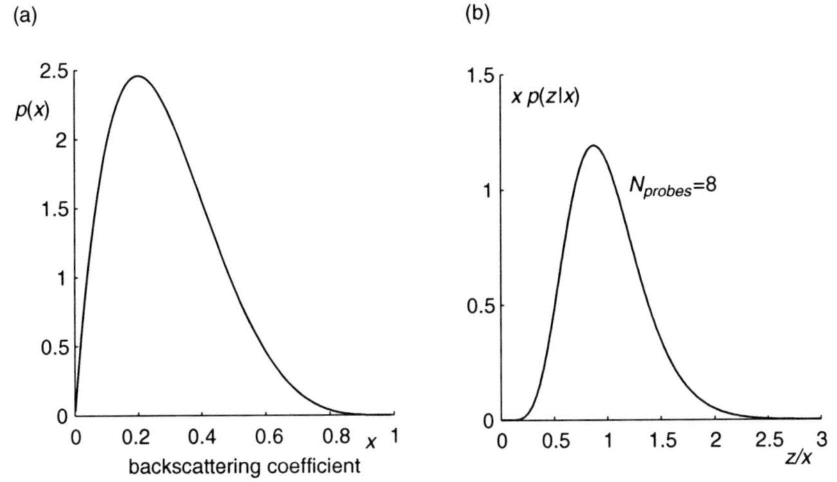

Figure 3.4 Probability densities for the backscattering coefficient. (a) Prior density $p(x)$. (b) Conditional density $p(z|x)$ with $N_{probes} = 8$. The two axes have been scaled with x and $1/x$, respectively, to obtain invariance with respect to x

values will be used throughout the examples in this chapter. Note that there is no physical evidence for the beta distribution of x. The assumption is a subjective result of our state of knowledge concerning the occurrence of x. If no such knowledge is available, a uniform distribution between 0 and 1 (i.e. all x are equally likely) would be more reasonable.

The measurement is denoted by z. The mathematical model for SAR measurements is that, with fixed x, the variable $N_{probes}z/x$ has a gamma distribution with parameter N_{probes} (the number of probes per measurement). The probability density associated with a gamma distribution is:

$$gamma\_pdf(u, \alpha) = \frac{U(u)}{\Gamma(\alpha)} u^{\alpha-1} \exp(-u) \tag{3.4}$$

where u is the independent variable, $\Gamma(\alpha)$ is the gamma function, a is the parameter of the distribution and $U(u)$ is the unit step function which returns 0 if u is negative and 1 otherwise. Since z can be regarded as a gamma-distributed random variable scaled by a factor x/N_{probes}, the conditional density of z becomes:

$$p(z|x) = U(z)\frac{N_{probes}}{x} gamma\_pdf\left(\frac{N_{probes}z}{x}, N_{probes}\right) \tag{3.5}$$

Figure 3.4(b) shows the conditional density.

Cost functions

The optimization criterion of Bayes, minimum risk, applies to statistical parameter estimation provided that two conditions are met. First, it must be possible to quantify the cost involved when the estimates differ from the true parameters. Second, the expectation of the cost, the risk, should be acceptable as an optimization criterion.

Suppose that the damage is quantified by a cost function $C(\hat{\mathbf{x}}|\mathbf{x})$: $\mathbb{R}^M \times \mathbb{R}^M \rightarrow \mathbb{R}$. Ideally, this function represents the true cost. In most applications, however, it is difficult to quantify the cost accurately. Therefore, it is common practice to choose a cost function whose mathematical treatment is not too complex. Often, the assumption is that the cost function only depends on the difference between estimated and true parameters: the estimation error $\mathbf{e} = \hat{\mathbf{x}} - \mathbf{x}$. With this assumption, the following cost functions are well known (see Table 3.1):

- *quadratic cost function*:

$$C(\hat{\mathbf{x}}|\mathbf{x}) = \|\hat{\mathbf{x}} - \mathbf{x}\|_2^2 = \sum_{m=0}^{M-1} (\hat{x}_m - x_m)^2 \tag{3.6}$$

- *absolute value cost function*:

$$C(\hat{\mathbf{x}}|\mathbf{x}) = \|\hat{\mathbf{x}} - \mathbf{x}\|_1 = \sum_{m=0}^{M-1} |\hat{x}_m - x_m| \tag{3.7}$$

Table 3.1 Three different Bayes estimators worked out for the scalar case

MMSE estimation Quadratic cost function	MMAE estimation Absolute cost function	MAP estimation Uniform cost function
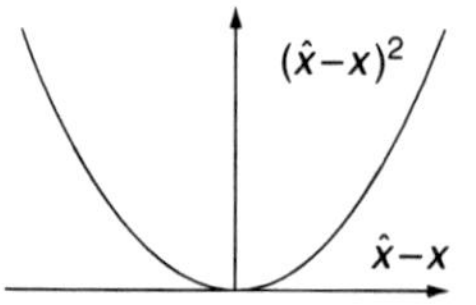	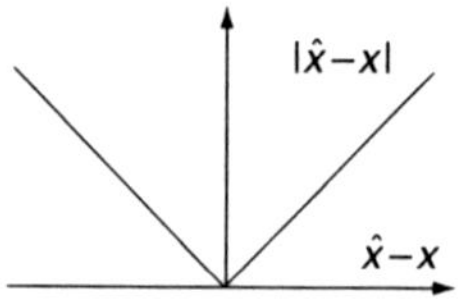	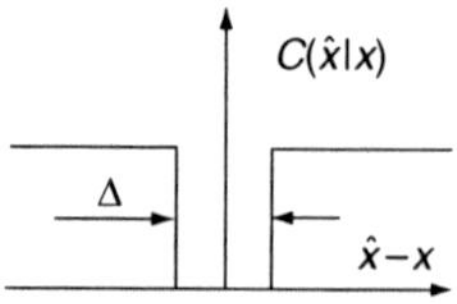
$\hat{x}_{MMSE}(\mathbf{z}) = E[x\|\mathbf{z}]$ $= \int_x xp(x\|\mathbf{z})dx$	$\hat{x}_{MMAE}(\mathbf{z}) = \hat{x}$ with $\int_{-\infty}^{\hat{x}} p(x\|\mathbf{z})dx = \frac{1}{2}$	$\hat{x}_{MAP}(\mathbf{z}) = \underset{x}{\text{argmax}}\{p(x\|\mathbf{z})\}$

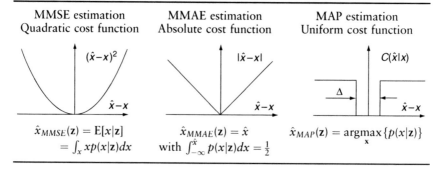

- *uniform cost function*:

$$C(\hat{\mathbf{x}}|\mathbf{x}) = \begin{cases} 1 & \text{if } \|\hat{\mathbf{x}} - \mathbf{x}\|_1 > \Delta \\ 0 & \text{if } \|\hat{\mathbf{x}} - \mathbf{x}\|_1 \leq \Delta \end{cases} \quad \text{with: } \Delta \to 0 \tag{3.8}$$

The first two cost functions are instances of the Minkowski distance measures (see Appendix A.2). The third cost function is an approximation of the distance measure mentioned in (a.22).

Risk minimization

With an arbitrarily selected estimate $\hat{\mathbf{x}}$ and a given measurement vector $\mathbf{z}$, the *conditional risk* of $\hat{\mathbf{x}}$ is defined as the expectation of the cost function:

$$R(\hat{\mathbf{x}}|\mathbf{z}) = \mathrm{E}[C(\hat{\mathbf{x}}|\mathbf{x})|\mathbf{z}] = \int_{\mathbf{x}} C(\hat{\mathbf{x}}|\mathbf{x})p(\mathbf{x}|\mathbf{z})d\mathbf{x} \tag{3.9}$$

In *Bayes estimation* (or *minimum risk estimation*) the estimate is the parameter vector that minimizes the risk:

$$\hat{\mathbf{x}}(\mathbf{z}) = \underset{\mathbf{x}}{\mathrm{argmin}}\{R(\mathbf{x}|\mathbf{z})\} \tag{3.10}$$

The minimization extends over the entire parameter space.

The *overall risk* (also called *average risk*) of an estimator $\hat{\mathbf{x}}(\mathbf{z})$ is the expected cost seen over the full set of possible measurements:

$$R = \mathrm{E}[R(\hat{\mathbf{x}}(\mathbf{z})|\mathbf{z})] = \int_{\mathbf{z}} R(\hat{\mathbf{x}}(\mathbf{z})|\mathbf{z})p(\mathbf{z})d\mathbf{z} \tag{3.11}$$

Minimization of the integral is accomplished by minimization of the integrand. However, since $p(\mathbf{z})$ is positive, it suffices to minimize $R(\hat{\mathbf{x}}(\mathbf{z})|\mathbf{z})$. Therefore, the Bayes estimator not only minimizes the conditional risk, but also the overall risk.

The Bayes solution is obtained by substituting the selected cost function in (3.9) and (3.10). Differentiating and equating it to zero yields for the three cost functions given in (3.6), (3.7) and (3.8):

- *MMSE* estimation (MMSE = minimum mean square error).
- *MMAE* estimation (MMAE = minimum mean absolute error).
- *MAP* estimation (MAP = maximum a posterior).

Table 3.1 gives the solutions that are obtained if x is a scalar. The MMSE and the MAP estimators will be worked out for the vectorial case in the next sections. But first, the scalar case will be illustrated by means of an example.

Example 3.3 Estimation of the backscattering coefficient
The estimators for the backscattering coefficient (see previous example) take the form as depicted in Figure 3.5. These estimators are found by substitution of (3.3) and (3.5) in the expressions in Table 3.1.

In this example, the three estimators do not differ much. Nevertheless their own typical behaviours manifest themselves clearly if we evaluate their results empirically. This can be done by means of the population of the $N_{pop} = 500$ realizations that are shown in the figure. For each sample z_i we calculate the average cost $1/N_{pop} \sum_{i=1}^{N_{pop}} C(\hat{x}(z_i)|x_i)$,

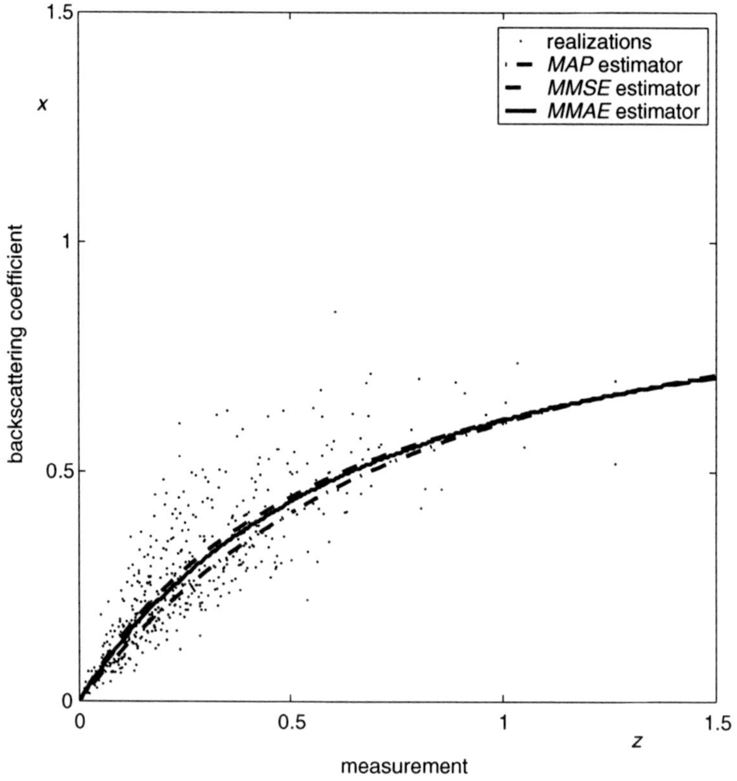

Figure 3.5 Three different Bayesian estimators

Table 3.2 Empirical evaluation of the three different Bayes estimators in Figure 3.5

Evaluation criterion	Type of estimation		
	MMSE estimation	*MMAE estimation*	*MAP estimation*
Quadratic cost function	**0.0067**	0.0069	0.0081
Absolute cost function	0.062	**0.060**	0.063
Uniform cost function (evaluated with $\Delta = 0.05$)	0.26	0.19	**0.10**

where x_i is the true value of the i-th sample and $\hat{x}(.)$ is the estimator under test. Table 3.2 gives the results of that evaluation for the three different estimators and the three different cost functions.

Not surprisingly, in Table 3.2 the MMSE, the MMAE and the MAP estimators are optimal with respect to their own criterion, i.e. the quadratic, the absolute value and the uniform cost criterion, respectively. It appears that the MMSE estimator is preferable if the cost of a large error is considerably higher than the one of a small error. The MAP estimator does not discriminate between small or large errors. The MMAE estimator takes its position in between.

MATLAB code for generating Figure 3.5 is given in Listing 3.1. It uses the Statistics toolbox for calculating the various probability density functions. Although the MAP solution can be found analytically, here we approximate all three solutions numerically. To avoid confusion, it is easy to create functions that calculate the various probabilities needed. Note how $p(z) = \int p(z|x)p(x)dx$ is approximated by a sum over a range of values of x, whereas $p(x|z)$ is found by Bayes' rule.

Listing 3.1
MATLAB code for MMSE, MMSA and MAP estimation in the scalar case.

```
function estimates
  global N Np a b xrange;

  N=500;                    % Number of samples
  Np=8;                     % Number of looks
  a=2; b=5;                 % Beta distribution parameters
  x=0.005:0.005:1;          % Interesting range of x
  z=0.005:0.005:1.5;        % Interesting range of z
  load scatter;             % Load set (for plotting only)
  xrange=x;
```

```
  for i=1:length(z)
    [dummy,ind]=max(px_z(x,z(i))); x_map(i)=x(ind);
    x_mse(i)=sum(pz_x(z(i),x).*px(x).*x)
      ./.sum(pz_x(z(i),x).*px(x));
    ind=find((cumsum(px_z(x,z(i))) ./ .sum(px_z(x,z(i))))>0.5);
    x_mae(i)=x(ind(1));
  end

  figure; clf; plot(zset,xset,'.'); hold on;
  plot(z,x_map,'k-.'); plot(z,x_mse,'k--');
  plot(z,x_mae,'k-');
  legend('realizations','MAP','MSE','MAE');
return

function ret=px(x)
  global a b; ret=betapdf(x,a,b);
return

function ret=pz_x(z,x)
  global Np; ret=(z>0).*(Np./x).*gampdf(Np*z./x,Np,1);
return

function ret=pz(z)
  global xrange; ret=sum(px(xrange).*pz_x(z,xrange));
return

function ret=px_z(x,z)
  ret=pz_x(z,x).*px(x)./pz(z);
return
```

3.1.1 MMSE estimation

The solution based on the quadratic cost function (3.6) is called the *minimum mean square error estimator*, also called the *minimum variance estimator* for reasons that will become clear in a moment. Substitution of (3.6) and (3.9) in (3.10) gives:

$$\hat{\mathbf{x}}_{MMSE}(\mathbf{z}) = \underset{\hat{\mathbf{x}}}{\operatorname{argmin}}\left\{\int_{\mathbf{x}} (\hat{\mathbf{x}} - \mathbf{x})^T (\hat{\mathbf{x}} - \mathbf{x}) p(\mathbf{x}|\mathbf{z}) d\mathbf{x}\right\} \tag{3.12}$$

Differentiating the function between braces with respect to $\hat{\mathbf{x}}$ (see Appendix B.4), and equating this result to zero yields a system of M linear equations, the solution of which is:

$$\hat{\mathbf{x}}_{MMSE}(\mathbf{z}) = \int_{\mathbf{x}} \mathbf{x} p(\mathbf{x}|\mathbf{z}) d\mathbf{x} = \mathrm{E}[\mathbf{x}|\mathbf{z}] \tag{3.13}$$

The conditional risk of this solution is the sum of the variances of the estimated parameters:

$$\begin{aligned} R(\hat{\mathbf{x}}_{MMSE}(\mathbf{z})|\mathbf{z}) &= \int_{\mathbf{x}} (\hat{\mathbf{x}}_{MMSE}(\mathbf{z}) - \mathbf{x})^T(\hat{\mathbf{x}}_{MMSE}(\mathbf{z}) - \mathbf{x})p(\mathbf{x}|\mathbf{z})d\mathbf{x} \\ &= \int_{\mathbf{x}} (\mathrm{E}[x|z] - \mathbf{x})^T(\mathrm{E}[x|z] - \mathbf{x})p(\mathbf{x}|\mathbf{z})d\mathbf{x} \\ &= \sum_{m=0}^{M-1} \mathrm{Var}[x_m|\mathbf{z}] \end{aligned} \tag{3.14}$$

Hence the name 'minimum variance estimator'.

3.1.2 MAP estimation

If the uniform cost function is chosen, the conditional risk (3.9) becomes:

$$\begin{aligned} R(\hat{\mathbf{x}}|\mathbf{z}) &= \int_{\mathbf{x}} p(\mathbf{x}|\mathbf{z})d\mathbf{x} - p(\hat{\mathbf{x}}|\mathbf{z})\Delta \\ &= 1 - p(\hat{\mathbf{x}}|\mathbf{z})\Delta \end{aligned} \tag{3.15}$$

The estimate which now minimizes the risk is called the maximum a posterior (MAP) estimate:

$$\hat{\mathbf{x}}_{MAP}(\mathbf{z}) = \underset{\mathbf{x}}{\mathrm{argmax}}\{p(\mathbf{x}|\mathbf{z})\} \tag{3.16}$$

This solution equals the mode (maximum) of the posterior probability. It can also be written entirely in terms of the prior probability densities and the conditional probabilities:

$$\hat{\mathbf{x}}_{MAP}(\mathbf{z}) = \underset{\mathbf{x}}{\mathrm{argmax}}\left\{\frac{p(\mathbf{z}|\mathbf{x})p(\mathbf{x})}{p(\mathbf{z})}\right\} = \underset{\mathbf{x}}{\mathrm{argmax}}\{p(\mathbf{z}|\mathbf{x})p(\mathbf{x})\} \tag{3.17}$$

This expression is similar to the one of MAP classification; see (2.12).

3.1.3 The Gaussian case with linear sensors

Suppose that the parameter vector $\mathbf{x}$ has a normal distribution with expectation vector $E[\mathbf{x}] = \boldsymbol{\mu}_x$ and covariance matrix $\mathbf{C}_x$. In addition, suppose that the measurement vector can be expressed as a linear combination of the parameter vector corrupted by additive Gaussian noise:

$$\mathbf{z} = \mathbf{H}\mathbf{x} + \mathbf{v} \tag{3.18}$$

where $\mathbf{v}$ is an N-dimensional random vector with zero expectation and covariance matrix $\mathbf{C}_v$. $\boldsymbol{x}$ and $\boldsymbol{v}$ are uncorrelated. $\mathbf{H}$ is an $N \times M$ matrix.

The assumption that both $\mathbf{x}$ and $\mathbf{v}$ are normal implies that the conditional probability density of $\mathbf{z}$ is also normal. The conditional expectation vector of $\mathbf{z}$ equals:

$$E[\mathbf{z}|\mathbf{x}] = \boldsymbol{\mu}_{z|x} = \mathbf{H}\boldsymbol{\mu}_x \tag{3.19}$$

The conditional covariance matrix of $\mathbf{z}$ is $\mathbf{C}_{z|x} = \mathbf{C}_v$.

Under these assumptions the posterior distribution of $\mathbf{x}$ is normal as well. Application of (3.2) yields the following expressions for the MMSE estimate and the corresponding covariance matrix:

$$\begin{aligned} \hat{\mathbf{x}}_{MMSE}(\mathbf{z}) &= \boldsymbol{\mu}_{x|z} = E[\mathbf{x}|\mathbf{z}] = \left(\mathbf{H}^T\mathbf{C}_v^{-1}\mathbf{H} + \mathbf{C}_x^{-1}\right)^{-1}\left(\mathbf{H}^T\mathbf{C}_v^{-1}\mathbf{z} + \mathbf{C}_x^{-1}\boldsymbol{\mu}_x\right) \\ \mathbf{C}_{\underline{x}|z} &= \left(\mathbf{H}^T\mathbf{C}_v^{-1}\mathbf{H} + \mathbf{C}_x^{-1}\right)^{-1} \end{aligned} \tag{3.20}$$

The proof is left as an exercise for the reader. See exercise 3. Note that $\boldsymbol{\mu}_{x|z}$, being the posterior expectation, is the MMSE estimate $\hat{\mathbf{x}}_{MMSE}(\mathbf{z})$.

The posterior expectation, $E[\mathbf{x}|\mathbf{z}]$, consists of two terms. The first term is linear in $\mathbf{z}$. It represents the information coming from the measurement. The second term is linear in $\boldsymbol{\mu}_x$, representing the prior knowledge. To show that this interpretation is correct it is instructive to see what happens at the extreme ends: either no information from the measurement, or no prior knowledge:

- The measurements are useless if the matrix $\mathbf{H}$ is virtually zero, or if the noise is too large, i.e. $\mathbf{C}_v^{-1}$ is too small. In both cases, the second term in (3.20) dominates. In the limit, the estimate becomes $\hat{\mathbf{x}}_{MMSE}(\mathbf{z}) = \boldsymbol{\mu}_x$ with covariance matrix $\mathbf{C}_x$, i.e. the estimate is purely based on prior knowledge.

- On the other hand, if the prior knowledge is weak, i.e. if the variances of the parameters are very large, the inverse covariance matrix $C_{\mathbf{x}}^{-1}$ tends to zero. In the limit, the estimate becomes:

$$\hat{\mathbf{x}}_{MMSE}(\mathbf{z}) = \left(\mathbf{H}^T\mathbf{C}_{\mathbf{v}}^{-1}\mathbf{H}\right)^{-1}\mathbf{H}^T\mathbf{C}_{\mathbf{v}}^{-1}\mathbf{z} \tag{3.21}$$

In this solution, the prior knowledge, i.e. $\boldsymbol{\mu}_{\mathbf{x}}$, is completely ruled out.

Note that the mode of a normal distribution coincides with the expectation. Therefore, in the linear-Gaussian case, MAP estimation and MMSE estimation coincide: $\hat{\mathbf{x}}_{MMSE}(\mathbf{z}) = \hat{\mathbf{x}}_{MAP}(\mathbf{z})$.

3.1.4 Maximum likelihood estimation

In many practical situations the prior knowledge needed in MAP estimation is not available. In these cases, an estimator which does not depend on prior knowledge is desirable. One attempt in that direction is the method referred to as *maximum likelihood estimation* (ML estimation). The method is based on the observation that in MAP estimation, (3.17), the peak of the first factor $p(\mathbf{z}|\mathbf{x})$ is often in an area of $\mathbf{x}$ in which the second factor $p(\mathbf{x})$ is almost constant. This holds true especially if little prior knowledge is available. In these cases, the prior density $p(\mathbf{x})$ does not affect the position of the maximum very much. Discarding the factor, and maximizing the function $p(\mathbf{z}|\mathbf{x})$ solely, gives the ML estimate:

$$\hat{\mathbf{x}}_{ML}(\mathbf{z}) = \underset{\mathbf{x}}{\operatorname{argmax}}\{p(\mathbf{z}|\mathbf{x})\} \tag{3.22}$$

Regarded as a function of $\mathbf{x}$ the conditional probability density is called the *likelihood function*. Hence the name 'maximum likelihood estimation'.

Another motivation for the ML estimator is when we change our viewpoint with respect to the nature of the parameter vector $\mathbf{x}$. In the Bayesian approach $\mathbf{x}$ is a random vector, statistically defined by means of probability densities. In contrast, we may also regard $\mathbf{x}$ as a non-random vector whose value is simply unknown. This is the so-called *Fisher approach*. In this view, there are no probability densities associated with $\mathbf{x}$. The only density in which $\mathbf{x}$ appears is $p(\mathbf{z}|\mathbf{x})$, but here $\mathbf{x}$ must be regarded as a parameter of the density of $\mathbf{z}$. From all estimators discussed so far, the only estimator that can handle this deterministic point of view on $\mathbf{x}$ is the ML estimator.

Example 3.4 Maximum likelihood estimation of the backscattering coefficient
The maximum likelihood estimator for the backscattering coefficient (see previous examples) is found by maximizing (3.5):

$$\frac{dp(z|x)}{dx} = 0 \quad \Rightarrow \quad \hat{x}_{ML}(z) = z \tag{3.23}$$

The estimator is depicted in Figure 3.6 together with the MAP estimator. The figure confirms the statement above that in areas of flat prior probability density the MAP estimator and the ML estimator coincide. However, the figure also reveals that the ML estimator can produce an estimate of the backscattering coefficient that is larger than one; a physical impossibility. This is the price that we have to pay for not using prior information about the physical process.

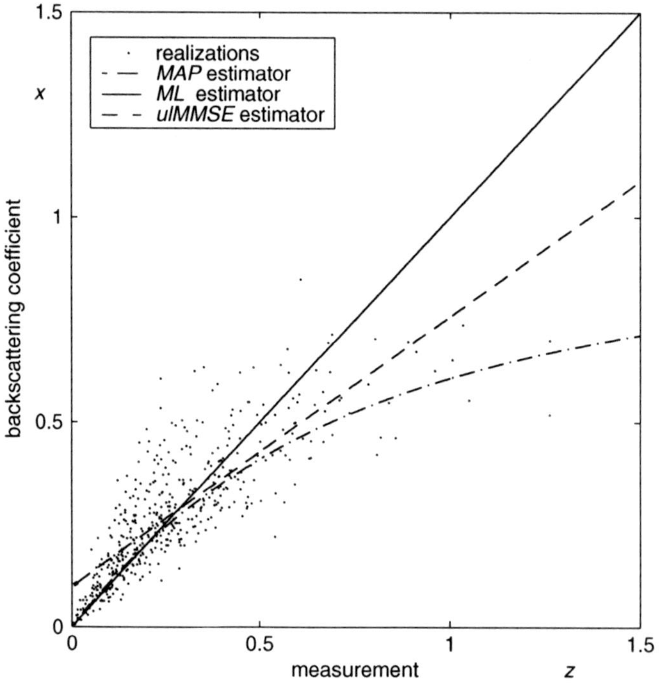

Figure 3.6 MAP estimation, ML estimation and linear MMSE estimation

If the measurement vector $\mathbf{z}$ is linear in $\mathbf{x}$ and corrupted by additive Gaussian noise, as given in equation (3.18), the likelihood of $\mathbf{x}$ is given in (3.21). Thus, in that case:

$$\hat{\mathbf{x}}_{ML}(\mathbf{z}) = \left(\mathbf{H}^T\mathbf{C}_\mathbf{v}^{-1}\mathbf{H}\right)^{-1}\mathbf{H}^T\mathbf{C}_\mathbf{v}^{-1}\mathbf{z} \tag{3.24}$$

A further simplification is obtained if we assume that the noise is white, i.e. $\mathbf{C}_\mathbf{v} \sim \mathbf{I}$:

$$\hat{\mathbf{x}}_{ML}(\mathbf{z}) = (\mathbf{H}^T\mathbf{H})^{-1}\mathbf{H}^T\mathbf{z} \tag{3.25}$$

The operation $(\mathbf{H}^T\mathbf{H})^{-1}\mathbf{H}^T$ is the *pseudo inverse* of $\mathbf{H}$. Of course, its validity depends on the existence of the inverse of $\mathbf{H}^T\mathbf{H}$. Usually, such is the case if the number of measurements exceeds the number of parameters, i.e. $N > M$. That is, if the system is overdetermined.

3.1.5 Unbiased linear MMSE estimation

The estimators discussed in the previous sections exploit full statistical knowledge of the problem. Designing such an estimator is often difficult. The first problem that arises is the adequate modelling of the probability densities. Such modelling requires detailed knowledge of the physical process and the sensors. Once we have the probability densities, the second problem is how to minimize the conditional risk. Analytic solutions are often hard to obtain. Numerical solutions are often burdensome.

If we constrain the expression for the estimator to some mathematical form, the problem of designing the estimator boils down to finding the suitable parameters of that form. An example is the *unbiased linear MMSE estimator* with the following form:[2]

$$\hat{\mathbf{x}}_{ulMMSE}(\mathbf{z}) = \mathbf{K}\mathbf{z} + \mathbf{a} \tag{3.26}$$

The matrix $\mathbf{K}$ and the vector $\mathbf{a}$ must be optimized during the design phase so as to match the behaviour of the estimator to the problem at hand. The estimator has the same optimization criterion as the MMSE

[2] The connotation of the term 'unbiased' becomes clear in Section 3.2.1. The *linear MMSE* (without the adjective 'unbiased') also exists. It has the form $\hat{\mathbf{x}}_{lMMSE}(\mathbf{z}) = \mathbf{K}\mathbf{z}$. See exercise 1.

estimator, i.e. a quadratic cost function. The constraint results in an estimator that is not as good as the (unconstrained) MMSE estimator, but it requires only knowledge of moments up to the order two, i.e. expectation vectors and covariance matrices.

The starting point is the overall risk expressed in (3.11). Together with the quadratic cost function we have:

$$R = \int_{\mathbf{z}} \int_{\mathbf{x}} (\mathbf{Kz} + \mathbf{a} - \mathbf{x})^T (\mathbf{Kz} + \mathbf{a} - \mathbf{x}) p(\mathbf{x}|\mathbf{z}) p(\mathbf{z}) d\mathbf{x} d\mathbf{z} \qquad (3.27)$$

The optimal unbiased linear MMSE estimator is found by minimizing R with respect to **K** and **a**. Hence we differentiate R with respect to **a** and equate the result to zero:

$$\begin{aligned} \frac{dR}{d\mathbf{a}} &= \int_{\mathbf{z}} \int_{\mathbf{x}} (2\mathbf{a} + 2\mathbf{Kz} - 2\mathbf{x}) p(\mathbf{x}|\mathbf{z}) p(\mathbf{z}) d\mathbf{x} d\mathbf{z} \\ &= 2\mathbf{a} + 2\mathbf{K}\boldsymbol{\mu}_{\mathbf{z}} - 2\boldsymbol{\mu}_{\mathbf{x}} = 0 \end{aligned}$$

yielding:

$$\mathbf{a} = \boldsymbol{\mu}_{\mathbf{x}} - \mathbf{K}\boldsymbol{\mu}_{\mathbf{z}} \qquad (3.28)$$

with $\boldsymbol{\mu}_{\mathbf{x}}$ and $\boldsymbol{\mu}_{\mathbf{z}}$ the expectations of **x** and **z**.

Substitution of **a** back into (3.27), differentiation with respect to **K**, and equating the result to zero (see also Appendix B.4):

$$\begin{aligned} R &= \int_{\mathbf{z}} \int_{\mathbf{x}} (\mathbf{K}(\mathbf{z} - \boldsymbol{\mu}_{\mathbf{z}}) - (\mathbf{x} - \boldsymbol{\mu}_{\mathbf{x}}))^T (\mathbf{K}(\mathbf{z} - \boldsymbol{\mu}_{\mathbf{z}}) - (\mathbf{x} - \boldsymbol{\mu}_{\mathbf{x}})) p(\mathbf{x}|\mathbf{z}) p(\mathbf{z}) d\mathbf{x} d\mathbf{z} \\ &= trace\left(\mathbf{K}\mathbf{C}_{\mathbf{z}}\mathbf{K}^T + \mathbf{C}_{\mathbf{x}} - 2\mathbf{K}\mathbf{C}_{\mathbf{zx}}\right) \end{aligned}$$

$$\frac{dR}{d\mathbf{K}} = 2\mathbf{K}\mathbf{C}_{\mathbf{z}} - 2\mathbf{C}_{\mathbf{xz}} = 0$$

yields:

$$\mathbf{K} = \mathbf{C}_{\mathbf{xz}}\mathbf{C}_{\mathbf{z}}^{-1} \qquad (3.29)$$

$\mathbf{C}_{\mathbf{z}}$ is the covariance matrix of **z**, and $\mathbf{C}_{\mathbf{xz}} = \mathrm{E}[(\mathbf{x} - \boldsymbol{\mu}_{\mathbf{x}})(\mathbf{z} - \boldsymbol{\mu}_{\mathbf{z}})^T]$ the cross-covariance matrix between **x** and **z**.

Example 3.5 Unbiased linear MMSE estimation of the backscattering coefficient
In the scalar case, the linear MMSE estimator takes the form:

$$\hat{x}_{ulMMSE}(z) = \frac{\text{cov}[x,z]}{\text{Var}[z]}(z - \text{E}[z]) + \text{E}[x] \tag{3.30}$$

where cov[x,z] is the covariance of x and z. In the backscattering problem, the required moments are difficult to obtain analytically. However, they are easily estimated from the population of the 500 realizations shown in Figure 3.6 using techniques from Chapter 5. The resulting estimator is shown in Figure 3.6. MATLAB code to plot the ML and unbiased linear MMSE estimates of the backscattering coefficient on a data set is given in Listing 3.2.

Listing 3.2
MATLAB code for unbiased linear MMSE estimation.

```
load scatter;                % Load dataset (zset,xset)
z=0.005:0.005:1.5;           % Interesting range of z
x_ml=z;                      % Maximum likelihood
mu_x=mean(xset); mu_z=mean(zset);
K=((xset-mu_x)'*(zset-mu_z))*inv((zset-mu_z)'*(zset-mu_z));
a = mu_x - K*mu_z;
x_ulmse = K*z + a;           % Unbiased linear MMSE
figure; clf; plot(zset,xset,'.'); hold on;
plot(z,x_ml,'k-'); plot(z,x_ulmse,'k--');
```

Linear sensors

The linear MMSE estimator takes a particular form if the sensory system is linear and the sensor noise is additive:

$$\mathbf{z} = \mathbf{H}\mathbf{x} + \mathbf{v} \tag{3.31}$$

This case is of special interest because of its crucial role in the *Kalman filter* (to be discussed in Chapter 4). Suppose that the noise has zero mean with covariance matrix $\mathbf{C_v}$. In addition, suppose that $\mathbf{x}$ and $\mathbf{v}$ are uncorrelated, i.e. $\mathbf{C_{xv}} = \mathbf{0}$. Under these assumptions the moments of $\mathbf{z}$ are as follows:

$$\begin{aligned} \boldsymbol{\mu}_\mathbf{z} &= \mathbf{H}\boldsymbol{\mu}_\mathbf{x} \\ \mathbf{C_z} &= \mathbf{H}\mathbf{C_x}\mathbf{H}^T + \mathbf{C_v} \\ \mathbf{C_{xz}} &= \mathbf{C_x}\mathbf{H}^T \\ \mathbf{C_{zx}} &= \mathbf{H}\mathbf{C_x} \end{aligned} \tag{3.32}$$

The proof is left as an exercise for the reader.

Substitution of (3.32), (3.28) and (3.29) in (3.26) gives rise to the following estimator:

$$\hat{\mathbf{x}}_{ulMMSE}(\mathbf{z}) = \boldsymbol{\mu}_{\mathbf{x}} + \mathbf{K}(\mathbf{z} - \mathbf{H}\boldsymbol{\mu}_{\mathbf{x}}) \quad \text{with} \quad \mathbf{K} = \mathbf{C}_{\mathbf{x}}\mathbf{H}^T\left(\mathbf{H}\mathbf{C}_{\mathbf{x}}\mathbf{H}^T + \mathbf{C}_{\mathbf{v}}\right)^{-1} \tag{3.33}$$

This version of the unbiased linear MMSE estimator is the so-called *Kalman form* of the estimator.

Examination of (3.20) reveals that the MMSE estimator in the Gaussian case with linear sensors is also expressed as a linear combination of $\boldsymbol{\mu}_{\mathbf{x}}$ and $\mathbf{z}$. Thus, in this special case (that is, Gaussian densities + linear sensors) $\hat{\mathbf{x}}_{MMSE}(\mathbf{z})$ is a linear estimator. Since $\hat{\mathbf{x}}_{ulMMSE}(\mathbf{z})$ and $\hat{\mathbf{x}}_{MMSE}(\mathbf{z})$ are based on the same optimization criterion, the two solutions must be identical here: $\hat{\mathbf{x}}_{ulMMSE}(\mathbf{z}) = \hat{\mathbf{x}}_{MMSE}(\mathbf{z})$. We conclude that (3.20) is an alternative form of (3.33). The forms are mathematically equivalent. See exercise 5.

The interpretation of $\hat{\mathbf{x}}_{ulMMSE}(\mathbf{z})$ is as follows. The term $\boldsymbol{\mu}_x$ represents the prior knowledge. The term $\mathbf{H}\boldsymbol{\mu}_{\mathbf{x}}$ is the prior knowledge that we have about the measurements. Therefore, the factor $\mathbf{z} - \mathbf{H}\boldsymbol{\mu}_{\mathbf{x}}$ is the informative part of the measurements (called the *innovation*). The so-called *Kalman gain matrix* $\mathbf{K}$ transforms the innovation into a *correction term* $\mathbf{K}(\mathbf{z} - \mathbf{H}\boldsymbol{\mu}_{\mathbf{x}})$ that represents the knowledge that we have gained from the measurements.

3.2 PERFORMANCE OF ESTIMATORS

No matter which precautions are taken, there will always be a difference between the estimate of a parameter and its true (but unknown) value. The difference is the *estimation error*. An estimate is useless without an indication of the magnitude of that error. Usually, such an indication is quantified in terms of the so-called *bias* of the estimator, and the *variance*. The main purpose of this section is to introduce these concepts.

Suppose that the true, but unknown value of a parameter is $\mathbf{x}$. An estimator $\hat{\mathbf{x}}(.)$ provides us with an estimate $\hat{\mathbf{x}} = \hat{\mathbf{x}}(\mathbf{z})$ based on measurements $\mathbf{z}$. The estimation error $\mathbf{e}$ is the difference between the estimate and the true value:

$$\mathbf{e} = \hat{\mathbf{x}} - \mathbf{x} \tag{3.34}$$

Since $\mathbf{x}$ is unknown, $\mathbf{e}$ is unknown as well.

3.2.1 Bias and covariance

The error $\mathbf{e}$ is composed of two parts. One part is the one that does not change value if we repeat the experiment over and over again. It is the expectation of the error, called the *bias*. The other part is the random part and is due to sensor noise and other random phenomena in the sensory system. Hence, we have:

$$\text{error} = \text{bias} + \text{random part}$$

If $\mathbf{x}$ is a scalar, the *variance* of an estimator is the variance of $\mathbf{e}$. As such the variance quantifies the magnitude of the random part. If $\mathbf{x}$ is a vector, each element of $\mathbf{e}$ has its own variance. These variances are captured in the covariance matrix of $\mathbf{e}$, which provides an economical and also a more complete way to represent the magnitude of the random error.

The application of the expectation and variance operators to $\mathbf{e}$ needs some discussion. Two cases must be distinguished. If $\mathbf{x}$ is regarded as a non-random, unknown parameter, then $\mathbf{x}$ is not associated with any probability density. The only randomness that enters the equations is due to the measurements $\mathbf{z}$ with density $p(\mathbf{z}|\mathbf{x})$. However, if $\mathbf{x}$ is regarded as random, it does have a probability density. We have two sources of randomness then, $\mathbf{x}$ and $\mathbf{z}$.

We start with the first case which applies to, for instance, the maximum likelihood estimator. Here, the bias $\mathbf{b}(\mathbf{x})$ is given by:

$$\begin{aligned} \mathbf{b}(\mathbf{x}) &\stackrel{def}{=} \mathrm{E}[\hat{\mathbf{x}} - \mathbf{x}|\mathbf{x}] \\ &= \int (\hat{\mathbf{x}}(\mathbf{z}) - \mathbf{x}) p(\mathbf{z}|\mathbf{x}) d\mathbf{z} \end{aligned} \tag{3.35}$$

The integral extends over the full space of $\mathbf{z}$. In general, the bias depends on $\mathbf{x}$. The bias of an estimator can be small or even zero in one area of $\mathbf{x}$, whereas in another area the bias of that same estimator might be large.

In the second case, both $\mathbf{x}$ and $\mathbf{z}$ are random. Therefore, we define an *overall bias* $\mathbf{b}$ by taking the expectation operator now with respect to both $\mathbf{x}$ and $\mathbf{z}$:

$$\begin{aligned} \mathbf{b} &\stackrel{def}{=} \mathrm{E}[\hat{\mathbf{x}} - \mathbf{x}] \\ &= \int \int (\hat{\mathbf{x}}(\mathbf{z}) - \mathbf{x}) p(\mathbf{x}, \mathbf{z}) d\mathbf{z} d\mathbf{x} \end{aligned} \tag{3.36}$$

The integrals extend over the full space of $\mathbf{x}$ and $\mathbf{z}$.

The overall bias must be considered as an average taken over the full range of $\mathbf{x}$. To see this, rewrite $p(\mathbf{x},\mathbf{z}) = p(\mathbf{z}|\mathbf{x})p(\mathbf{x})$ to yield:

$$\mathbf{b} = \int \mathbf{b}(\mathbf{x})p(\mathbf{x})d\mathbf{x} \tag{3.37}$$

where $\mathbf{b}(\mathbf{x})$ is given in (3.35).

If the overall bias of an estimator is zero, then the estimator is said to be *unbiased*. Suppose that in two different areas of $\mathbf{x}$ the biases of an estimator have opposite sign, then these two opposite biases may cancel out. We conclude that, even if an estimator is unbiased (i.e. its overall bias is zero), then this does not imply that the bias for a specific value of $\mathbf{x}$ is zero. Estimators that are unbiased for every $\mathbf{x}$ are called *absolutely unbiased.*

The variance of the error, which serves to quantify the random fluctuations, follows the same line of reasoning as the one of the bias. First we determine the covariance matrix of the error with non-random $\mathbf{x}$:

$$\begin{aligned} \mathbf{C}_\mathbf{e}(\mathbf{x}) &\stackrel{def}{=} \mathrm{E}\left[(\mathbf{e} - \mathrm{E}[\mathbf{e}])(\mathbf{e} - \mathrm{E}[\mathbf{e}])^T|\mathbf{x}\right] \\ &= \int (\hat{\mathbf{x}}(\mathbf{z}) - \mathbf{x} - \mathbf{b}(\mathbf{x}))(\hat{\mathbf{x}}(\mathbf{z}) - \mathbf{x} - \mathbf{b}(\mathbf{x}))^T p(\mathbf{z}|\mathbf{x})d\mathbf{z} \end{aligned} \tag{3.38}$$

As before, the integral extends over the full space of $\mathbf{z}$. The variances of the elements of $\mathbf{e}$ are at the diagonal of $\mathbf{C}_\mathbf{e}(\mathbf{x})$.

The magnitude of the full error (bias + random part) is quantified by the so-called *mean square error* (the second order moment matrix of the error):

$$\begin{aligned} \mathbf{M}_\mathbf{e}(\mathbf{x}) &\stackrel{def}{=} \mathbf{E}\left[\mathbf{e}\mathbf{e}^T|\mathbf{x}\right] \\ &= \int (\hat{\mathbf{x}}(\mathbf{z}) - \mathbf{x})(\hat{\mathbf{x}}(\mathbf{z}) - \mathbf{x})^T p(\mathbf{z}|\mathbf{x})d\mathbf{z} \end{aligned} \tag{3.39}$$

It is straightforward to prove that:

$$\mathbf{M}_\mathbf{e}(\mathbf{x}) = \mathbf{b}(\mathbf{x})\mathbf{b}^T(\mathbf{x}) + \mathbf{C}_\mathbf{e}(\mathbf{x}) \tag{3.40}$$

This expression underlines the fact that the error is composed of a bias and a random part.

The overall mean square error $\mathbf{M}_\mathbf{e}$ is found by averaging $\mathbf{M}_\mathbf{e}(\mathbf{x})$ over all possible values of $\mathbf{x}$:

$$\mathbf{M_e} \stackrel{def}{=} \mathrm{E}\left[\mathbf{e}\mathbf{e}^T\right] = \int \mathbf{M_e}(\mathbf{x})p(\mathbf{x})d\mathbf{x} \tag{3.41}$$

Finally, the *overall covariance matrix* of the estimation error is found as:

$$\begin{aligned} \mathbf{C_e} &= \mathrm{E}\left[(\mathbf{e}-\mathbf{b})(\mathbf{e}-\mathbf{b})^T\right] \\ &= \mathbf{M_e} - \mathbf{b}\mathbf{b}^T \end{aligned} \tag{3.42}$$

The diagonal elements of this matrix are the overall variances of the estimation errors.

The MMSE estimator and the unbiased linear MMSE estimator are always unbiased. To see this, rewrite (3.36) as follows:

$$\begin{aligned} \mathbf{b} &= \int\int (\hat{\mathbf{x}}_{MMSE}(\mathbf{z}) - \mathbf{x})p(\mathbf{x}|\mathbf{z})p(\mathbf{z})d\mathbf{x}d\mathbf{z} \\ &= \int\int (\mathrm{E}[\mathbf{x}|\mathbf{z}] - \mathbf{x})p(\mathbf{x}|\mathbf{z})p(\mathbf{z})d\mathbf{x}d\mathbf{z} \end{aligned} \tag{3.43}$$

The inner integral is identical to zero, and thus **b** must be zero. The proof of the unbiasedness of the unbiased linear MMSE estimator is left as an exercise.

Other properties related to the quality of an estimator are *stability* and *robustness*. In this context, stability refers to the property of being insensitive to small random errors in the measurement vector. Robustness is the property of being insensitive to large errors in a few elements of the measurements (outliers); see Section 3.3.2. Often, the enlargement of prior knowledge increases both the stability and the robustness.

Example 3.6 Bias and variance in the backscattering application
Figure 3.7 shows the bias and variance of the various estimators discussed in the previous examples. To enable a fair comparison between bias and variance in comparable units, the square root of the latter, i.e. the standard deviation, has been plotted. Numerical evaluation of (3.37), (3.41) and (3.42) yields:[3]

[3] In this example, the vector $\mathbf{b}(\mathbf{x})$ and the matrix $\mathbf{C}(\mathbf{x})$ turn into scalars because here $\mathbf{x}$ is a scalar.

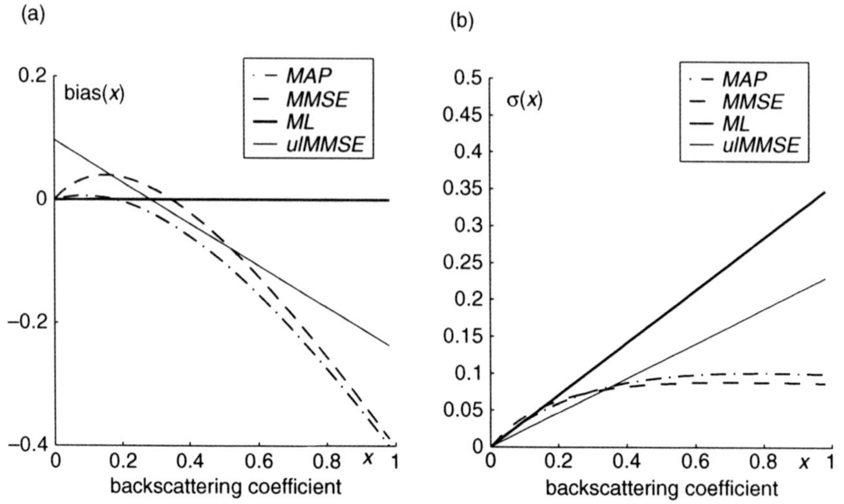

Figure 3.7 The bias and the variance of the various estimators in the backscattering problem

$$\begin{aligned} b_{MMSE} &= 0 & \sigma_{MMSE} &= \sqrt{C_{MMSE}} = 0.086 \\ b_{ulMMSE} &= 0 & \sigma_{ulMMSE} &= \sqrt{C_{ulMMSE}} = 0.094 \\ b_{ML} &= 0 & \sigma_{ML} &= \sqrt{C_{ML}} = 0.116 \\ b_{MAP} &= -0.036 & \sigma_{MAP} &= \sqrt{C_{MAP}} = 0.087 \end{aligned}$$

From this, and from Figure 3.7, we observe that:

- The overall bias of the ML estimator appears to be zero. So, in this example, the ML estimator is unbiased (together with the two MMSE estimators which are intrinsically unbiased). The MAP estimator is biased.
- Figure 3.7 shows that for some ranges of x the bias of the MMSE estimator is larger than its standard deviation. Nevertheless, the MMSE estimator outperforms all other estimators with respect to overall bias and variance. Hence, although a small bias is a desirable property, sometimes the overall performance of an estimator can be improved by allowing a larger bias.
- The ML estimator appears to be linear here. As such, it is comparable with the unbiased linear MMSE estimator. Of these two linear estimators, the unbiased linear MMSE estimator outperforms the ML estimator. The reason is that – unlike the ML estimator – the ulMMSE

estimator exploits prior knowledge about the parameter. In addition, the ulMMSE estimator is more apt to the evaluation criterion.

- Of the two nonlinear estimators, the MMSE estimator outperforms the MAP estimator. The obvious reason is that the cost function of the MMSE estimator matches the evaluation criterion.
- Of the two MMSE estimators, the nonlinear MMSE estimator outperforms the linear one. Both estimators have the same optimization criterion, but the constraint of the ulMMSE degrades its performance.

3.2.2 The error covariance of the unbiased linear MMSE estimator

We now return to the case of having linear sensors, $\mathbf{z} = \mathbf{Hx} + \mathbf{v}$, as discussed in Section 3.1.5. The unbiased linear MMSE estimator appeared to be (see eq. (3.33)):

$$\hat{\mathbf{x}}_{ulMMSE}(\mathbf{z}) = \boldsymbol{\mu}_{\mathbf{x}} + \mathbf{K}(\mathbf{z} - \mathbf{H}\boldsymbol{\mu}_{\mathbf{x}}) \quad \text{with} \quad \mathbf{K} = \mathbf{C_x H}^T\left(\mathbf{HC_x H}^T + \mathbf{C_v}\right)^{-1}$$

where $\mathbf{C_v}$ and $\mathbf{C_x}$ are the covariance matrices of $\mathbf{v}$ and $\mathbf{x}$. $\boldsymbol{\mu}_{\mathbf{x}}$ is the (prior) expectation vector of $\mathbf{x}$. As said before, the $\hat{\mathbf{x}}_{ulMMSE}(.)$ is unbiased.

Due to the unbiasedness of $\hat{\mathbf{x}}_{ulMMSE}(.)$, the mean of the estimation error $\mathbf{e} = \hat{\mathbf{x}}_{ulMMSE}(.) - \mathbf{x}$ is zero. The error covariance matrix, $\mathbf{C_e}$, of $\mathbf{e}$ expresses the uncertainty that remains after having processed the measurements. Therefore, $\mathbf{C_e}$ is identical to the covariance matrix associated with the posterior probability density. It is given by (3.20):

$$\mathbf{C_e} = \mathbf{C_{x|z}} = \left(\mathbf{C_x}^{-1} + \mathbf{H}^T\mathbf{C_v}^{-1}\mathbf{H}\right)^{-1} \tag{3.44}$$

The inverse of a covariance matrix is called an *information matrix*. For instance, $\mathbf{C_e}^{-1}$ is a measure of the information provided by the estimate $\hat{\mathbf{x}}_{ulMMSE}(.)$. If the norm of $\mathbf{C_e}^{-1}$ is large, then the norm of $\mathbf{C_e}$ must be small implying that the uncertainty in $\hat{\mathbf{x}}_{ulMMSE}(.)$ is small as well. Equation (3.44) shows that $\mathbf{C_e}^{-1}$ is made up of two terms. The term $\mathbf{C_x}^{-1}$ represents the prior information provided by $\boldsymbol{\mu}_{\mathbf{x}}$. The matrix $\mathbf{C_v}^{-1}$ represents the information that is given by $\mathbf{z}$ about the vector $\mathbf{H_x}$. Therefore, the matrix $\mathbf{H}^T\mathbf{C_v}^{-1}\mathbf{H}$ represents the information about $\mathbf{x}$ provided by $\mathbf{z}$. The two sources of information add up. So, the information about $\mathbf{x}$ provided by $\hat{\mathbf{x}}_{ulMMSE}(.)$ is $\mathbf{C_e}^{-1} = \mathbf{C_x}^{-1} + \mathbf{H}^T\mathbf{C_v}^{-1}\mathbf{H}$.

Using the matrix inversion lemma (b.10) the expression for the error covariance matrix can be given in an alternative form:

$$\mathbf{C}_e = \mathbf{C}_\mathbf{x} - \mathbf{K}\mathbf{S}\mathbf{K}^T \quad \text{with:} \ \mathbf{S} = \mathbf{H}\mathbf{C}_\mathbf{x}\mathbf{H}^T + \mathbf{C}_\mathbf{n} \tag{3.45}$$

The matrix $\mathbf{S}$ is called the *innovation matrix* because it is the covariance matrix of the innovations $\mathbf{z} - \mathbf{H}\boldsymbol{\mu}_\mathbf{x}$. The factor $\mathbf{K}(\mathbf{z} - \mathbf{H}\boldsymbol{\mu}_\mathbf{x})$ is a correction term for the prior expectation vector $\boldsymbol{\mu}_\mathbf{x}$. Equation (3.45) shows that the prior covariance matrix $\mathbf{C}_\mathbf{x}$ is reduced by the covariance matrix $\mathbf{K}\mathbf{S}\mathbf{K}^T$ of the correction term.

3.3 DATA FITTING

In data-fitting techniques, the measurement process is modelled as:

$$\mathbf{z} = \mathbf{h}(\mathbf{x}) + \mathbf{v} \tag{3.46}$$

where $\mathbf{h}(.)$ is the *measurement function* that models the sensory system, and $\mathbf{v}$ are disturbing factors, such as sensor noise and modelling errors. The purpose of fitting is to find the parameter vector $\mathbf{x}$ which 'best' fits the measurements $\mathbf{z}$.

Suppose that $\hat{\mathbf{x}}$ is an estimate of $\mathbf{x}$. Such an estimate is able to 'predict' the modelled part of $\mathbf{z}$, but it cannot predict the disturbing factors. Note that $\mathbf{v}$ represents both the noise and the unknown modelling errors. The *prediction* of the estimate $\hat{\mathbf{x}}$ is given by $\mathbf{h}(\hat{\mathbf{x}})$. The *residuals* $\boldsymbol{\varepsilon}$ are defined as the difference between observed and predicted measurements:

$$\varepsilon = \mathbf{z} - \mathbf{h}(\hat{\mathbf{x}}) \tag{3.47}$$

Data fitting is the process of finding the estimate $\hat{\mathbf{x}}$ that minimizes some error norm $\|\boldsymbol{\varepsilon}\|$ of the residuals. Different error norms (see Appendix A.1.1) lead to different data fits. We will shortly discuss two error norms.

3.3.1 Least squares fitting

The most common error norm is the squared Euclidean norm, also called the *sum of squared differences* (SSD), or simply the LS norm (*least squared error norm*):

$$\|\boldsymbol{\varepsilon}\|_2^2 = \sum_{n=0}^{N-1} \varepsilon_n^2 = \sum_{n=0}^{N-1} (z_n - h_n(\hat{\mathbf{x}}))^2 = (\mathbf{z} - \mathbf{h}(\hat{\mathbf{x}}))^T(\mathbf{z} - \mathbf{h}(\hat{\mathbf{x}})) \tag{3.48}$$

The *least squares fit*, or *least squares estimate* (LS) is the parameter vector which minimizes this norm:

$$\hat{\mathbf{x}}_{LS}(\mathbf{z}) = \underset{\mathbf{x}}{\operatorname{argmin}}\left\{(\mathbf{z} - \mathbf{h}(\mathbf{x}))^T(\mathbf{z} - \mathbf{h}(\mathbf{x}))\right\} \tag{3.49}$$

If $\mathbf{v}$ is random with a normal distribution, zero mean and covariance matrix $\mathbf{C_v} = \sigma_v^2\mathbf{I}$, the LS estimate is identical to the ML estimate: $\hat{\mathbf{x}}_{LS} \equiv \hat{\mathbf{x}}_{ML}$. To see this, it suffices to note that in the Gaussian case the likelihood takes the form

$$p(\mathbf{z}|\mathbf{x}) = \frac{1}{\sqrt{(2\pi)^N \sigma_v^{2N}}} \exp\left(-\frac{(\mathbf{z} - \mathbf{h}(\mathbf{x}))^T(\mathbf{z} - \mathbf{h}(\mathbf{x}))}{2\sigma_v^2}\right) \tag{3.50}$$

The minimization of (3.48) is equivalent to the minimization of (3.50). If the measurement function is linear, that is $\mathbf{z} = \mathbf{Hx} + \mathbf{v}$, and $\mathbf{H}$ is an $N \times M$ matrix having a rank M with $M < N$, then according to (3.25):

$$\hat{\mathbf{x}}_{LS}(\mathbf{z}) = \hat{\mathbf{x}}_{ML}(\mathbf{z}) = (\mathbf{H}^T\mathbf{H})^{-1}\mathbf{H}^T\mathbf{z} \tag{3.51}$$

Example 3.7 Repeated measurements
Suppose that a scalar parameter x is N times repeatedly measured using a calibrated measurement device: $z_n = x + v_n$. These repeated measurements can be represented by a vector $\mathbf{z} = [z_1 \ldots z_N]^T$. The corresponding measurement matrix is $\mathbf{H} = [\,1 \ldots 1\,]^T$. Since $(\mathbf{H}^T\mathbf{H})^{-1} = 1/N$, the resulting least squares fit is:

$$\hat{x}_{LS} = \frac{1}{N}\mathbf{H}^T\mathbf{z} = \frac{1}{N}\sum_{n=1}^{N} z_n$$

In other words, the best fit is found by averaging the measurements.

Nonlinear sensors

If $\mathbf{h}(.)$ is nonlinear, an analytic solution of (3.49) is often difficult. One is compelled to use a numerical solution. For that, several algorithms exist, such as '*Gauss–Newton*', '*Newton–Raphson*', '*steepest descent*' and many others. Many of these algorithms are implemented within MATLAB's optimization toolbox. The 'Gauss–Newton' method will be explained shortly.

Assuming that some initial estimate $\mathbf{x}_{ref}$ is available we expand (3.46) in a Taylor series and neglect all terms of order higher than two:

$$\begin{aligned} \mathbf{z} &= \mathbf{h}(\mathbf{x}) + \mathbf{v} \\ &\approx \mathbf{h}(\mathbf{x}_{ref}) + \mathbf{H}_{ref}\left(\mathbf{x} - \mathbf{x}_{ref}\right) + \mathbf{v} \quad \text{with:} \quad \mathbf{H}_{ref} = \left.\frac{\partial \mathbf{h}(\mathbf{x})}{\partial \mathbf{x}}\right|_{\mathbf{x}=\mathbf{x}_{ref}} \end{aligned} \tag{3.52}$$

where $\mathbf{H}_{ref}$ is the Jacobian matrix of $\mathbf{h}(.)$ evaluated at $\mathbf{x}_{ref}$, see Appendix B.4. With such a linearization, (3.51) applies. Therefore, the following approximate value of the LS estimate is obtained:

$$\hat{\mathbf{x}}_{LS} \approx \mathbf{x}_{ref} + \left(\mathbf{H}_{ref}^T \mathbf{H}_{ref}\right)^{-1} \mathbf{H}_{ref}\left(\mathbf{z} - \mathbf{h}(\mathbf{x}_{ref})\right) \tag{3.53}$$

A refinement of the estimate could be achieved by repeating the procedure with the approximate value as reference. This suggests an iterative approach. Starting with some initial guess $\hat{\mathbf{x}}(0)$, the procedure becomes as follows:

$$\begin{aligned} \hat{\mathbf{x}}(i+1) &= \hat{\mathbf{x}}(i) + \left(\mathbf{H}^T(i)\mathbf{H}(i)\right)^{-1}\mathbf{H}(i)(\mathbf{z} - \mathbf{h}(\hat{\mathbf{x}}(i))) \\ \text{with:} \quad \mathbf{H}(i) &= \left.\frac{\partial \mathbf{h}(\mathbf{x})}{\partial \mathbf{x}}\right|_{\mathbf{x}=\hat{\mathbf{x}}(i)} \end{aligned} \tag{3.54}$$

In each iteration, the variable i is incremented. The iterative process stops if the difference between $\mathbf{x}(i+1)$ and $\mathbf{x}(i)$ is smaller than some predefined threshold. The success of the method depends on whether the first initial guess is already close enough to the global minimum. If not, the process will either diverge, or get stuck in a local minimum.

Example 3.8 Estimation of the diameter of a blood vessel
In vascular X-ray imaging, one of the interesting parameters is the diameter of blood vessels. This parameter provides information about a possible constriction. As such, it is an important aspect in cardiologic diagnosis.

Figure 3.8(a) is a (simulated) X-ray image of a blood vessel of the coronary circulation. The image quality depends on many factors. Most important are the low-pass filtered noise (called quantum mottle) and the image blurring due to the image intensifier.

Figure 3.8(b) shows the one-dimensional, vertical cross-section of the image at a location as indicated by the two black arrows in Figure 3.8(a). Suppose that our task is to estimate the diameter of the imaged blood vessel from the given cross-section. Hence, we define

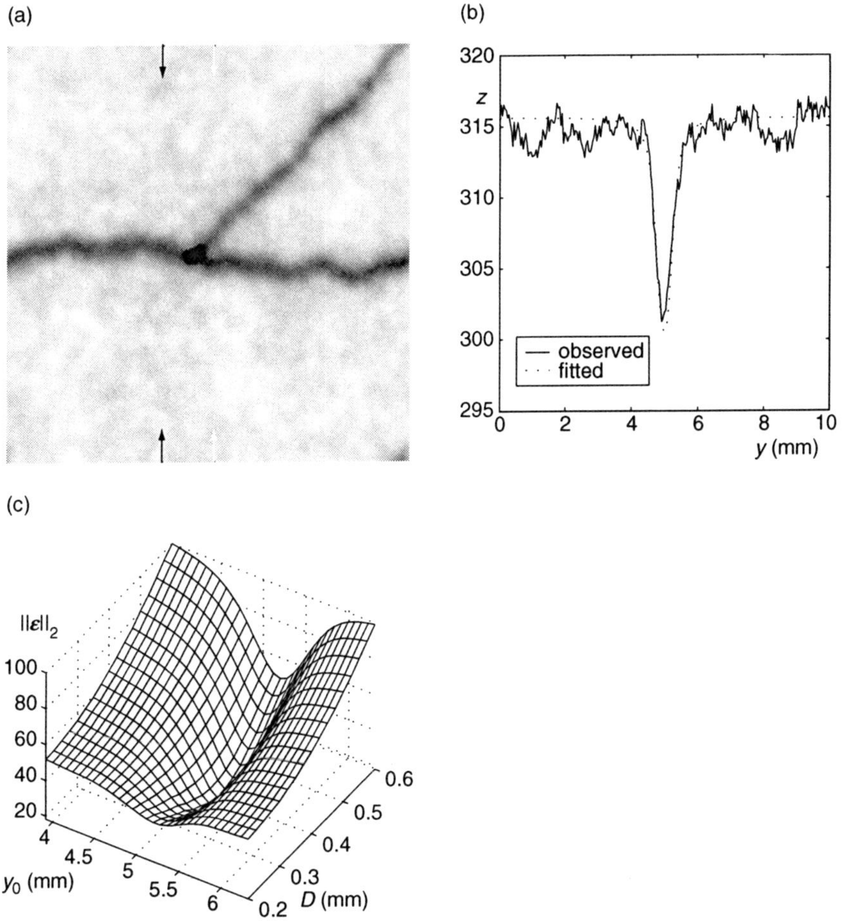

Figure 3.8 LS estimation of the diameter D and the position y_0 of a blood vessel. (a) X-ray image of the blood vessel. (b) Cross-section of the image together with fitted profile. (c) The sum of least squared errors as a function of the diameter and the position

a measurement vector **z** whose elements consist of the pixel grey values along the cross-section.

The parameter of interest is the diameter D. However, other parameters might be unknown as well, e.g. the position and orientation of the blood vessel, the attenuation coefficient or the intensity of the X-ray source. This example will be confined to the case where the only unknown parameters are the diameter D and the position y_0 of the image blood vessel in the cross-section. Thus, the parameter vector is two-dimensional: $\mathbf{x}^T = [\,D \quad y_0\,]$. Since all other parameters are assumed to be known, a radiometric model can be worked out to a measurement function $\mathbf{h}(\mathbf{x})$ which quantitatively predicts the cross-section, and thus also the measurement vector $\mathbf{z}$ for any value of the parameter vector $\mathbf{x}$.

With this measurement function it is straightforward to calculate the LS norm $\|\boldsymbol{\varepsilon}\|_2^2$ for a couple of values of $\mathbf{x}$. Figure 3.8(c) is a graphical representation of that. It appears that the minimum of $\|\boldsymbol{\varepsilon}\|_2^2$ is obtained if $\hat{D}_{LS} = 0.42$ mm. The true diameter is $D = 0.40$ mm. The thus obtained fitted cross-section is also shown in Figure 3.8(b).

Note that the LS norm in Figure 3.8(c) is a smooth function of $\mathbf{x}$. Hence, the convergence region of a numerical optimizer will be large.

3.3.2 Fitting using a robust error norm

Suppose that the measurement vector in an LS estimator has a few number of elements with large measurement errors, the so-called *outliers*. The influence of an outlier is much larger than the one of the others because the LS estimator weights the errors quadraticly. Consequently, the robustness of LS estimation is poor.

Much can be improved if the influence is bounded in one way or another. This is exactly the general idea of applying a *robust error norm*. Instead of using the sum of squared differences, the objective function of (3.48) becomes:

$$\|\boldsymbol{\varepsilon}\|_{robust} = \sum_{n=0}^{N-1} \rho(\varepsilon_n) = \sum_{n=0}^{N-1} \rho(z_n - h_n(\hat{\mathbf{x}})) \tag{3.55}$$

$\rho(.)$ measures the size of each individual residual $z_n - h_n(\hat{\mathbf{x}})$. This measure should be selected such that above a given level of ε_n its influence is ruled out. In addition, one would like to have $\rho(.)$ being smooth so that numerical optimization of $\|\boldsymbol{\varepsilon}\|_{robust}$ is not too difficult. A suitable choice (among others) is the so-called Geman–McClure error norm:

$$\rho(\varepsilon) = \frac{\varepsilon^2}{\varepsilon^2 + \sigma^2} \tag{3.56}$$

A graphical representation of this function and its derivative is shown in Figure 3.9. The parameter σ is a soft threshold value. For values of $\boldsymbol{\varepsilon}$ smaller than about σ, the function follows the LS norm. For values larger than σ, the function gets saturated. Consequently, for small values of $\boldsymbol{\varepsilon}$ the

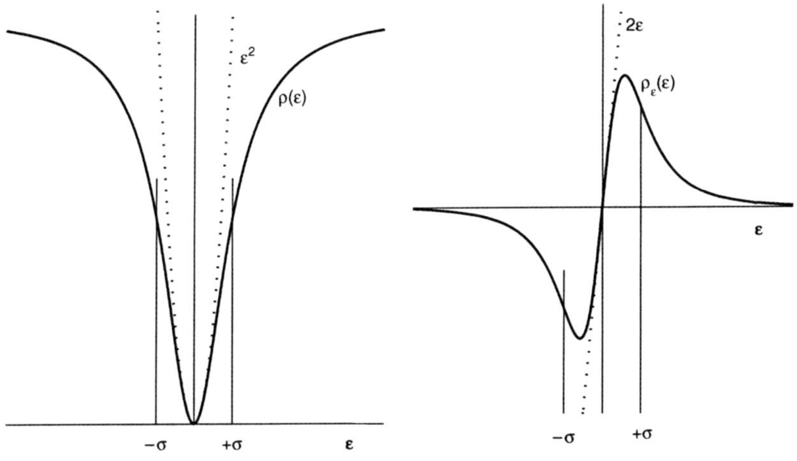

Figure 3.9 A robust error norm and its derivative

derivative $\rho_{\boldsymbol{\varepsilon}}(\boldsymbol{\varepsilon}) = \partial \|\boldsymbol{\varepsilon}\|_{robust}/\partial \mathbf{x}$ of $\rho(.)$ is nearly a constant. But for large values of $\boldsymbol{\varepsilon}$, i.e. for outliers, it becomes nearly zero. Therefore, in a Gauss–Newton style of optimization, the Jacobian matrix is virtually zero for outliers. Only residuals that are about as large as σ or smaller than that play a role.

Example 3.9 Robust estimation of the diameter of a blood vessel
If in example 3.8 the diameter must be estimated near the bifurcation (as indicated in Figure 3.8(a) by the white arrows) a large modelling error occurs because of the branching vessel. See the cross-section in Figure 3.10(a). These modelling errors are large compared to the noise and they should be considered as outliers. However, Figure 3.10(b) shows that the error landscape $\|\boldsymbol{\varepsilon}\|_2^2$ has its minimum at $\hat{D}_{LS} = 0.50\,\text{mm}$. The true value is $D = 0.40\,\text{mm}$. Furthermore, the minimum is less pronounced than the one in Figure 3.8(c), and therefore also less stable.

Note also that in Figure 3.10(a) the position found by the LS estimator is in the middle between the two true positions of the two vessels.

Figure 3.11 shows the improvements that are obtained by applying a robust error norm. The threshold σ is selected just above the noise level. For this setting, the error landscape clearly shows two pronounced minima corresponding to the two blood vessels. The global minimum is reached at $\hat{D}_{robust} = 0.44\,\text{mm}$. The estimated position now clearly corresponds to one of the two blood vessels as shown in Figure 3.11(a).

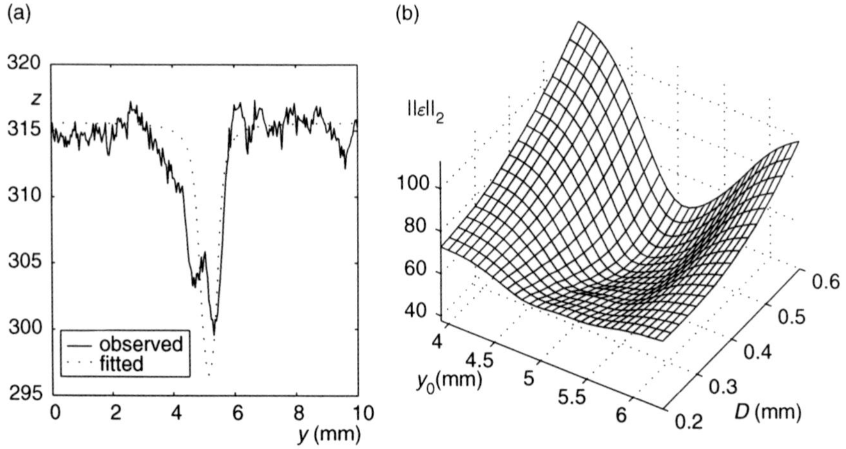

Figure 3.10 LS estimation of the diameter D and the position y_0. (a) Cross-section of the image together with a profile fitted with the LS norm. (b) The LS norm as a function of the diameter and the position

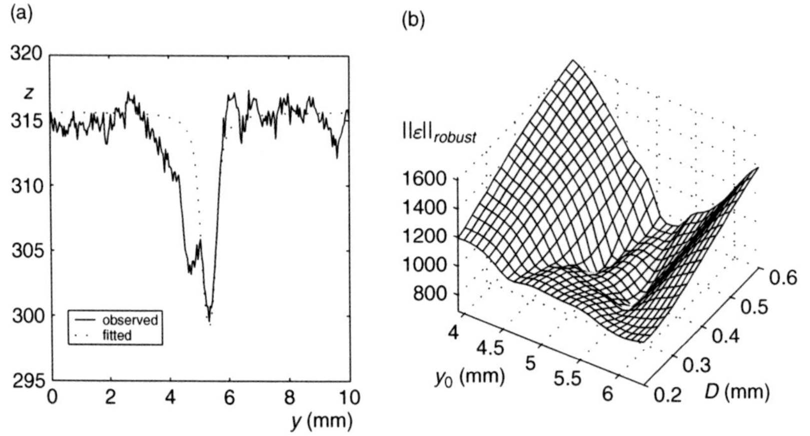

Figure 3.11 Robust estimation of the diameter D and the position y_0. (a) Cross-section of the image together with a profile fitted with a robust error norm. (b) The robust error norm as a function of the diameter and the position

3.3.3 Regression

Regression is the act of deriving an empirical function from a set of experimental data. Regression analysis considers the situation involving pairs of measurements (t, z). The variable t is regarded as a measurement without any appreciable error. t is called the *independent variable*.

We assume that some empirical function $f(.)$ is chosen that (hopefully) can predict z from the independent variable t. Furthermore, a parameter vector $\mathbf{x}$ can be used to control the behaviour of $f(.)$. Hence, the model is:

$$z = f(t, \mathbf{x}) + \varepsilon \tag{3.57}$$

$f(.,.)$ is the *regression curve*, and ε represents the residual, i.e. the part of z that cannot be predicted by $f(.,.)$. Such a residual can originate from sensor noise (or other sources of randomness) which makes the prediction uncertain, but it can also be caused by an inadequate choice of the regression curve.

The goal of regression is to determine an estimate $\hat{\mathbf{x}}$ of the parameter vector $\mathbf{x}$ based on N observations (t_n, z_n), $n = 0, \ldots, N-1$ such that the residuals ε_n are as small as possible. We can stack the observations z_n in a vector $\mathbf{z}$. Using (3.57), the problem of finding $\hat{\mathbf{x}}$ can be transformed to the standard form of (3.47):

$$\mathbf{z} = \mathbf{h}(\hat{\mathbf{x}}) + \boldsymbol{\varepsilon} \quad \text{with:} \quad \mathbf{z} \stackrel{def}{=} \begin{bmatrix} z_0 \\ \vdots \\ z_{N-1} \end{bmatrix} \quad \mathbf{h}(\mathbf{x}) \stackrel{def}{=} \begin{bmatrix} f(t_0, \mathbf{x}) \\ \vdots \\ f(t_{N-1}, \mathbf{x}) \end{bmatrix} \quad \boldsymbol{\varepsilon} \stackrel{def}{=} \begin{bmatrix} \varepsilon_0 \\ \vdots \\ \varepsilon_{N-1} \end{bmatrix} \tag{3.58}$$

where $\boldsymbol{\varepsilon}$ is the vector that embodies the residuals ε_n.

Since the model is in the standard form, $\mathbf{x}$ can be estimated with a least squares approach as in Section 3.3.1. Alternatively, we use a robust error norm as defined in Section 3.3.2. The minimization of such a norm is called *robust regression analysis*.

In the simplest case, the regression curve $f(t, \mathbf{x})$ is linear in $\mathbf{x}$. With that, the model becomes of the form $\mathbf{z} = \mathbf{H}\hat{\mathbf{x}} + \boldsymbol{\varepsilon}$, and thus, the solution of (3.51) applies. As an example, we consider *polynomial regression* for which the regression curve is a polynomial of order $M-1$:

$$f(t, \mathbf{x}) = x_0 + x_1 t + \cdots + x_{M-1} t^{M-1} \tag{3.59}$$

If, for instance, $M = 3$, then the regression curve is a parabola described by three parameters. These parameters can be found by least squares estimation using the following model:

$$\mathbf{z} = \begin{bmatrix} 1 & t_0 & t_0^2 \\ 1 & t_1 & t_1^2 \\ \vdots & \vdots & \vdots \\ 1 & t_{N-1} & t_{N-1}^2 \end{bmatrix} \hat{\mathbf{x}} + \boldsymbol{\varepsilon} \tag{3.60}$$

Example 3.10 The calibration curve of a level sensor
In this example, the goal is to determine a calibration curve of a level sensor to be used in a water tank. For that purpose, a second measurement system is available with a much higher precision than the 'sensor under test'. The measurement results of the second system serve as a reference. Figure 3.12 shows the observed errors of the sensor versus the reference values. Here, 46 pairs of measurements are shown. A zero order (fit with a constant), a first order (linear fit) and a tenth order polynomial are fitted to the data. As can be seen, the constant fit appears to be inadequate for describing the data (the model is too simple). The first order polynomial describes the data reasonably well, and is also suitable for extrapolation. The tenth order polynomial follows the measurement points better, but also the noise. It cannot be used for extrapolation because it is an example of *overfitting* the data. This occurs whenever the model has too many degrees of freedom compared to the number of data samples.

Listing 3.3 illustrates how to fit and evaluate polynomials using MATLAB's `polyfit()` and `polyval()` routines.

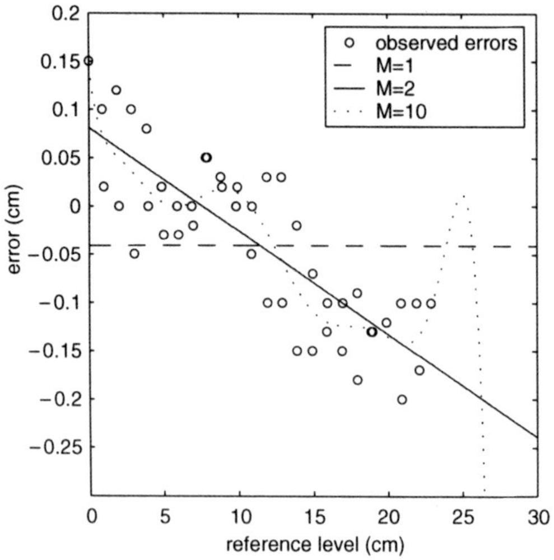

Figure 3.12 Determination of a calibration curve by means of polynomial regression

Listing 3.3
MATLAB code for polynomial regression.

```
load levelsensor;                          % Load dataset (t,z)
figure; clf; plot(t,z,'k.'); hold on;      % Plot it
y=0:0.2:30; M=[ 1 2 10 ]; plotstring={'k--','k-','k:'};
for m=1:3
  p=polyfit(t,z,M(m)-1);                   % Fit polynomial
  z_hat=polyval(p,y);                      % Calculate plot points
  plot(y,z_hat,plotstring{m});             % and plot them
end;
axis([0 30 -0.3 0.2]);
```

3.4 OVERVIEW OF THE FAMILY OF ESTIMATORS

The chapter concludes with the overview shown in Figure 3.13. Two main approaches have been discussed, the Bayes estimators and the fitting techniques. Both approaches are based on the minimization of an objective function. The difference is that with Bayes, the objective function is defined in the parameter domain, whereas with fitting techniques, the objective function is defined in the measurement domain. Another difference is that the Bayes approach has a probabilistic context, whereas the approach of fitting lacks such a context.

Within the family of Bayes estimators we have discussed two estimators derived from two cost functions. The quadratic cost function leads to MMSE (minimum variance) estimation. The cost function is such that small errors are regarded as unimportant, while larger errors are considered more and more serious. The solution is found as the conditional mean, i.e. the expectation of the posterior probability density. The estimator is unbiased.

If the MMSE estimator is constrained to be linear, the solution can be expressed entirely in terms of first and second order moments. If, in addition, the sensory system is linear with additive, uncorrelated noise, a simple form of the estimator appears that is used in Kalman filtering. This form is sometimes referred to as the Kalman form.

The other Bayes estimator is based on the uniform cost function. This cost function is such that the damage of small and large errors are equally weighted. It leads to MAP estimation. The solution appears to be the mode of the posterior probability density. The estimator is not guaranteed to be unbiased.

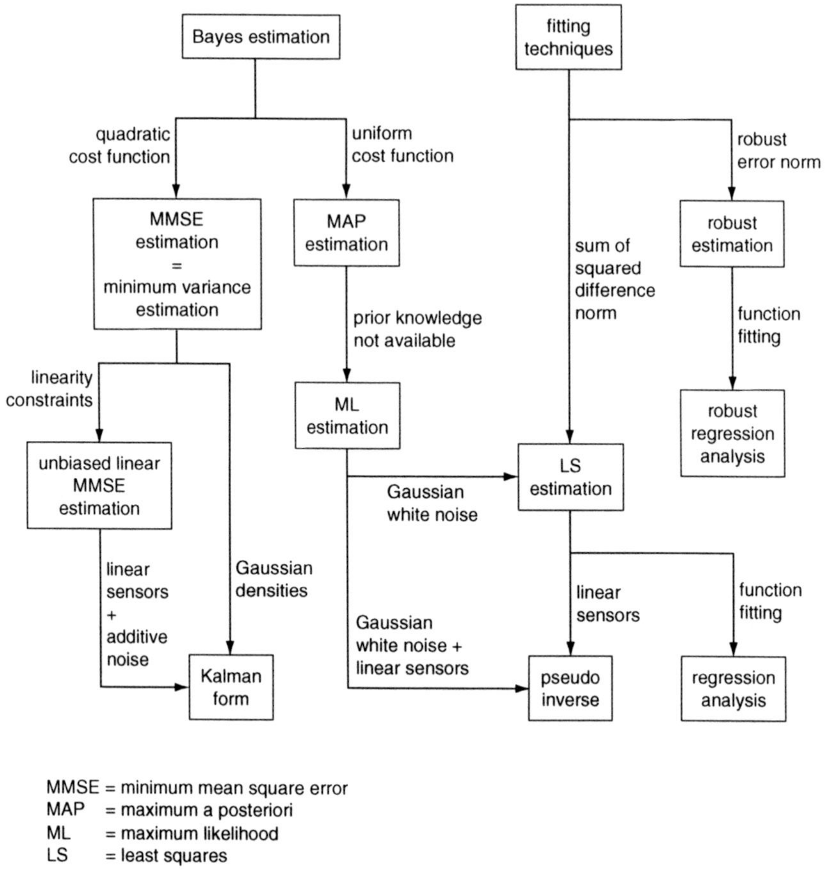

Figure 3.13 A family tree of estimators

It is remarkable that although the quadratic cost function and the unit cost function differ a lot, the solutions are identical provided that the posterior density is uni-modal and symmetric. An example of this occurs when the prior probability density and conditional probability density are both Gaussian. In that case the posterior probability density is Gaussian too.

If no prior knowledge about the parameter is available, one can use the ML estimator. Another possibility is to resort to fitting techniques, of which the LS estimator is most popular. The ML estimator is essentially a MAP estimator with uniform prior probability. Under the assumptions of normal distributed sensor noise the ML solution and the LS solution are identical. If, in addition, the sensors are linear, the ML and LS estimator become the pseudo inverse.

A robust estimator is one which can cope with outliers in the measurements. Such an estimator can be achieved by application of a robust error norm.

3.5 SELECTED BIBLIOGRAPHY

Kay, S.M., *Fundamentals of Statistical Signal Processing – Estimation Theory*, Prentice Hall, New Jersey, 1993.

Papoulis, A., *Probability, Random Variables and Stochastic Processes*, Third Edition, McGraw-Hill, New York, 1991.

Sorenson, H.W., *Parameter Estimation: Principles and Problems*, Marcel Dekker, New York, 1980.

Tarantola, A., *Inverse Problem Theory: Methods for Data Fitting and Model Parameter Estimation*, Elsevier, New York, 1987.

Van Trees, H.L., *Detection, Estimation and Modulation Theory*, Vol. 1, Wiley, New York, 1968.

3.6 EXERCISES

1. Prove that the linear MMSE estimator, whose form is $\hat{\mathbf{x}}_{lMMSE}(\mathbf{z}) = \mathbf{Kz}$, is found as:

$$\mathbf{K} = \mathbf{M}_{\mathbf{xz}}\mathbf{M}_{\mathbf{z}}^{-1} \quad \text{with} \quad \mathbf{M}_{\mathbf{xz}} \stackrel{def}{=} \mathrm{E}[\mathbf{xz}^T] \quad \text{and} \quad \mathbf{M}_{\mathbf{z}} \stackrel{def}{=} \mathrm{E}[\mathbf{zz}^T] \qquad (*)$$

2. In the Gaussian case, in Section 3.1.3, we silently assumed that the covariance matrices $\mathbf{C}_\mathbf{x}$ and $\mathbf{C}_\mathbf{v}$ are invertible. What can be said about the elements of $\mathbf{x}$ if $\mathbf{C}_\mathbf{x}$ is singular? And what about the elements of $\mathbf{v}$ if $\mathbf{C}_\mathbf{v}$ is singular? What must be done to avoid such a situation? (*)
3. Prove that, in Section 3.1.3, the posterior density is Gaussian, and prove equation (3.20). (*) Hint: use equation (3.2), and expand the argument of the exponential.
4. Prove equation (3.32). (0)
5. Use the matrix inversion lemma (b.10) to prove that the form given in (3.20):

$$\hat{\mathbf{x}}_{MMSE}(\mathbf{z}) = \mathbf{C}_e\left(\mathbf{C}_\mathbf{x}^{-1}\boldsymbol{\mu}_\mathbf{x} + \mathbf{H}^T\mathbf{C}_\mathbf{v}^{-1}\mathbf{z}\right) \quad \text{with} \quad \mathbf{C}_e = \left(\mathbf{H}^T\mathbf{C}_\mathbf{v}^{-1}\mathbf{H} + \mathbf{C}_\mathbf{x}^{-1}\right)^{-1}$$

   is equivalent to the Kalman form given in (3.33). (*)
6. Explain why (3.42) cannot be replaced by $\mathbf{C}_e = \int \mathbf{C}_e(\mathbf{x})p(\mathbf{x})d\mathbf{x}$. (*)
7. Prove that the unbiased linear MMSE estimator is indeed unbiased. (0)
8. Given that the random variable z is binominally distributed (Appendix C.1.3) with parameters (x,M). x is the probability of success of a single trial. M is the number of independent trials. z is the number of successes in M trials. The parameter x must be estimated from an observed value of z.

   - Develop the ML estimator for x. (0)
   - What is the bias and the variance of this estimator? (0)

9. If, without having observed z, the parameter x in exercise 8 is uniformly distributed between 0 and 1, what will be the posterior density $p(x|z)$ of x? Develop the MMSE estimator and the MAP estimator for this case. What will be the bias and the variance of these estimators? (*)
10. A Geiger counter is an instrument that measures radioactivity. Essentially, it counts the number of events (arrival of nuclear particles) within a given period of time. These numbers are Poisson distributed with expectation λ, i.e. the mean number of events within the period. z is the counted number of events within a period. We assume that λ is uniform distributed between 0 and L.

    - Develop the ML estimator for λ. (0)
    - Develop the MAP estimator for λ. What is the bias and the variance of the ML estimator? (*)
    - Show that the ML estimator is absolutely unbiased, and that the MAP estimator is biased. (*)
    - Give an expression for the (overall) variance of the ML estimator. (0)

4

State Estimation

The theme of the previous two chapters will now be extended to the case in which the variables of interest change over time. These variables can be either real-valued vectors (as in Chapter 3), or discrete class variables that only cover a finite number of symbols (as in Chapter 2). In both cases, the variables of interest are called *state variables*.

The design of a state estimator is based on a *state space model* that describes the underlying physical process of the application. For instance, in a tracking application, the variables of interest are the position and velocity of a moving object. The state space model gives the connection between the velocity and the position (which, in this case, is a kinematical relation). Variables, like position and velocity, are real numbers. Such variables are called *continuous states*.

The design of a state estimator is also based on a *measurement model* that describes how the data of a sensory system depend on the state variables. For instance, in a radar tracking system, the measurements are the azimuth and range of the object. Here, the measurements are directly related to the two-dimensional position of the object if represented in polar coordinates.

The estimation of a dynamic class variable, i.e. a *discrete state* variable is sometimes called *mode estimation* or *labelling*. An example is in speech recognition where – for the recognition of a word – a sequence of phonetic classes must be estimated from a sequence of acoustic features. Here too, the analysis is based on a state space model and a

Classification, Parameter Estimation and State Estimation: An Engineering Approach using MATLAB
F. van der Heijden, R.P.W. Duin, D. de Ridder and D.M.J. Tax
 ISBN: 0-470-09013-8

measurement model (in fact, each possible word has its own state space model).

The outline of the chapter is as follows. Section 4.1 gives a framework for estimation in dynamic systems. It introduces the various concepts, notations and mathematical models. Next, it presents a general scheme to obtain the optimal solution. In practice, however, such a general scheme is of less value because of the computational complexity involved when trying to implement the solution directly. Therefore, the general approach needs to be worked out for different cases. Section 4.2 is devoted to the case of continuous state variables. Practical solutions are feasible if the models are linear-Gaussian (Section 4.2.1). If the model is not linear, one can resort to suboptimal methods (Section 4.2.2). Section 4.3 deals with the discrete state case. The chapter finalizes with Section 4.4 which contains an introduction to particle filtering. This technique can handle nonlinear and non-Gaussian models covering the continuous and the discrete case, and even mixed cases (i.e. combinations of continuous and discrete states).

The chapter confines itself to the theoretical aspects of state estimation. Practical issues, like implementations, deployment, consistency checks are dealt with in Chapter 8. The use of MATLAB is also deferred to that chapter.

4.1 A GENERAL FRAMEWORK FOR ONLINE ESTIMATION

Usually, the estimation problem is divided into three paradigms:

- online estimation (optimal filtering)
- prediction
- retrodiction (smoothing, offline estimation).

Online estimation is the estimation of the present state using all the measurements that are available, i.e. all measurements up to the present time. *Prediction* is the estimation of future states. *Retrodiction* is the estimation of past states.

This section sets up a framework for the online estimation of the states of *time-discrete* processes. Of course, most physical processes evolve in the continuous time. Nevertheless, we will assume that these systems can be described adequately by a model where the continuous time is reduced to a sequence of specific times. Methods for the conversion from time-continuous to time-discrete models are described in many textbooks, for instance, on control engineering.

4.1.1 Models

We assume that the continuous time t is equidistantly sampled with period Δ. The discrete time index is denoted by an integer variable i. Hence, the moments of sampling are $t_i = i\Delta$. Furthermore, we assume that the estimation problem starts at $t = 0$. Thus, i is a non-negative integer denoting the discrete time.

The state space model

The state at time i is denoted by $\mathbf{x}(i) \in X$ where X is the state space. For discrete states, $X = \Omega = \{\omega_1, \ldots, \omega_K\}$ where ω_k is the k-th symbol (label, or class) out of K possible classes. For real-valued vectors with dimension M, we have $X = \mathbb{R}^M$.

Suppose for a moment that we have observed the state of a process during its whole history, i.e. from the beginning of time up to the present. In other words, the sequence of states $\mathbf{x}(0), \mathbf{x}(1), \ldots, \mathbf{x}(i)$ are observed and as such fully known. i denotes the present time. In addition, suppose that – using this sequence – we want to estimate (predict) the next state $\mathbf{x}(i+1)$. Assuming that the states can be modelled as random variables, we need to evaluate the conditional probability density[1] $p(\mathbf{x}(i+1)|\mathbf{x}(0), \mathbf{x}(1), \ldots, \mathbf{x}(i))$. Once this probability density is known, the application of the theory in Chapters 2 and 3 will provide the optimal estimate of $\mathbf{x}(i+1)$. For instance, if X is a real-valued vector space, the Bayes estimator from Chapter 3 provides the best prediction of the next state (the density $p(\mathbf{x}(i+1)|\mathbf{x}(0), \mathbf{x}(1), \ldots, \mathbf{x}(i))$ must be used instead of the posterior density).

Unfortunately, the evaluation of $p(\mathbf{x}(i+1)|\mathbf{x}(0), \mathbf{x}(1), \ldots, \mathbf{x}(i))$ is a nasty task because it is not clear how to establish such a density in real-world problems. Things become much easier if we succeed to define the state such that the so-called *Markov condition* applies:

$$p(\mathbf{x}(i+1)|\mathbf{x}(0), \mathbf{x}(1), \ldots, \mathbf{x}(i)) = p(\mathbf{x}(i+1)|\mathbf{x}(i)) \qquad (4.1)$$

---

[1] For the finite-state case, the probability densities transform into probabilities, and appropriate summations replace the integrals.

The probability of $\mathbf{x}(i+1)$ depends solely on $\mathbf{x}(i)$ and not on the past states. In order to predict $\mathbf{x}(i+1)$, the knowledge of the full history is not needed. It suffices to know the present state. If the Markov condition applies, the state of a physical process is a summary of the history of the process.

Example 4.1 The density of a substance mixed with a liquid
Mixing and diluting are tasks frequently encountered in the food industries, paper industry, cement industry, and so. One of the parameters of interest during the production process is the density $D(t)$ of some substance. It is defined as the fraction of the volume of the mix that is made up by the substance.

Accurate models for these production processes soon involve a large number of state variables. Figure 4.1 is a simplified view of the process. It is made up by two real-valued state variables and one discrete state. The volume $V(t)$ of the liquid in the barrel is regulated by an on/off feedback control of the input flow $f_1(t)$ of the liquid: $f_1(t) = f_0 x(t)$. The on/off switch is represented by the discrete state variable $x(t) \in \{0,1\}$. A hysteresis mechanism using a level detector (LT) prevents jitter of the switch. $x(t)$ switches to the 'on' state (=1) if $V(t) < V_{low}$, and switches back to the 'off' state (=0) if $V(t) > V_{high}$.

The rate of change of the volume is $\dot{V}(t) = f_1(t) + f_2(t) - f_3(t)$ with $f_2(t)$ the volume flow of the substance, and $f_3(t)$ the output volume flow of the mix. We assume that the output flow is governed by Torricelli's law: $f_3(t) = c\sqrt{V(t)/V_{ref}}$. The density is defined as $D(t) = V_S(t)/V(t)$ where $V_S(t)$ is the volume of the substance. The rate of change of $V_S(t)$ is: $\dot{V}_S(t) = f_2(t) - D(t)f_3(t)$. After some

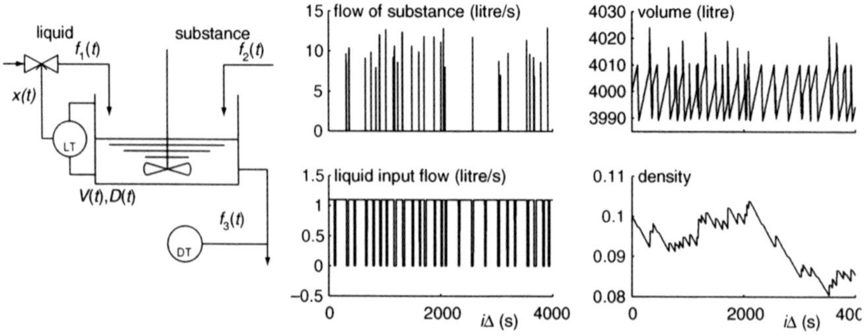

Figure 4.1 A density control system for the process industry

manipulations the following system of differential equations appears:

$$\begin{aligned}\dot{V}(t) &= f_1(t) + f_2(t) - c\sqrt{V(t)/V_{ref}} \\ \dot{D}(t) &= \frac{f_2(t)(1 - D(t)) - f_1(t)D(t)}{V(t)}\end{aligned} \tag{4.2}$$

A discrete time approximation[2] (notation: $V(i) \equiv V(i\Delta)$, $D(i) \equiv D(i\Delta)$, and so on) is:

$$\begin{aligned}V(i+1) &\simeq V(i) + \Delta\left(f_0 x(i) + f_2(i) - c\sqrt{V(i)/V_{ref}}\right) \\ D(i+1) &\simeq D(i) - \Delta\frac{f_0 x(i)D(i) - f_2(i)(1 - D(i))}{V(i)} \\ x(i+1) &= \left(x(i) \wedge \neg(V(i) > V_{high})\right) \vee (V(i) < V_{low})\end{aligned} \tag{4.3}$$

This equation is of the type $\mathbf{x}(i+1) = \mathbf{f}(\mathbf{x}(i), \mathbf{u}(i), \mathbf{w}(i))$ with $\mathbf{x}(i) = [\, V(i)\ D(i)\ x(i)\,]^T$. The elements of the vector $\mathbf{u}(i)$ are the known input variables, i.e. the non-random part of $f_2(i)$. The vector $\mathbf{w}(i)$ contains the random input, i.e. the random part of $f_2(i)$. The probability density of $\mathbf{x}(i+1)$ depends on the present state $\mathbf{x}(i)$, but not on the past states.

Figure 4.1 shows a realization of the process. Here, the substance is added to the volume in chunks with an average volume of 10 litre and at random points in time.

If the *transition probability density* $p(\mathbf{x}(i+1)|\mathbf{x}(i))$ is known together with the initial probability density $p(\mathbf{x}(0))$, then the probability density at an arbitrary time can be determined recursively:

$$p(\mathbf{x}(i+1)) = \int_{\mathbf{x}(i)\in X} p(\mathbf{x}(i+1)|\mathbf{x}(i))p(\mathbf{x}(i))d\mathbf{x} \text{ for } i = 0, 1, \ldots \tag{4.4}$$

---

[2] The approximation that is used here is $\dot{V}(t)\Delta \simeq V((i+1)\Delta) - V(i\Delta)$. The approximation is only close if Δ is sufficiently small. Other approximations may be more accurate, but this subject is outside the scope of the book.

The joint probability density of the sequence $\mathbf{x}(0), \mathbf{x}(1), \ldots, \mathbf{x}(i)$ follows readily

$$\begin{aligned} p(\mathbf{x}(0), \ldots, \mathbf{x}(i)) &= p(\mathbf{x}(i)|\mathbf{x}(i-1)|)p(\mathbf{x}(0), \ldots, \mathbf{x}(i-1)) \\ &= p(\mathbf{x}(0)) \prod_{j=1}^{i} p(\mathbf{x}(j)|x(j-1)) \end{aligned} \tag{4.5}$$

The measurement model

In addition to the state space model, we also need a *measurement model* that describes the data from the sensor in relation with the state. Suppose that at moment i the measurement data is $\mathbf{z}(i) \in Z$ where Z is the *measurement space*. For the real-valued state variables, the measurement space is often a real-valued vector space, i.e. $Z = \mathbb{R}^N$. For the discrete case, one often assumes that the measurement space is also finite, i.e. $Z = \{\vartheta_1, \ldots, \vartheta_N\}$.

The probabilistic model of the sensory system is fully defined by the *conditional probability density* $p(\mathbf{z}(i)|\mathbf{x}(0), \ldots, \mathbf{x}(i), \mathbf{z}(0), \ldots, \mathbf{z}(i-1))$. We assume that the sequence of measurements starts at time $i = 0$. In order to shorten the notation, the sequence of all measurements up to the present will be denoted by:

$$\mathbf{Z}(i) = \{\mathbf{z}(0), \ldots, \mathbf{z}(i)\} \tag{4.6}$$

We restrict ourselves to *memoryless* sensory systems, i.e. systems where $\mathbf{z}(i)$ depends on the value of $\mathbf{x}(i)$, but not on previous states nor on previous measurements. In other words:

$$p(\mathbf{z}(i)|\mathbf{x}(0), \ldots, \mathbf{x}(i), \mathbf{Z}(i-1)) = p(\mathbf{z}(i)|\mathbf{x}(i)) \tag{4.7}$$

4.1.2 Optimal online estimation

Figure 4.2 presents an overview of the scheme for the online estimation of the state. The connotation of the phrase *online* is that for each time index i an estimate $\hat{\mathbf{x}}(i)$ of $\mathbf{x}(i)$ is produced based on $\mathbf{Z}(i)$, i.e. based on all measurements that are available at that time. The crux of optimal online estimation is to maintain the posterior density $p(\mathbf{x}(i)|\mathbf{Z}(i))$ for

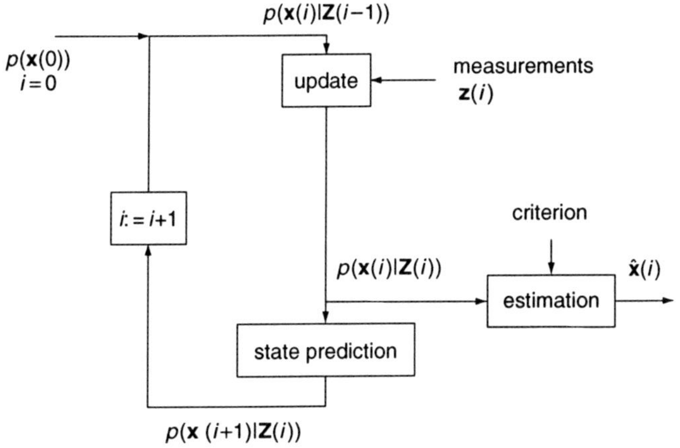

Figure 4.2 An overview of online estimation

running values of i. This density captures all the available information of the current state $\mathbf{x}(i)$ after having observed the current measurement and all previous ones. With the availability of the posterior density, the methods discussed in Chapters 2 and 3 become applicable. The only work to be done, then, is to adopt an optimality criterion and to work this out using the posterior density to get the optimal estimate of the current state.

The maintenance of the posterior density is done efficiently by means of a recursion. From the posterior density $p(\mathbf{x}(i)|\mathbf{Z}(i))$, valid for the current period i, the density $p\big(\mathbf{x}(i+1)|\mathbf{Z}(i+1)\big)$, valid for the next period $i+1$, is derived. The first step of the recursion cycle is a *prediction step*. The knowledge about $\mathbf{x}(i)$ is extrapolated to knowledge about $\mathbf{x}(i+1)$. Using Bayes' theorem for conditional probabilities in combination with the Markov condition (4.1), we have:

$$\begin{aligned} p(\mathbf{x}(i+1)|\mathbf{Z}(i)) &= \int_{\mathbf{x}(i)\in X} p\big(\mathbf{x}(i+1),\mathbf{x}(i)|\mathbf{Z}(i)\big)d\mathbf{x}(i) \\ &= \int_{\mathbf{x}(i)\in X} p\big(\mathbf{x}(i+1)|\mathbf{x}(i),\mathbf{Z}(i)\big)p(\mathbf{x}(i)|\mathbf{Z}(i))d\mathbf{x}(i) \qquad (4.8) \\ &= \int_{\mathbf{x}(i)\in X} p\big(\mathbf{x}(i+1)|\mathbf{x}(i)\big)p(\mathbf{x}(i)|\mathbf{Z}(i))d\mathbf{x}(i) \end{aligned}$$

At this point, we increment the counter i, so that $p(\mathbf{x}(i+1)|\mathbf{Z}(i))$ now becomes $p\big(\mathbf{x}(i)|\mathbf{Z}(i-1)\big)$. The increment can be done anywhere in the loop, but the choice to do it at this point leads to a shorter notation of the second step.

The second step is an *update step*. The knowledge gained from observing the measurement $\mathbf{z}(i)$ is used to refine the density. Using – once again – the theorem of Bayes, now in combination with the conditional density for memoryless sensory systems (4.7):

$$\begin{aligned} p(\mathbf{x}(i)|\mathbf{Z}(i)) &= p(\mathbf{x}(i)|\mathbf{Z}(i-1),\mathbf{z}(i)) \\ &= \frac{1}{c}p(\mathbf{z}(i)|\mathbf{x}(i),\mathbf{Z}(i-1))p(\mathbf{x}(i)|\mathbf{Z}(i-1)) \\ &= \frac{1}{c}p(\mathbf{z}(i)|\mathbf{x}(i))p(\mathbf{x}(i)|\mathbf{Z}(i-1)) \end{aligned} \tag{4.9}$$

where c is a normalization constant:

$$c = \int_{\mathbf{x}(i)\in X} p(\mathbf{z}(i)|\mathbf{x}(i))p(\mathbf{x}(i)|\mathbf{Z}(i-1))d\mathbf{x}(i) \tag{4.10}$$

The recursion starts with the processing of the first measurement $\mathbf{z}(0)$. The posterior density $p(\mathbf{x}(0)|\mathbf{Z}(0))$ is obtained using $p(\mathbf{x}(0))$ as the prior.

The outline for optimal estimation, expressed in (4.8), (4.9) and (4.10), is useful in the discrete case (where integrals turn into summations). For continuous states, a direct implementation is difficult for two reasons:

- It requires efficient representations for the N- and M-dimensional density functions.
- It requires efficient algorithms for the integrations over an M-dimensional space.

Both requirements are hard to fulfil, especially if M is large. Nonetheless, many researchers have tried to implement the general scheme. One of the most successful endeavours has resulted in what is called *particle filtering* (see Section 4.4). But first, the discussion will be focused on special cases.

4.2 CONTINUOUS STATE VARIABLES

This session addresses the case when both the state and the measurements are real-valued vectors with dimensions of M and N, respectively. The starting point is the general scheme for online estimation discussed in the previous section, and illustrated in Figure 4.2. As said before, in general, a direct implementation of the scheme is difficult. Fortunately, there are circumstances which allow a fast implementation. For instance, in the special case, where the models are linear and the disturbances have a normal distribution, an implementation based on an 'expectation and covariance matrix' representation of the probability densities is feasible (Section 4.2.1). If the models are nonlinear, but the nonlinearity is smooth, linearization techniques can be applied (Section 4.2.2). If the models are highly nonlinear, but the dimensions N and M are not too large, numerical methods are possible (Section 4.2.3 and 4.4).

4.2.1 Optimal online estimation in linear-Gaussian systems

Most literature in optimal estimation in dynamic systems deals with the particular case in which both the state model and the measurement model are linear, and the disturbances are Gaussian (the *linear-Gaussian systems*). Perhaps the main reason for the popularity is the mathematical tractability of this case.

Linear-Gaussian state space models

The state model is said to be linear if the transition from one state to the next can be expressed by a so-called *linear system equation* (or: *linear state equation, linear plant equation, linear dynamic equation*):

$$\mathbf{x}(i+1) = \mathbf{F}(i)\mathbf{x}(i) + \mathbf{L}(i)\mathbf{u}(i) + \mathbf{w}(i) \tag{4.11}$$

$\mathbf{F}(i)$ is the *system matrix*. It is an $M \times M$ matrix where M is the dimension of the state vector. M is called the *order* of the system. The vector $\mathbf{u}(i)$ is the *control vector* (*input vector*) of dimension L. Usually, the vector is generated by a controller according to some control law. As such the input vector is a deterministic signal that is fully known, at least up to the present. $\mathbf{L}(i)$ is the *gain matrix* of dimension $M \times L$. Sometimes the matrix is called the *distribution matrix* as it distributes the control vector across the elements of the state vector.

$\mathbf{w}(i)$ is the *process noise* (*system noise, plant noise*). It is a sequence of random vectors of dimension[3] M. The process noise represents the

[3] Sometimes the process noise is represented by $\mathbf{G}(i)\mathbf{w}(i)$ where $\mathbf{G}(i)$ is the *noise gain matrix*. With that, $\mathbf{w}(i)$ is not restricted to have dimension M. Of course, the dimension of $\mathbf{G}(i)$ must be appropriate.

unknown influences on the system, for instance, formed by disturbances from the environment. The process noise can also represent an unknown input/control signal. Sometimes process noise is also used to take care of modelling errors. The general assumption is that the process noise is a *white random sequence* with normal distribution. The term 'white' is used here to indicate that the expectation is zero and the autocorrelation is governed by the Kronecker delta function:

$$\begin{aligned} \mathrm{E}[\mathbf{w}(i)] &= 0 \\ \mathrm{E}\left[\mathbf{w}(i)\mathbf{w}^T(j)\right] &= \mathbf{C}_\mathbf{w}(i)\delta(i,j) \end{aligned} \tag{4.12}$$

$\mathbf{C}_\mathbf{w}(i)$ is the covariance matrix of $\mathbf{w}(i)$. Since $\mathbf{w}(i)$ is supposed to have a normal distribution with zero mean, $\mathbf{C}_\mathbf{w}(i)$ defines the density of $\mathbf{w}(i)$ in full.

The initial condition of the state model is given in terms of the expectation $\mathrm{E}[\mathbf{x}(0)]$ and the covariance matrix $\mathbf{C}_\mathbf{x}(0)$. In order to find out how these parameters of the process propagate to an arbitrary time i, the state equation (4.11) must be used recursively:

$$\begin{aligned} \mathrm{E}[\mathbf{x}(i+1)] &= \mathbf{F}(i)\mathrm{E}[\mathbf{x}(i)] + \mathbf{L}(i)\mathbf{u}(i) \\ \mathbf{C}_\mathbf{x}(i+1) &= \mathbf{F}(i)\mathbf{C}_\mathbf{x}(i)\mathbf{F}^T(i) + \mathbf{C}_\mathbf{w}(i) \end{aligned} \tag{4.13}$$

The first equation follows from $\mathrm{E}[\mathbf{w}(i)] = \mathbf{0}$. The second equation uses the fact that the process noise is white, i.e. $\mathrm{E}[\mathbf{w}(i)\mathbf{w}^T(j)] = 0$ for $i \neq j$. See (4.12).

If $\mathrm{E}[\mathbf{x}(0)]$ and $\mathbf{C}_\mathbf{x}(0)$ are known, then equation (4.13) can be used to calculate $\mathrm{E}[\mathbf{x}(1)]$ and $\mathbf{C}_\mathbf{x}(1)$. From that, by reapplying the (4.13), the next values, $\mathrm{E}[\mathbf{x}(2)]$ and $\mathbf{C}_\mathbf{x}(2)$, can be found, and so on. Thus, the iterative use of equation (4.13) gives us $\mathrm{E}[\mathbf{x}(i)]$ and $\mathbf{C}_\mathbf{x}(i)$ for arbitrary $i > 0$.

In the special case, where neither $\mathbf{F}(i)$ nor $\mathbf{C}_\mathbf{w}(i)$ depend on i, the state space model is *time invariant*. The notation can be shortened then by dropping the index, i.e. $\mathbf{F}$ and $\mathbf{C}_\mathbf{w}$. If, in addition, $\mathbf{F}$ is stable (the magnitudes of the eigenvalues of $\mathbf{F}$ are all less than one; Appendix D.3.2), the sequence $\mathbf{C}_\mathbf{x}(i)$, $i = 0, 1, \ldots$ converges to a constant matrix. The balance in (4.13) is reached when the decrease of $\mathbf{C}_\mathbf{x}(i)$ due to $\mathbf{F}$ compensates the increase due to $\mathbf{C}_\mathbf{w}$. If such is the case, then:

$$\mathbf{C}_\mathbf{x} = \mathbf{F}\mathbf{C}_\mathbf{x}\mathbf{F}^T + \mathbf{C}_\mathbf{w} \tag{4.14}$$

This is the *discrete Lyapunov equation*.

Some special state space models

In this section, we introduce some elementary random processes. They are presented here not only to illustrate the properties of state models with random inputs, but also because they are often used as building blocks for models of complicated physical processes.

Random constants Sometimes it is useful to model static parameters as states in a dynamic system. In that case, the states do not change in time:

$$\mathbf{x}(i+1) = \mathbf{x}(i) \tag{4.15}$$

Such a model is useful when the sequential measurements $\mathbf{z}(i)$ of $\mathbf{x}$ are processed online so that the estimate of $\mathbf{x}$ becomes increasingly accurate as time proceeds.

First order autoregressive models A first order autoregressive (AR) model is of the type

$$x(i+1) = \alpha x(i) + w(i) \tag{4.16}$$

where $w(i)$ is a white, zero mean, normally distributed sequence with variance σ_w^2. In this particular example, $\sigma_x^2(i) \equiv \mathrm{C_x(i)}$ since $x(i)$ is a scalar. $\sigma_x^2(i)$ can be expressed in closed form: $\sigma_x^2(\infty) = \alpha^{2i}\sigma_x^2(0)+ (1-\alpha^{2i+2})\sigma_w^2/(1-\alpha^2)$. The equation holds if $\alpha \neq 1$. The system is stable provided that $|\alpha| < 1$. In that case, the term $\alpha^{2i}\sigma_x^2(0)$ exponentially fades out. The second term asymptotically reaches the steady state, i.e. the solution of the Lyapunov equation:

$$\sigma_x^2(\infty) = \frac{1}{1-\alpha^2}\sigma_w^2 \tag{4.17}$$

If $|\alpha| > 0$, the system is not stable, and both terms grow exponentially.

First order AR models are used to describe slowly fluctuating phenomena. Physically, such phenomena occur when broadband noise is dampened by a first order system, e.g. mechanical shocks damped by a mass/dampener system. Processes that involve exponential growth are also modelled by first AR models. Figure 4.3 shows a realization of a stable and an unstable AR process.

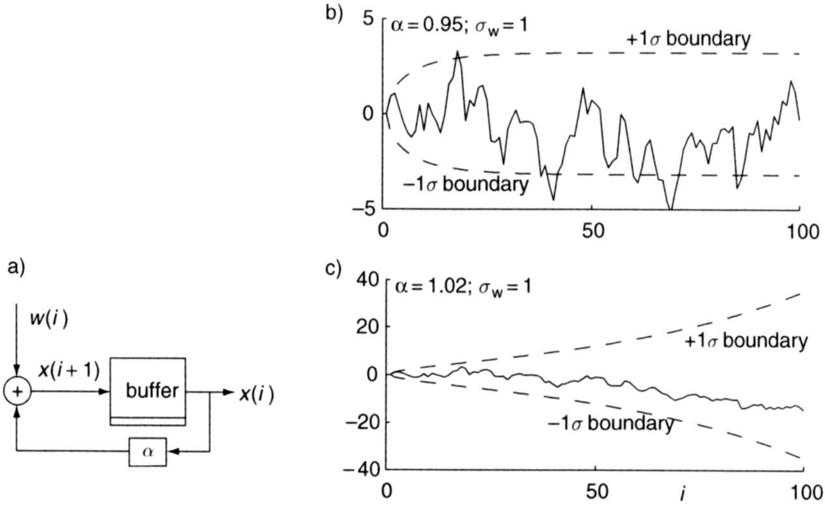

Figure 4.3 First order autoregressive models. (a) Schematic diagram of the model. (b) A stable realization. (c) An unstable realization

Random walk Consider the process:

$$x(i+1) = x(i) + w(i) \quad \text{with} \quad x(0) = 0 \tag{4.18}$$

$w(i)$ is a random sequence of independent increments, $+d$, and decrements, $-d$; each occurs with probability $\frac{1}{2}$. Suppose that after i time steps, the number of increments is $n(i)$, then the number of decrements is $i - n(i)$. Thus, $x(i)/d = 2n(i) - i$. The variable $n(i)$ has a binomial distribution (Appendix C.1.3) with parameters $(i, \frac{1}{2})$. Its mean value is $\frac{1}{2}i$; hence $\mathrm{E}[x(i)] = 0$. The variance of $n(i)$ is $\frac{1}{4}i$. Therefore, $\sigma_x^2(i) = id^2$. Clearly, $\sigma_x^2(i)$ is not limited, and the solution of the Lyapunov equation does not exist.

According to the central limit theorem (Appendix C.1.4), after about 20 time steps the distribution of $x(i)$ is reasonably well approximated by a normal distribution. Figure 4.4 shows a realization of a random walk process. Random walk processes find application in navigation problems.

Second order autoregressive models Second order autoregressive models are of the type:

$$x(i+1) = \alpha x(i) + \beta x(i-1) + w(i) \tag{4.19}$$

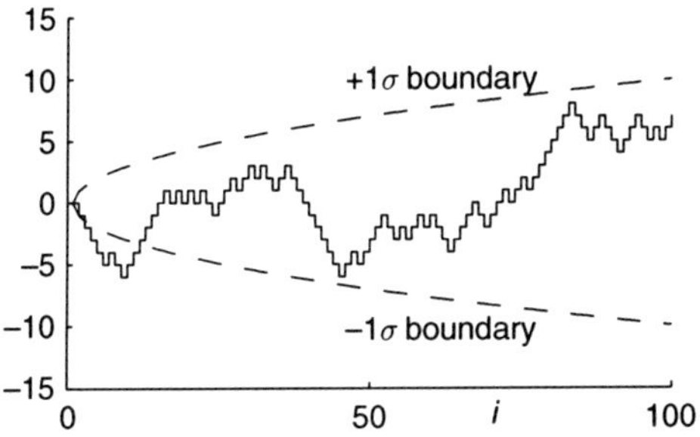

Figure 4.4 Random walk

The model can be cast into a state space model by defining $\mathbf{x}(i) \stackrel{def}{=} [x(i)\ x(i-1)]^T$:

$$\mathbf{x}(i+1) = \begin{bmatrix} x(i+1) \\ x(i) \end{bmatrix} = \begin{bmatrix} \alpha & \beta \\ 1 & 0 \end{bmatrix} \mathbf{x}(i) + \begin{bmatrix} w(i) \\ 0 \end{bmatrix} \tag{4.20}$$

The eigenvalues of this system are $\frac{1}{2}\alpha \pm \frac{1}{2}\sqrt{\alpha^2 + 4\beta}$. If $\alpha^2 > -4\beta$, the system can be regarded as a cascade of two first order AR processes with two real eigenvalues. However, if $\alpha^2 < -4\beta$, the eigenvalues become complex and can be written as $de^{\pm 2\pi jf}$ with $j = \sqrt{-1}$. The magnitude of the eigenvalues, i.e. the damping, is $d = \sqrt{-\beta}$. The frequency f is found by the relation $\cos 2\pi f = |\alpha|/(2d)$. The solution of the Lyapunov equation is obtained by multiplying (4.19) on both sides by $x(i+1)$, $x(i)$ and $x(i-1)$, and taking the expectation:

$$\begin{aligned}
\mathrm{E}[x(i+1)x(i+1)] &= \mathrm{E}[\alpha x(i)x(i+1) + \beta x(i-1)x(i+1) + w(i)x(i+1)] \\
\mathrm{E}[x(i+1)x(i)] &= \mathrm{E}[\alpha x(i)x(i) + \beta x(i-1)x(i) + w(i)x(i)] \\
\mathrm{E}[x(i+1)x(i-1)] &= \mathrm{E}[\alpha x(i)x(i-1) + \beta x(i-1)x(i-1) + w(i)x(i-1)]
\end{aligned}$$

$$\Downarrow$$

$$\begin{aligned}
\sigma_x^2 &= \alpha\sigma_x^2 r_1 + \beta\sigma_x^2 r_2 + \sigma_w^2 \\
\sigma_x^2 r_1 &= \alpha\sigma_x^2 + \beta\sigma_x^2 r_1 \\
\sigma_x^2 r_2 &= \alpha\sigma_x^2 r_1 + \beta\sigma_x^2
\end{aligned} \tag{4.21}$$

$$\Downarrow$$

$$r_1 = \frac{\alpha}{1-\beta} \quad r_2 = \frac{\alpha^2 + \beta - \beta^2}{1-\beta} \quad \sigma_x^2 = \frac{\sigma_w^2}{1 - ar_1 - \beta r_2}$$

The equations are valid if the system is in the steady state, i.e. when $\sigma_x^2(i) = \sigma_x^2(i+1)$ and $\mathrm{E}[x(i+1)x(i)] = \mathrm{E}[x(i)x(i-1)]$. For this situation the abbreviated notation $\sigma_x^2 \equiv \sigma_x^2(\infty)$ is used. Furthermore, r_k denotes the autocorrelation between $x(i)$ and $x(i+k)$. That is, $\mathrm{E}[x(i)x(i+k)] = \mathrm{Cov}[x(i)x(i+k)] = \sigma_x^2 r_k$ (only valid in the steady state). See also Section 8.1.5 and Appendix C.2.

Second order AR models are the time-discrete counterparts of second order differential equations describing physical processes that behave like a damped oscillator, e.g. a mass/spring/dampener system, a swinging pendulum, an electrical *LCR*-circuit, and so on. Figure 4.5 shows a realization of a second order AR process.

Prediction

Equation (4.13) is the basis for prediction. Suppose that at time i an unbiased estimate $\hat{\mathbf{x}}(i)$ is known together with the associated error covariance $\mathbf{C}_e(i)$. The best predicted value (MMSE) of the state for ℓ samples ahead of i is obtained by the recursive application of (4.13). The recursion starts with $\mathrm{E}[\mathbf{x}(i)] = \hat{\mathbf{x}}(i)$ and terminates when $\mathrm{E}[\mathbf{x}(i+\ell)]$ is obtained. The covariance matrix $\mathbf{C}_e(i)$ is a measure of the magnitudes of the random fluctuations of $\mathbf{x}(i)$ around $\hat{\mathbf{x}}(i)$. As such it is also a measure of uncertainty. Therefore, the recursive usage of (4.13) applied to $\mathbf{C}_e(i)$ gives $\mathbf{C}_e(i+\ell)$, i.e. the uncertainty of the prediction. With that, the recursive equations for the prediction become:

$$\begin{aligned} \hat{\mathbf{x}}(i+\ell+1) &= \mathbf{F}(i+\ell)\hat{\mathbf{x}}(i+\ell) + \mathbf{L}(i+\ell)\mathbf{u}(i+\ell) \\ \mathbf{C}_e(i+\ell+1) &= \mathbf{F}(i+\ell)\mathbf{C}_e(i+\ell)\mathbf{F}^T(i+\ell) + \mathbf{C}_w(i+\ell) \end{aligned} \tag{4.22}$$

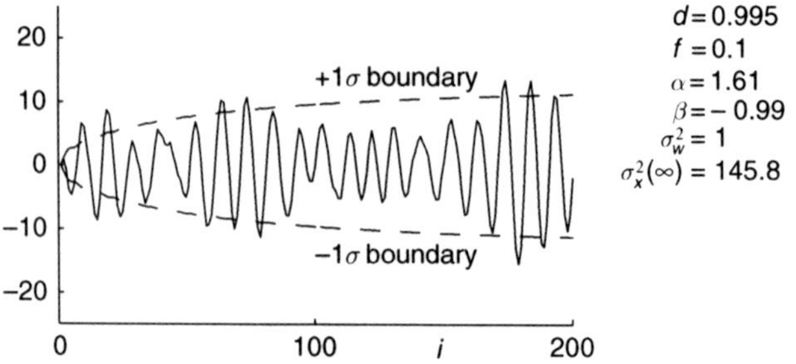

Figure 4.5 Second order autoregressive process

Example 4.2 Prediction of a swinging pendulum
The mechanical system shown in Figure 4.6 is a pendulum whose position is described by the angle $\theta(t)$ and the position of the hinge. The length R of the arm is constant. The mass m is concentrated at the end. The hinge moves randomly in the horizontal direction with an acceleration given by $a(t)$. Newton's law, applied to the geometrical set up, gives:

$$ma(t)\cos\theta(t) + mR\ddot{\theta}(t) = -mg\sin\theta(t) - \frac{mk}{R}\dot{\theta}(t) \tag{4.23}$$

k is a viscous friction constant; g is the gravitation constant. If the sampling period Δ, and $\max(|\theta|)$ is sufficiently small, the equation can be transformed to a second order AR process. The following state model, with $x_1(i) = \theta(i\Delta)$ and $x_2(i) = \dot{\theta}(i\Delta)$, is equivalent to that AR process:

$$\begin{aligned} x_1(i+1) &= x_1(i) + \Delta x_2(i) \\ x_2(i+1) &= x_2(i) - \frac{\Delta}{R}\left(gx_1(i) + \frac{k}{R}x_2(i) + a(i)\right) \end{aligned} \tag{4.24}$$

Figure 4.7(a) shows the result of a so-called *fixed interval prediction*. The prediction is performed from a fixed point in time (i is fixed), and with a running lead, that is $\ell = 1, 2, 3, \ldots$. In Figure 4.7(a), the fixed point is $i \sim 10$(s). Assuming that for that i the state is fully known, $\hat{\mathbf{x}}(i) = \mathbf{x}(i)$ and $\mathbf{C}_e(i) = 0$, predictions for the next states are calculated and plotted. It can be seen that the prediction error increases with the lead. For larger leads, the prediction covariance matrix approaches the state covariance matrix, i.e. $\mathbf{C}_e(\infty) = \mathbf{C}_x(\infty)$.

Figure 4.7(b) shows the results from *fixed lead prediction*. Here, the recursions are reinitiated for each i. The lead is fixed and chosen such that the relative prediction error is 36%.

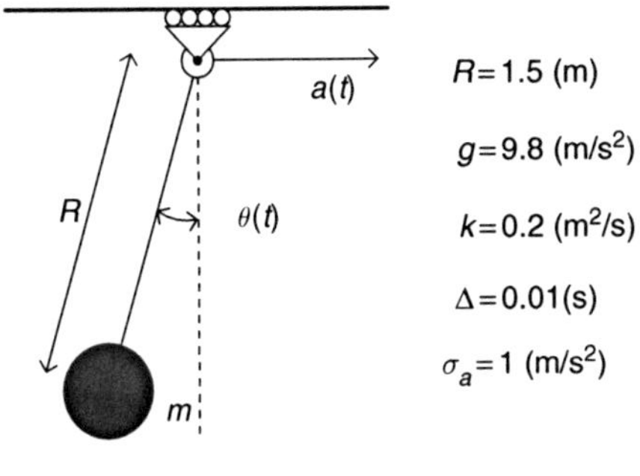

Figure 4.6 A swinging pendulum

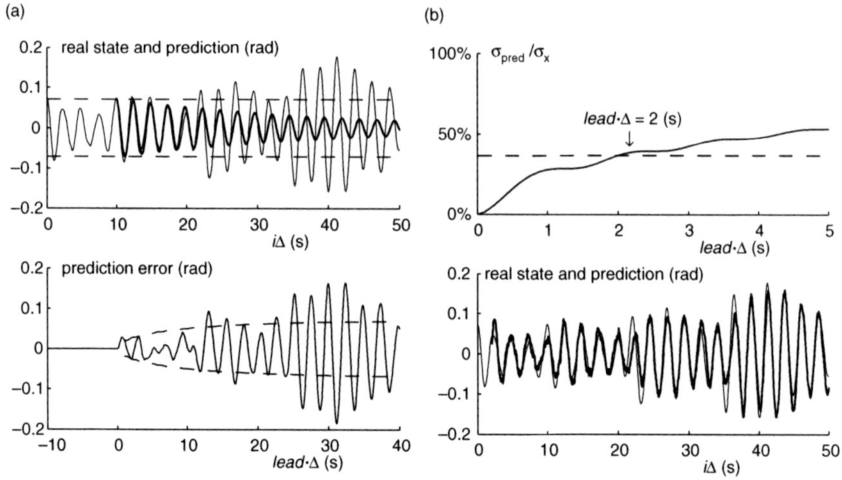

Figure 4.7 Prediction. (a) Fixed interval prediction. (b) Fixed lead prediction

Linear-Gaussian measurement models

A linear measurement model takes the following form:

$$\mathbf{z}(i) = \mathbf{H}(i)\mathbf{x}(i) + \mathbf{v}(i) \tag{4.25}$$

$\mathbf{H}(i)$ is the so-called *measurement matrix*. It is an $N \times M$ matrix. $\mathbf{v}(i)$ is the *measurement noise*. It is a sequence of random vectors of dimension N. Obviously, the measurement noise represents the noise sources in the sensory system. Examples are thermal noise in a sensor and the quantization errors of an AD converter.

The general assumption is that the measurement noise is a zero mean, white random sequence with normal distribution. In addition, the sequence is supposed to have no correlation between the measurement noise and the process noise:

$$\begin{aligned} &\mathrm{E}[\mathbf{v}(i)] = \mathbf{0} \\ &\mathrm{E}\left[\mathbf{v}(i)\mathbf{v}^T(j)\right] = \mathbf{C}_v(i)\delta(i,j) \\ &\mathrm{E}\left[\mathbf{v}(i)\mathbf{w}^T(j)\right] = \mathbf{0} \end{aligned} \tag{4.26}$$

$\mathbf{C}_v(i)$ is the covariance matrix of $\mathbf{v}(i)$. $\mathbf{C}_v(i)$ specifies the density of $\mathbf{v}(i)$ in full.

The sensory system is time invariant if neither $\mathbf{H}$ nor $\mathbf{C}_v$ depends on i.

The discrete Kalman filter

The concepts developed in the previous section are sufficient to transform the general scheme presented in Section 4.1 into a practical solution. In order to develop the estimator, first the initial condition valid for $i = 0$ must be established. In the general case, this condition is defined in terms of the probability density $p(\mathbf{x}(0))$ for $\mathbf{x}(0)$. Assuming a normal distribution for $\mathbf{x}(0)$ it suffices to specify only the expectation $\mathrm{E}[\mathbf{x}(0)]$ and the covariance matrix $\mathbf{C}_\mathbf{x}(0)$. Hence, the assumption is that these parameters are available. If not, we can set $\mathrm{E}[\mathbf{x}(0)] = \mathbf{0}$ and let $\mathbf{C}_\mathbf{x}(0)$ approach to infinity, i.e. $\mathbf{C}_\mathbf{x}(0) \to \infty\mathbf{I}$. Such a large covariance matrix represents the lack of prior knowledge.

The next step is to establish the posterior density $p(\mathbf{x}(0)|\mathbf{z}(0))$ from which the optimal estimate for $\mathbf{x}(0)$ follows. At this point, we enter the loop of Figure 4.2. Hence, we calculate the density $p(\mathbf{x}(1)|\mathbf{z}(0))$ of the next state, and process the measurement $\mathbf{z}(1)$ resulting in the updated density $p(\mathbf{x}(1)|\mathbf{z}(0), \mathbf{z}(1)) = p(\mathbf{x}(1)|\mathbf{Z}(1))$. From that, the optimal estimate for $\mathbf{x}(1)$ follows. This procedure has to be iterated for all the next time cycles.

The representation of all the densities that are involved can be given in terms of expectations and covariances. The reason is that any linear combination of Gaussian random vectors yields a vector that is also Gaussian. Therefore, both $p(\mathbf{x}(i+1)|\mathbf{Z}(i))$ and $p((\mathbf{x}(i)|\mathbf{Z}(i))$ are fully represented by their expectations and covariances. In order to discriminate between the two situations a new notation is needed. From now on, the conditional expectation $\mathrm{E}[\mathbf{x}(i)|\mathbf{Z}(j)]$ will be denoted by $\bar{\mathbf{x}}(i|j)$. It is the expectation associated with the conditional density $p(\mathbf{x}(i)|\mathbf{Z}(j))$. The covariance matrix associated with this density is denoted by $\mathbf{C}(i|j)$.

The update, i.e. the determination of $p((\mathbf{x}(i)|\mathbf{Z}(i))$ given $p(\mathbf{x}(i)|\mathbf{Z}(i-1))$, follows from Section 3.1.5 where it has been shown that the unbiased linear MMSE estimate in the linear-Gaussian case equals the MMSE estimate, and that this estimate is the conditional expectation. Application of (3.33) and (3.45) to (4.25) and (4.26) gives:

$$
\begin{aligned}
\hat{\mathbf{z}}(i) &= \mathbf{H}(i)\bar{\mathbf{x}}(i|i-1) \\
\mathbf{S}(i) &= \mathbf{H}(i)\mathbf{C}(i|i-1)\mathbf{H}^T(i) + \mathbf{C}_\mathbf{v}(i) \\
\mathbf{K}(i) &= \mathbf{C}(i|i-1)\mathbf{H}^T(i)\mathbf{S}^{-1}(i) \\
\bar{\mathbf{x}}(i|i) &= \bar{\mathbf{x}}(i|i-1) + \mathbf{K}(i)(\mathbf{z}(i) - \hat{\mathbf{z}}(i)) \\
\mathbf{C}(i|i) &= \mathbf{C}(i|i-1) - \mathbf{K}(i)\mathbf{S}(i)\mathbf{K}^T(i)
\end{aligned}
\tag{4.27}
$$

The interpretation is as follows. $\hat{\mathbf{z}}(i)$ is the *predicted measurement*. It is an unbiased estimate of $\mathbf{z}(i)$ using all information from the past. The so-called *innovation matrix* $\mathbf{S}(i)$ represents the uncertainty of the predicted measurement. The uncertainty is due to two factors: the uncertainty of $\mathbf{x}(i)$ as expressed by $\mathbf{C}(i|i-1)$, and the uncertainty due to the measurement noise $\mathbf{v}(i)$ as expressed by $\mathbf{C}_\mathbf{v}(i)$. The matrix $\mathbf{K}(i)$ is the *Kalman gain matrix*. This matrix has large, when $\mathbf{S}(i)$ is small and $\mathbf{C}(i|i-1)\mathbf{H}^T(i)$ is large, that is, when the measurements are relatively accurate. When this is the case, the values in the error covariance matrix $\mathbf{C}(i|i)$ will be much smaller than $\mathbf{C}(i|i-1)$.

The prediction, i.e. the determination of $p(\mathbf{x}(i+1)|\mathbf{Z}(i))$ given $p((\mathbf{x}(i)|\mathbf{Z}(i))$, boils down to finding out how the expectation $\overline{\mathbf{x}}(i|i)$ and the covariance matrix $\mathbf{C}(i|i)$ propagate to the next state. Using (4.11) and (4.13) we have:

$$\begin{aligned} \overline{\mathbf{x}}(i+1|i) &= \mathbf{F}(i)\overline{\mathbf{x}}(i|i) + \mathbf{L}(i)\mathbf{u}(i) \\ \mathbf{C}(i+1|i) &= \mathbf{F}(i)\mathbf{C}(i|i)\mathbf{F}^T(i) + \mathbf{C}_\mathbf{w}(i) \end{aligned} \tag{4.28}$$

At this point, we increment the counter, and $\overline{\mathbf{x}}(i+1|i)$ and $\mathbf{C}(i+1|i)$ become $\overline{\mathbf{x}}(i|i-1)$ and $\mathbf{C}(i|i-1)$. These recursive equations are generally referred to as the *discrete Kalman filter* (DKF).

In the Gaussian case, it does not matter much which optimality criterion we select. MMSE estimation, MMAE estimation and MAP estimation yield the same result, i.e. the conditional mean. Hence, the final estimate is found as $\hat{\mathbf{x}}(i) = \overline{\mathbf{x}}(i|i)$. It is an absolute unbiased estimate, and its covariance matrix is $\mathbf{C}(i|i)$. Therefore, this matrix is often called the *error covariance matrix*.

In the time invariant case, and assuming that the Kalman filter is stable, the error covariance matrix converges to a constant matrix. In that case, the innovation matrix and the Kalman gain matrix become constant as well. The filter is said to be in the *steady state*. The steady state condition simply implies that $\mathbf{C}(i+1|i) = \mathbf{C}(i|i-1)$. If the notation for this matrix is shortened to $\mathbf{P}$, then (4.28) and (4.27) lead to the following equation:

$$\mathbf{P} = \mathbf{F}\mathbf{P}\mathbf{F}^T + \mathbf{C}_\mathbf{w} - \mathbf{F}\mathbf{P}\mathbf{H}^T\left(\mathbf{H}\mathbf{P}\mathbf{H}^T + \mathbf{C}_\mathbf{v}\right)^{-1}\mathbf{H}\mathbf{P}\mathbf{F}^T \tag{4.29}$$

The equation is known as the *discrete algebraic Ricatti equation*. Usually, it is solved numerically. Its solution implicitly defines the steady state solution for $\mathbf{S}$, $\mathbf{K}$ and $\mathbf{C}$.

Example 4.3 Application to the swinging pendulum
In this example we reconsider the mechanical system shown in Figure 4.6, and described in Example 4.2. Suppose a gyroscope measures the angular speed $\dot{\theta}(t)$ at regular intervals of 0.4 s. The discrete model in Example 4.2 uses a sampling period of $\Delta = 0.01$ s. We could increase the sampling period to 0.4 s in order to match it with the sampling period of the measurements, but then the applied discrete approximation would be poor. Instead, we model the measurements with a time variant model: $z(i) = \mathbf{H}(i)\mathbf{x}(i) + v(i)$ where both $\mathbf{H}(i)$ and $\mathbf{C}_v(i)$ are always zero except for those i that are multiples of 40:

$$\mathbf{H}(i) = \begin{cases} [0\ 1] & \text{if } \mathrm{mod}(i, 40) = 0 \\ \mathbf{0} & \text{elsewhere} \end{cases} \tag{4.30}$$

The effect of such a measurement matrix is that during 39 consecutive cycles of the loop only predictions take place. During these cycles $\mathbf{H}(i) = 0$, and consequently the Kalman gains are zero. The corresponding updates would have no effect, and can be skipped. Only during each 40th cycle $\mathbf{H}(i) \neq 0$, and a useful update takes place.

Figure 4.8 shows measurements obtained from the swinging pendulum. The variance of the measurement noise is $\sigma_v^2 \equiv \mathbf{C}_v(i) = 0.1^2$ (rad^2/s^2). The filter is initiated with $\bar{\mathbf{x}}(0) = 0$ and with $\mathbf{C}_\mathbf{x}(0) \to \infty$. The figure also shows the second element of the Kalman gain matrix, e.g. $\mathbf{K}_{2,1}(i)$ (the first

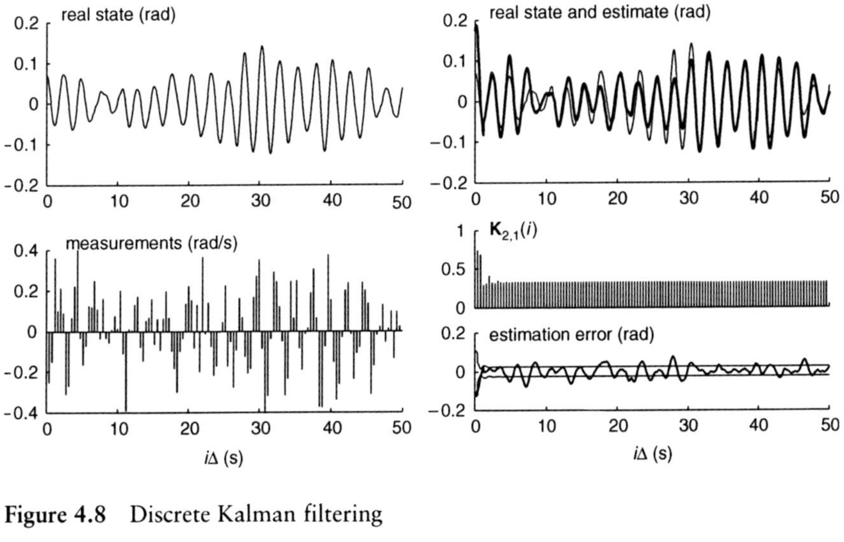

Figure 4.8 Discrete Kalman filtering

one is much smaller, but follows the same pattern). It can be seen that after about 3 (s) the Kalman gain (and the error covariance matrix) remain constant. The filter has reached its steady state.

4.2.2 Suboptimal solutions for nonlinear systems

This section extends the discussion on state estimation to the more general case of nonlinear systems and nonlinear measurement functions:

$$\begin{aligned} \mathbf{x}(i+1) &= \mathbf{f}(\mathbf{x}(i), \mathbf{u}(i), i) + \mathbf{w}(i) \\ \mathbf{z}(i) &= \mathbf{h}(\mathbf{x}(i), i) + \mathbf{v}(i) \end{aligned} \tag{4.31}$$

The vector $\mathbf{f}(\cdot, \cdot, \cdot)$ is a nonlinear, time variant function of the state $\mathbf{x}(i)$ and the control vector $\mathbf{u}(i)$. The control vector is a deterministic signal. Since $\mathbf{u}(i)$ is fully known, it only causes an explicit time dependency in $\mathbf{f}(\cdot, \cdot, \cdot)$. Without loss of generality, the notation can be shortened to $\mathbf{f}(\mathbf{x}(i),i)$, because such an explicit time dependency is already implied in that shorter notation. If no confusion can occur, the abbreviation $\mathbf{f}(\mathbf{x}(i))$ will be used instead of $\mathbf{f}(\mathbf{x}(i), i)$ even if the system does depend on time. As before, $\mathbf{w}(i)$ is the process noise. It is modelled as zero mean, Gaussian white noise with covariance matrix $\mathbf{C}_\mathbf{w}(i)$. The vector $\mathbf{h}(\cdot,\cdot)$ is a nonlinear measurement function. Here too, if no confusion is possible, the abbreviation $\mathbf{h}(\mathbf{x}(i))$ will be used covering both the time variant and the time invariant case. $\mathbf{v}(i)$ represents the measurement noise, modelled as zero mean, Gaussian white noise with covariance matrix $\mathbf{C}_\mathbf{v}(i)$.

Any Gaussian random vector that undergoes a *linear* operation retains its Gaussian distribution. A linear operator only affects the expectation and the covariance matrix of that vector. This property is the basis of the Kalman filter. It is applicable to linear-Gaussian systems, and it permits a solution that is entirely expressed in terms of expectations and covariance matrices. However, the property does not hold for nonlinear operations. In nonlinear systems, the state vectors and the measurement vectors are not Gaussian distributed, even though the process noise and the measurement noise might be. Consequently, the expectation and the covariance matrix do not fully specify the probability density of the state vector. The question is then how to determine this non-Gaussian density, and how to represent it in an economical way. Unfortunately, no general answer exists to this question.

This section seeks the answer by assuming that the nonlinearities of the system are smooth enough to allow linear or quadratic approximations.

Using these approximations, Kalman-like filters become within reach. These solutions are *suboptimal* since there is no guarantee that the approximations are close.

An obvious way to get the approximations is by application of a Taylor series expansion of the functions. Ignorance of the higher order terms of the expansion gives the desired approximation. The Taylor series exists by virtue of the assumed smoothness of the nonlinear function; it does not work out if the nonlinearity is a discontinuity, i.e. saturation, dead zone, hysteresis, and so on. The Taylor series expansions of the system equations are as follows:

$$\begin{aligned}\mathbf{f}(\mathbf{x}+\boldsymbol{\varepsilon}) &= \mathbf{f}(\mathbf{x}) + \mathbf{F}(\mathbf{x})\boldsymbol{\varepsilon} + \frac{1}{2}\sum_{m=0}^{M-1} \mathbf{e}_m \boldsymbol{\varepsilon}^T \mathbf{F}_{xx}^m(\mathbf{x})\boldsymbol{\varepsilon} + HOT \\ \mathbf{h}(\mathbf{x}+\boldsymbol{\varepsilon}) &= \mathbf{h}(\mathbf{x}) + \mathbf{H}(\mathbf{x})\boldsymbol{\varepsilon} + \frac{1}{2}\sum_{n=0}^{N-1} \mathbf{e}_n \boldsymbol{\varepsilon}^T \mathbf{H}_{xx}^n(\mathbf{x})\boldsymbol{\varepsilon} + HOT\end{aligned} \tag{4.32}$$

$\mathbf{e}_m$ is the Cartesian basis vector with appropriate dimension. The m-th element of $\mathbf{e}_m$ is one; the other are zeros. $\mathbf{e}_m$ can be used to select the m-th element of a vector: $\mathbf{e}_m^T\mathbf{x} = x_m$. $\mathbf{F}(\mathbf{x})$ and $\mathbf{H}(\mathbf{x})$ are Jacobian matrices. $\mathbf{F}_{xx}^m(\mathbf{x})$ and $\mathbf{H}_{xx}^n(\mathbf{x})$ are Hessian matrices. These matrices are defined in Appendix B.4. *HOT* are the higher order terms. The quadratic approximation arises if the higher order terms are ignored. If, in addition, the quadratic term is ignored, the approximation becomes linear, e.g. $\mathbf{f}(\mathbf{x}+\boldsymbol{\varepsilon}) \cong \mathbf{f}(\mathbf{x}) + \mathbf{F}(\mathbf{x})\boldsymbol{\varepsilon}$.

The linearized Kalman filter

The simplest approximation occurs when the system equations are linearized around some fixed value $\bar{\bar{\mathbf{x}}}$ of $\mathbf{x}(i)$. This approximation is useful if the system is time invariant and stable, and if the states swing around an equilibrium state. Such a state is the solution of:

$$\bar{\bar{\mathbf{x}}} = \mathbf{f}(\bar{\bar{\mathbf{x}}}) \tag{4.33}$$

Defining $\boldsymbol{\varepsilon}(i) = \mathbf{x}(i) - \bar{\bar{\mathbf{x}}}$, the linear approximation of the state equation (4.31) becomes $\mathbf{x}(i+1) \cong \mathbf{f}(\bar{\bar{\mathbf{x}}}) + \mathbf{F}(\bar{\bar{\mathbf{x}}})\boldsymbol{\varepsilon}(i) + \mathbf{w}(i)$. After some manipulations:

$$\begin{aligned}\mathbf{x}(i+1) &\cong \mathbf{F}(\bar{\bar{\mathbf{x}}})\mathbf{x}(i) + \left(\mathbf{I} - \mathbf{F}(\bar{\bar{\mathbf{x}}})\right)\bar{\bar{\mathbf{x}}} + \mathbf{w}(i) \\ \mathbf{z}(i) &\cong \mathbf{h}(\bar{\bar{\mathbf{x}}}) + \mathbf{H}(\bar{\bar{\mathbf{x}}})\left(\mathbf{x}(i) - \bar{\bar{\mathbf{x}}}\right) + \mathbf{v}(i)\end{aligned} \tag{4.34}$$

By interpreting the term $(\mathbf{I} - \mathbf{F}(\bar{\bar{\mathbf{x}}}))\bar{\bar{\mathbf{x}}}$ as a constant control input, and by compensating the offset term $\mathbf{h}(\bar{\bar{\mathbf{x}}}) - \mathbf{H}(\bar{\bar{\mathbf{x}}})\bar{\bar{\mathbf{x}}}$ in the measurement vector, these equations become equivalent to (4.11) and (4.25). This allows the direct application of the DKF as given in (4.28) and (4.27).

Many practical implementations of the discrete Kalman filter are inherently linearized Kalman filters because physical processes are seldom exactly linear, and often a linear model is only an approximation of the real process.

Example 4.4 The swinging pendulum

The swinging pendulum from Example 4.2 is described by (4.23):

$$ma(t)\cos\theta(t) + mR\ddot{\theta}(t) = -mg\sin\theta(t) - \frac{mk}{R}\dot{\theta}(t)$$

This equation is transformed in the linear model given in (4.24). In fact, this is a linearized model derived from:

$$\begin{aligned} x_1(i+1) &= x_1(i) + \Delta x_2(i) \\ x_2(i+1) &= x_2(i) - \frac{\Delta}{R}\left(g\sin x_1(i) + \frac{k}{R}x_2(i) + a(i)\cos x_1(i)\right) \end{aligned} \tag{4.35}$$

The equilibrium for $a(i) = 0$ is $\bar{\bar{\mathbf{x}}} = \mathbf{0}$. The linearized model is obtained by equating $\sin x_1(i) \cong x_1(i)$ and $\cos x_1(i) \cong 1$.

Example 4.5 A linearized model for volume density estimation

In Section 4.1.1 we introduced the nonlinear, non-Gaussian problem of the volume density estimation of a mix in the process industry (Example 4.1). The model included a discrete state variable to describe the on/off regulation of the input flow. We will now replace this model by a linear feedback mechanism:

$$\begin{aligned} V(i+1) &\simeq V(i) + \Delta\Big(\alpha(V_0 - V(i)) + w_1(i) + \bar{f}_2 \\ &\qquad + w_2(i) - c\sqrt{V(i)/V_{ref}}\Big) \\ D(i+1) &\simeq D(i) \\ &\qquad - \Delta\frac{(\alpha(V_0 - V(i)) + w_1(i))D(i) - \left(\bar{f}_2 + w_2(i)\right)(1 - D(i))}{V(i)} \end{aligned} \tag{4.36}$$

The liquid input flow has now been modelled by $f_1(i) = \alpha(V_0 - V(i)) + w_1(i)$. The constant $V_0 = V_{ref} + (c - \overline{f}_2)/\alpha$ (with $V_{ref} = ½(V_{high} + V_{low})$) realizes the correct mean value of $V(i)$. The random part $w_1(i)$ of $f_1(i)$ establishes a first order AR model of $V(i)$ which is used as a rough approximation of the randomness of $f_1(i)$. The substance input flow $f_2(i)$, which in Example 4.1 appears as chunks at some discrete points of time, is now modelled by $\overline{f}_2 + w_2(i)$, i.e. a continuous flow with some randomness.

The equilibrium is found as the solution of $V(i+1) = V(i)$ and $D(i+1) = D(i)$. The results are:

$$\bar{\bar{\mathbf{x}}} = \begin{bmatrix} \bar{\bar{V}} \\ \bar{\bar{D}} \end{bmatrix} = \begin{bmatrix} V_{ref} \\ \overline{f}_2/(\overline{f}_2 + \alpha(V_0 - V_{ref})) \end{bmatrix} \tag{4.37}$$

The expressions for the Jacobian matrices are:

$$\mathbf{F}(\mathbf{x}) = \begin{bmatrix} 1 - \Delta\left(\alpha + \dfrac{c}{2\sqrt{VV_{ref}}}\right) & 0 \\ \Delta\left(\dfrac{\alpha D}{V} - \dfrac{\overline{f}_2(1-D) - \alpha(V_0 - V)D}{V^2}\right) & 1 - \Delta\dfrac{\overline{f}_2 + \alpha(V_0 - V)}{V} \end{bmatrix} \tag{4.38}$$

$$\mathbf{G}(\mathbf{x}) = \frac{\partial \mathbf{f}(\mathbf{x}, \mathbf{w})}{\partial \mathbf{w}} = \begin{bmatrix} \Delta & \Delta \\ -\Delta\dfrac{D}{V} & \Delta\dfrac{1-D}{V} \end{bmatrix} \tag{4.39}$$

The considered measurement system consists of two sensors:

- A level sensor that measures the volume $V(i)$ of the barrel.
- A radiation sensor that measures the density $D(i)$ of the output flow.

The latter uses the radiation of some source, e.g. X-rays, that is absorbed by the fluid. According to Beer–Lambert's law, $P_{out} = P_{in} \exp(-\mu D(i))$ where μ is a constant depending on the path length of the ray and on the material. Using an optical detector the measurement function becomes $z = U \exp(-\mu D) + v$ with U a constant voltage. With that, the model of the two sensors becomes:

$$\mathbf{z}(i) = \mathbf{h}(\mathbf{x}(i)) + \mathbf{v}(i) = \begin{bmatrix} V(i) + v_1(i) \\ U \exp(-\mu D(i)) + v_2(i) \end{bmatrix} \tag{4.40}$$

with the Jacobian matrix:

$$\mathbf{H}(\mathbf{x}) = \begin{bmatrix} 1 & 0 \\ 0 & -U\mu \exp(-\mu D) \end{bmatrix} \tag{4.41}$$

The best fitted parameters of this model are as follows:

Volume control	Substance flow	Output flow	Measurement system
$\Delta = 1$ (s)	$\bar{f}_2 = 0.1$ (lit/s)	$V_{ref} = 4000$ (lit)	$\sigma_{v_1} = 16$ (lit)
$V_0 = 4001$ (lit)	$\sigma_{w_2} = 0.9$ (lit/s)	$c = 1$ (lit/s)	$U = 1000$ (V)
$\alpha = 0.95$ (1/s)			$\mu = 100$
$\sigma_{w_1} = 0.1225$ (lit/s)			$\sigma_{v_2} = 0.02$ (V)

Figure 4.9 shows the real states (obtained from a simulation using the model from Example 4.1), observed measurements, estimated states and estimation errors. It can be seen that:

- The density can only be estimated if the real density is close to the equilibrium. In every other region, the linearization of the measurement is not accurate enough.
- The estimator is able to estimate the mean volume, but cannot keep track of the fluctuations. The estimation error of the volume is much larger than indicated by the 1σ boundaries (obtained from the error covariance matrix). The reason for the inconsistent behaviour is that the linear-Gaussian AR model does not fit well enough.

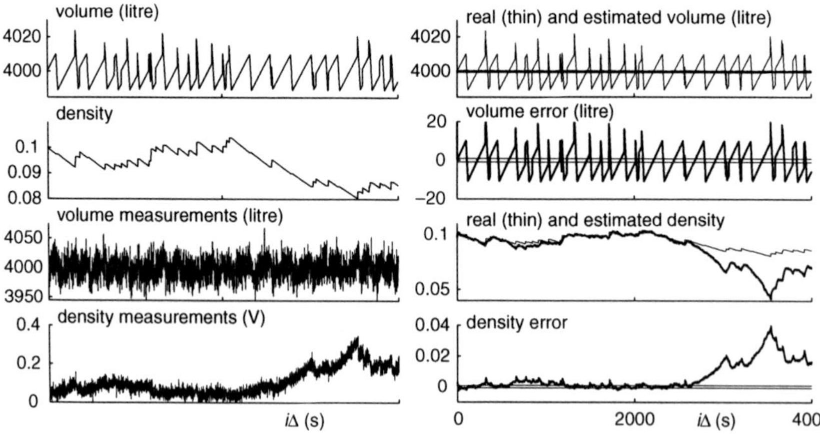

Figure 4.9 Linearized Kalman filtering applied to the volume density estimation problem

The extended Kalman filter

A straightforward generalization of the linearized Kalman filter occurs when the equilibrium point $\bar{\bar{\mathbf{x}}}$ is replaced with a *nominal trajectory* $\bar{\bar{\mathbf{x}}}(i)$, recursively defined as:

$$\bar{\bar{\mathbf{x}}}(i+1) = \mathbf{f}(\bar{\bar{\mathbf{x}}}(i)) \text{ with } \bar{\bar{\mathbf{x}}}(0) = \mathrm{E}[\mathbf{x}(0)] \tag{4.42}$$

Although the approach is suitable for time variant systems, it is not often used. There is another approach with almost the same computational complexity, but with better performance. That approach is the *extended Kalman filter* (EKF).

Again, the intention is to keep track of the conditional expectation $\bar{\mathbf{x}}(i|i)$ and the covariance matrix $\mathbf{C}(i|i)$. In the linear-Gaussian case, where all distributions are Gaussian, the conditional mean is identical to both the MMSE estimate (= minimum variance estimate), the MMAE estimate, and the MAP estimate; see Section 3.1.3. In the present case, the distributions are not necessarily Gaussian, and the solutions of the three estimators do not coincide. The extended Kalman filter provides only an approximation of the MMSE estimate.

Each cycle of the extended Kalman filter consists of a 'one step ahead' prediction and an update, as before. However, the tasks are much more difficult now, because the calculation of, for instance, the 'one step ahead' expectation:

$$\bar{\mathbf{x}}(i+1|i) = \mathrm{E}[\mathbf{x}(i+1)|\mathbf{Z}(i)] = \int \mathbf{x}(i+1)p(\mathbf{x}(i+1)|\mathbf{Z}(i))d\mathbf{x}(i+1) \tag{4.43}$$

requires the probability density $p(\mathbf{x}(i+1)|\mathbf{Z}(i))$; see (4.8). But, as said before, it is not clear how to represent this density. The solution of the EKF is to apply a linear approximation of the system function. With that, the 'one step ahead' expectation can be expressed entirely in terms of the moments of $p(\mathbf{x}(i)|\mathbf{Z}(i))$.

The EKF uses linear approximations of the system functions using the first two terms of the Taylor series expansions.[4] Suppose that at time i we have the updated estimate $\hat{\mathbf{x}}(i) \cong \bar{\mathbf{x}}(i|i)$ and the associated approximate

[4] With more terms in the Taylor series expansion, the approximation becomes more accurate. For instance, the *second order extended Kalman filter* uses quadratic approximations based on the first three terms of the Taylor series expansions. The discussion on the extensions of this type is beyond the scope of this book. See Bar-Shalom (1993)

error covariance matrix $\mathbf{C}(i|i)$. The word 'approximate' expresses the fact that our estimates are not guaranteed to be unbiased due to the linear approximations. However, we do assume that the influence of the errors induced by the linear approximations is small. If the estimation error is denoted by $\mathbf{e}(i)$, then:

$$\begin{aligned} \mathbf{x}(i) &= \bar{\mathbf{x}}(i|i) - \mathbf{e}(i) \\ \mathbf{x}(i+1) &= \mathbf{f}(\mathbf{x}(i)) + \mathbf{w}(i) \\ &= \mathbf{f}(\bar{\mathbf{x}}(i|i) - \mathbf{e}(i)) + \mathbf{w}(i) \\ &\cong \mathbf{f}(\bar{\mathbf{x}}(i|i)) - \mathbf{F}(\bar{\mathbf{x}}(i|i))\mathbf{e}(i) + \mathbf{w}(i) \end{aligned} \tag{4.44}$$

In these expressions, $\bar{\mathbf{x}}(i|i)$ is our estimate.[5] It is available at time i, and as such, it is deterministic. Only $\mathbf{e}(i)$ and $\mathbf{w}(i)$ are random. Taking the expectation on both sides of (4.44), we obtain approximate values for the 'one step ahead' prediction:

$$\begin{aligned} \bar{\mathbf{x}}(i+1|i) &\cong \mathbf{f}(\bar{\mathbf{x}}(i|i)) \\ \mathbf{C}(i+1|i) &\cong \mathbf{F}(\bar{\mathbf{x}}(i|i))\mathbf{C}(i|i)\mathbf{F}^T(\bar{\mathbf{x}}(i|i)) + \mathbf{C}_\mathbf{w}(i) \end{aligned} \tag{4.45}$$

We have approximations instead of equalities for two reasons. First, we neglect the possible bias of $\bar{\mathbf{x}}(i|i)$, i.e. a nonzero mean of $\mathbf{e}(i)$. Second, we ignore the higher order terms of the Taylor series expansion.

Upon incrementing the counter, $\bar{\mathbf{x}}(i+1|i)$ becomes $\bar{\mathbf{x}}(i|i-1)$, and we now have to update the prediction $\bar{\mathbf{x}}(i|i-1)$ by using a new measurement $\mathbf{z}(i)$ in order to get an approximation of the conditional mean $\bar{\mathbf{x}}(i|i)$. First we calculate the predicted measurement $\hat{\mathbf{z}}(i)$ based on $\bar{\mathbf{x}}(i|i-1)$ using a linear approximation of the measurement function, that is $\mathbf{h}(\hat{\mathbf{x}} - \mathbf{e}) \cong \mathbf{h}(\hat{\mathbf{x}}) - \mathbf{H}(\hat{\mathbf{x}})\mathbf{e}$. Next, we calculate the innovation matrix $\mathbf{S}(i)$ using that same approximation. Then, we apply the update according to the same equations as in the linear-Gaussian case.

[5] Up to this point $\bar{\mathbf{x}}(i|i)$ has been the exact expectation of $\mathbf{x}(i)$ given all measurements up to $\mathbf{z}(i)$. From now on, in this section, $\bar{\mathbf{x}}(i|i)$ will denote an approximation of that.

The predicted measurements are:

$$
\begin{aligned}
\mathbf{x}(i) &= \bar{\mathbf{x}}(i|i-1) - \mathbf{e}(i|i-1) \\
\hat{\mathbf{z}}(i) &= \mathrm{E}[\mathbf{z}(i)|\bar{\mathbf{x}}(i|i-1)] \\
&= \mathrm{E}[\mathbf{h}(\mathbf{x}(i)) + \mathbf{v}(i)|\bar{\mathbf{x}}(i|i-1)] \\
&\cong \mathrm{E}[\mathbf{h}(\bar{\mathbf{x}}(i|i-1)) - \mathbf{H}(\bar{\mathbf{x}}(i|i-1))\mathbf{e}(i|i-1) + \mathbf{v}(i)] \\
&\cong \mathbf{h}(\bar{\mathbf{x}}(i|i-1))
\end{aligned}
\tag{4.46}
$$

The approximation is based on the assumption that $\mathrm{E}[\mathbf{e}(i|i-1)] \cong \mathbf{0}$, and on the Taylor series expansion of $\mathbf{h}(\cdot)$. The innovation matrix becomes:

$$
\mathbf{S}(i) = \mathbf{H}(\bar{\mathbf{x}}(i|i-1))\mathbf{C}(i|i-1)\mathbf{H}^T(\bar{\mathbf{x}}(i|i-1)) + \mathbf{C}_\mathbf{v}(i) \tag{4.47}
$$

From this point on, the update continues as in the linear-Gaussian case; see (4.27):

$$
\begin{aligned}
\mathbf{K}(i) &= \mathbf{C}(i|i-1)\mathbf{H}^T(\bar{\mathbf{x}}(i|i-1))\mathbf{S}^{-1}(i) \\
\bar{\mathbf{x}}(i|i) &= \bar{\mathbf{x}}(i|i-1) + \mathbf{K}(i)(\mathbf{z}(i) - \hat{\mathbf{z}}(i)) \\
\mathbf{C}(i|i) &= \mathbf{C}(i|i-1) - \mathbf{K}(i)\mathbf{S}(i)\mathbf{K}^T(i)
\end{aligned}
\tag{4.48}
$$

Despite the similarity of the last equation with respect to the linear case, there is an important difference. In the linear case, the Kalman gain $\mathbf{K}(i)$ depends solely on deterministic parameters: $\mathbf{H}(i)$, $\mathbf{F}(i)$, $\mathbf{C}_\mathbf{w}(i)$, $\mathbf{C}_\mathbf{v}(i)$ and $\mathbf{C}_\mathbf{x}(0)$. It does not depend on the data. Therefore, $\mathbf{K}(i)$ is fully deterministic. It could be calculated in advance instead of online. In the EKF, the gains depend upon the estimated states $\bar{\mathbf{x}}(i|i-1)$ through $\mathbf{H}(\bar{\mathbf{x}}(i|i-1))$, and thus also upon the measurements $\mathbf{z}(i)$. As such, the Kalman gains are random matrices. Two runs of the extended Kalman filter in two repeated experiments lead to two different sequences of the Kalman gains. In fact, this randomness of $\mathbf{K}(i)$ can cause instable behaviour.

Example 4.6 The extended Kalman filter for volume density estimation

Application of the EKF to the density estimation problem introduced in Example 4.1 and represented by a linear-Gaussian model in Example 4.5 gives the results as shown in Figure 4.10. Compared with the results of the linearized KF (Figure 4.9) the density errors are now much better consistent with the 1σ boundaries obtained from the

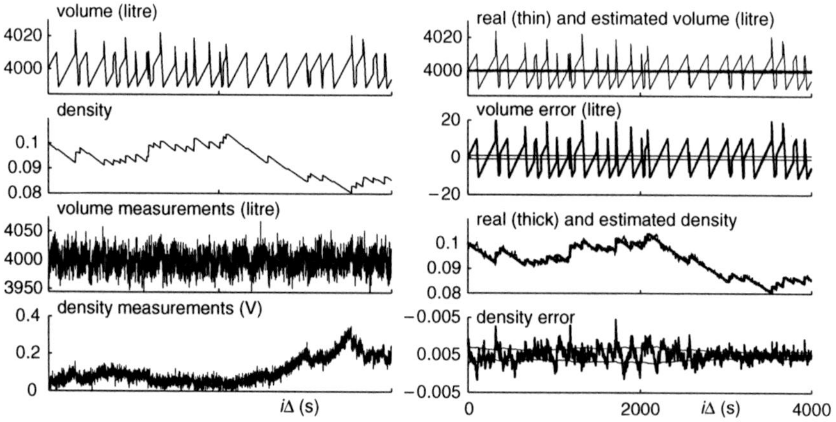

Figure 4.10 Extended Kalman filtering for the volume density estimation problem

error covariance matrix. However, the EKF is still not able to cope with the non-Gaussian disturbances of the volume.

Note also that the 1σ boundaries do not reach a steady state. The filter remains time variant, even in the long term.

The iterated extended Kalman filter

A further improvement of the update step in the extended Kalman filter is within reach if the current estimate $\bar{\mathbf{x}}(i|i)$ is used to get an improved linear approximation of the measurement function yielding an improved predicted measurement $\hat{\mathbf{z}}(i)$. In turn, such an improved predicted measurement can improve the current estimate. This suggests an iterative approach.

Let $\hat{\mathbf{z}}_\ell(i)$ be the predicted measurement in the ℓ-th iteration, and let $\bar{\mathbf{x}}_\ell(i)$ be the ℓ-th improvement of $\bar{\mathbf{x}}(i|i)$. The iteration is initiated with $\bar{\mathbf{x}}_0(i) = \bar{\mathbf{x}}(i|i-1)$. A naive approach for the calculation of $\bar{\mathbf{x}}_{\ell+1}(i)$ simply uses a relinearization of $\mathbf{h}(\cdot)$ based on $\bar{\mathbf{x}}_\ell(i)$.

$$\begin{aligned}
\mathbf{H}_{\ell+1} &= \mathbf{H}(\bar{\mathbf{x}}_\ell(i)) \\
\mathbf{S}_{\ell+1} &= \mathbf{H}_{\ell+1}\mathbf{C}(i|i-1)\mathbf{H}_{\ell+1}^T + \mathbf{C}_\mathbf{v}(i) \\
\mathbf{K}_{\ell+1} &= \mathbf{C}(i|i-1)\mathbf{H}_{\ell+1}^T\mathbf{S}_{\ell+1}^{-1} \\
\bar{\mathbf{x}}_{\ell+1}(i) &= \bar{\mathbf{x}}(i|i-1) + \mathbf{K}_{\ell+1}(\mathbf{z}(i) - \mathbf{h}(\bar{\mathbf{x}}(i|i-1)))
\end{aligned} \tag{4.49}$$

Hopefully, the sequence $\bar{\mathbf{x}}_\ell(i)$, with $\ell = 0, 1, 2, \ldots$, converges to a final solution.

A better approach is the so-called *iterated extended Kalman filter* (IEKF). Here, the approximation is made that both the predicted state and the measurement noise are normally distributed. With that, the posterior probability density (4.9)

$$p(\mathbf{x}(i)|\mathbf{Z}(i)) = \frac{1}{c}p(\mathbf{x}(i)|\mathbf{Z}(i-1))p(\mathbf{z}(i)|\mathbf{x}(i)) \tag{4.50}$$

being the product of two Gaussians, is also a Gaussian. Thus, the MMSE estimate coincides with the MAP estimate, and the task is now to find the maximum of $p(\mathbf{x}(i)|\mathbf{Z}(i))$. Equivalently, we maximize its logarithm. After the elimination of the irrelevant constants and factors, it all boils down to minimizing the following function w.r.t. $\mathbf{x}$:

$$f(\mathbf{x}) = \underbrace{\frac{1}{2}(\mathbf{x}-\bar{\mathbf{x}}_p)^T \mathbf{C}_p^{-1}(\mathbf{x}-\bar{\mathbf{x}}_p)}_{\text{comes from } p(\mathbf{x}(i)|\mathbf{Z}(i-1))} + \underbrace{\frac{1}{2}(\mathbf{z}-\mathbf{h}(\mathbf{x}))^T \mathbf{C}_\mathbf{v}^{-1}(\mathbf{z}-\mathbf{h}(\mathbf{x}))}_{\text{comes from } p(\mathbf{z}(i)|\mathbf{x}(i))} \tag{4.51}$$

For brevity, the following notation has been used:

$$\begin{aligned}
\bar{\mathbf{x}}_p &= \bar{\mathbf{x}}(i|i-1) \\
\mathbf{C}_p &= \mathbf{C}(i|i-1) \\
\mathbf{z} &= \mathbf{z}(i) \\
\mathbf{C}_\mathbf{v} &= \mathbf{C}_\mathbf{v}(i)
\end{aligned}$$

The strategy to find the minimum is to use Newton–Raphson iteration starting from $\bar{\mathbf{x}}_0 = \bar{\mathbf{x}}(i|i-1)$. In the ℓ-th iteration step, we have already an estimate $\bar{\mathbf{x}}_{\ell-1}$ obtained from the previous step. We expand $f(\mathbf{x})$ in a second order Taylor series approximation:

$$\begin{aligned}
f(\mathbf{x}) \cong{} & f(\bar{\mathbf{x}}_{\ell-1}) + (\mathbf{x}-\bar{\mathbf{x}}_{\ell-1})^T \frac{\partial f(\bar{\mathbf{x}}_{\ell-1})}{\partial \mathbf{x}} \\
& + \frac{1}{2}(\mathbf{x}-\bar{\mathbf{x}}_{\ell-1})^T \frac{\partial^2 f(\bar{\mathbf{x}}_{\ell-1})}{\partial \mathbf{x}^2}(\mathbf{x}-\bar{\mathbf{x}}_{\ell-1})
\end{aligned} \tag{4.52}$$

where $\partial f/\partial \mathbf{x}$ is the gradient and $\partial^2 f/\partial \mathbf{x}^2$ is the Hessian of $f(\mathbf{x})$. See Appendix B.4. The estimate $\bar{\mathbf{x}}_\ell$ is the minimum of the approximation.

It is found by equating the gradient of the approximation to zero. Differentiation of (4.52) w.r.t. $\mathbf{x}$ gives:

$$\begin{gathered}\frac{\partial f(\bar{\mathbf{x}}_{\ell-1})}{\partial \mathbf{x}} + \frac{\partial^2 f(\bar{\mathbf{x}}_{\ell-1})}{\partial \mathbf{x}^2}(\mathbf{x} - \bar{\mathbf{x}}_{\ell-1}) = \mathbf{0} \\ \Downarrow \\ \bar{\mathbf{x}}_\ell = \bar{\mathbf{x}}_{\ell-1} - \left(\frac{\partial^2 f(\bar{\mathbf{x}}_{\ell-1})}{\partial \mathbf{x}^2}\right)^{-1} \frac{\partial f(\bar{\mathbf{x}}_{\ell-1})}{\partial \mathbf{x}}\end{gathered} \tag{4.53}$$

The Jacobian and Hessian of (4.51), in explicit form, are:

$$\begin{aligned}\frac{\partial f(\bar{\mathbf{x}}_{\ell-1})}{\partial \mathbf{x}} &= \mathbf{C}_p^{-1}(\bar{\mathbf{x}}_{\ell-1} - \mathbf{x}_p) - \mathbf{H}_\ell^T \mathbf{C}_\mathbf{v}^{-1}(\mathbf{z} - \mathbf{h}(\bar{\mathbf{x}}_{\ell-1})) \\ \frac{\partial^2 f(\bar{\mathbf{x}}_{\ell-1})}{\partial \mathbf{x}^2} &= \mathbf{C}_p^{-1} + \mathbf{H}_\ell^T \mathbf{C}_\mathbf{v}^{-1}\mathbf{H}_\ell\end{aligned} \tag{4.54}$$

where $\mathbf{H}_\ell = \mathbf{H}(\bar{\mathbf{x}}_{\ell-1})$ is the Jacobian matrix of $\mathbf{h}(\mathbf{x})$ evaluated at $\bar{\mathbf{x}}_{\ell-1}$. Substitution of (4.54) in (4.53) yields the following iteration scheme:

$$\begin{aligned}\bar{\mathbf{x}}_\ell = \bar{\mathbf{x}}_{\ell-1} - \left(\mathbf{C}_p^{-1} + \mathbf{H}_\ell^T \mathbf{C}_\mathbf{v}^{-1}\mathbf{H}_\ell\right)^{-1} \Big[\mathbf{C}_p^{-1}(\bar{\mathbf{x}}_{\ell-1} - \mathbf{x}_p) \\ - \mathbf{H}_\ell^T \mathbf{C}_\mathbf{v}^{-1}(\mathbf{z} - \mathbf{h}(\bar{\mathbf{x}}_{\ell-1}))\Big]\end{aligned} \tag{4.55}$$

The result after one iteration, i.e. $\bar{\mathbf{x}}_1(i)$, is identical to the ordinary extended Kalman filter. The required number of further iterations depends on how fast $\bar{\mathbf{x}}_\ell(i)$ converges. Convergence is not guaranteed, but if the algorithm converges, usually a small number of iterations suffices. Therefore, it is common practice to fix the number of iterations to some practical number L. The final result is set to the last iteration, i.e. $\bar{\mathbf{x}}(i|i) = \bar{\mathbf{x}}_L$.

Equation (3.44) shows that the factor $(\mathbf{C}_p^{-1} + \mathbf{H}_\ell^T \mathbf{C}_\mathbf{v}^{-1}\mathbf{H}_\ell)^{-1}$ is the error covariance matrix associated with $\bar{\mathbf{x}}(i|i)$:

$$\mathbf{C}(i|i) \cong \left(\mathbf{C}_p^{-1} + \mathbf{H}_{L+1}^T \mathbf{C}_\mathbf{v}^{-1}\mathbf{H}_{L+1}\right)^{-1} \tag{4.56}$$

This insight gives another connotation to the last term in (4.55) because, in fact, the term $\mathbf{C}(i|i)\mathbf{H}_\ell^T \mathbf{C}_\mathbf{v}^{-1}$ can be regarded as the Kalman gain matrix $\mathbf{K}_\ell$ during the ℓ-th iteration; see (3.20).

Example 4.7 The iterated EKF for volume density estimation
In the previous example, the EKF was applied to the density estimation problem introduced in Example 4.1. The filter was initiated with the equilibrium state as prior knowledge, i.e. $E[\mathbf{x}(0)] = \bar{\bar{\mathbf{x}}} = [4000\ 0.1]^T$. Figure 4.11(b) shows the transient which occurs if the EKF is initiated with $E[\mathbf{x}(0)] = [2000\ 0]^T$. It takes about 40 (s) before the estimated density reaches the true densities. This slow transient is due to the fact that in the beginning the linearization is poor. The iterated EKF is of much help here. Figure 4.11(c) shows the results. From the first measurement on the estimated density is close to the real density. There is no transient.

The extended Kalman filter is widely used because for a long period of time no viable alternative solution existed. Nevertheless, it has numerous disadvantages:

- It only works well if the various random vectors are approximately Gaussian distributed. For complicated densities, the expectation-covariance representation does not suffice.
- It only works well if the nonlinearities of the system are not too severe because otherwise the Taylor series approximations fail. Discontinuities are deadly for the EKF's proper functioning.
- Recalculating the Jacobian matrices at every time step is computationally expensive.

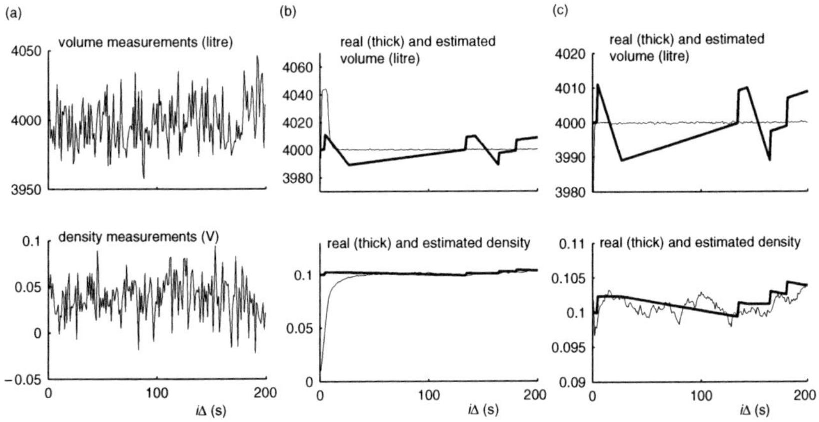

Figure 4.11 Iterated extended Kalman filtering for the volume density estimation problem. (a) Measurements (b) Results from the EKF (c) Results from the iterated EKF (no. of iterations = 20)

- In some applications, it is too difficult to find the Jacobian matrix analytically. In these cases, numerical approximations of the Jacobian matrix are needed. However, this introduces other types of problems because now the influence of having approximations rather than the true values comes in.
- In the EKF, the Kalman gain matrix depends on the data. With that, the stability of the filter is not assured anymore. Moreover, it is very hard to analyse the behaviour of the filter.
- The EKF does not guarantee unbiased estimates. In addition, the calculated error covariance matrices do not necessarily represent the true error covariances. The analysis of these effects is also hard.

4.2.3 Other filters for nonlinear systems

Besides the extended Kalman filter there are many more types of estimators for nonlinear systems. Particle filtering is a relatively new approach for the implementation of the scheme depicted in Figure 4.2. The discussion about particle filtering will be deferred to Section 4.4 because it not only applies to continuous states. Particle filtering is generally applicable; it covers the nonlinear, non-Gaussian continuous systems, but also discrete systems and mixed systems.

Statistical linearization is a method comparable with the extended Kalman filter. But, instead of using a truncated Taylor series approximation for the nonlinear system functions, a linear approximation $\mathbf{f}(\mathbf{x}+\boldsymbol{\varepsilon}) \cong \mathbf{f}(\mathbf{x}) + \mathbf{F}\boldsymbol{\varepsilon}$ is used such that the deviation $\mathbf{f}(\mathbf{x}+\boldsymbol{\varepsilon}) - \mathbf{f}(\mathbf{x}) - \mathbf{F}\boldsymbol{\varepsilon}$ is minimized according to a statistical criterion. For instance, one could try to determine $\mathbf{F}$ such that $\mathrm{E}\left[\|\mathbf{f}(\mathbf{x}+\boldsymbol{\varepsilon}) - \mathbf{f}(\mathbf{x}) - \mathbf{F}\boldsymbol{\varepsilon}\|^2\right]$ is minimal.

Another method is the *unscented Kalman filter*. This is a filter midway between the extended Kalman filter and the particle filter. Assuming Gaussian densities for $\mathbf{x}$ (as in the Kalman filter), the expectation and the error covariance matrix is represented by means of a number of samples $\mathbf{x}^{(k)}$, that are used to calculate the effects of a nonlinear system function on the expectation and the error covariance matrix. Unlike the particle filter, these samples are not randomly selected. Instead the filter uses a small amount of samples that are carefully selected and that uniquely represent the covariance matrix. The transformed points, i.e. $\mathbf{f}(\mathbf{x}^{(k)})$ are used to reconstruct the covariance matrix of $\mathbf{f}(\mathbf{x})$. Such a reconstruction is much more accurate than the approximation that is obtained by means of the truncated Taylor series expansion.

4.3 DISCRETE STATE VARIABLES

We consider physical processes that are described at any time as being in one of a finite number of states. Examples of such processes are:

- The sequence of strokes of a tennis player during a game, e.g. service, backhand-volley, smash, etc.
- The sequence of actions that a tennis player performs during a particular stroke.
- The different types of manoeuvres of an airplane, e.g. a linear flight, a turn, a nose dive, etc.
- The sequence of characters in a word, and the sequence of words in a sentence.
- The sequence of tones of a melody as part of a musical piece.
- The emotional modes of a person: angry, happy, astonished, etc.

These situations are described by a state variable $x(i)$ that can only take a value from a finite set of states $\Omega = \{\omega_1, \ldots, \omega_K\}$.

The task is to determine the sequence of states that a particular process goes through (or has gone through). For that purpose, at any time measurements $z(i)$ are available. Often, the output of the sensors is real-valued. But nevertheless we will assume that the measurements take their values from a finite set. Thus, some sort of discretization must take place that maps the range of the sensor data onto a finite set $Z = \{\vartheta_1, \ldots, \vartheta_N\}$.

This section first introduces a state space model that is often used for discrete state variables, i.e. the hidden Markov model. This model will be used in the next subsections for online and offline estimation of the states.

4.3.1 Hidden Markov models

A *hidden Markov model* (HMM) is an instance of the state space model discussed in Section 4.1.1. It describes a sequence $x(i)$ of discrete states starting at time $i = 0$. The sequence is observed by means of measurements $z(i)$ that can only take values from a finite set. The model consists of the following ingredients:

- The set Ω containing the K states ω_k that $x(i)$ can take.
- The set Z containing the N symbols ϑ_n that $z(i)$ can take.
- The *initial state probability* $P_0(x(0))$.

- The *state transition probability* $P_t(x(i)|x(i-1))$.
- The observation probability $P_z(z(i)|x(i))$.

The expression $P_0(x(0))$ with $x(0) \in \{1, \ldots, K\}$ denotes the probability that the random state variable $\underline{x}(0)$ takes the value $\omega_{x(0)}$. Thus $P_0(k) \stackrel{def}{=} P_0(\underline{x}(0) = \omega_k)$. Similar conventions hold for other expressions, like $P_t(x(i)|x(i-1))$ and $P_z(z(i)|x(i))$.

The Markov condition of an HMM states that $P(x(i)|x(0), \ldots, x(i-1))$, i.e. the probability of $x(i)$ under the condition of all previous states, equals the transition probability. The assumption of the validity of the Markov condition leads to a simple, yet powerful model. Another assumption of the HMM is that the measurements are memoryless. In other words, $z(i)$ only depends on $x(i)$ and not on the states at other time points: $P(z(j)|x(0), \ldots, x(i)) = P(z(j)|x(j))$.

An *ergodic* Markov model is one for which the observation of a single sequence $x(0), x(1), \ldots, x(\infty)$ suffices to determine all the state transition probabilities. A suitable technique for that is histogramming, i.e. the determination of the relative frequency with which a transition occurs; see Section 5.2.5. A sufficient condition for ergodicity is that all state probabilities are nonzero. In that case, all states are reachable from everywhere within one time step. Figure 4.12 is an illustration of an ergodic model.

Another type is the so-called *left–right model*. See Figure 4.13. This model has the property that the state index k of a sequence is nondecreasing as time proceeds. Such is the case when $P_t(k|\ell) = 0$ for all $k < \ell$. In addition, the sequence always starts with ω_1 and terminates

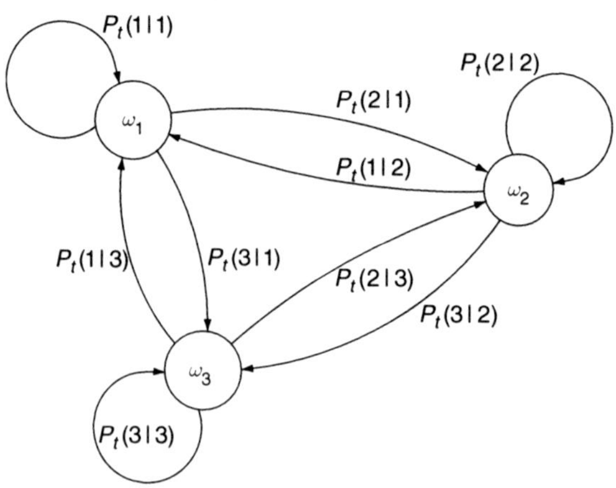

Figure 4.12 A three-state ergodic Markov model

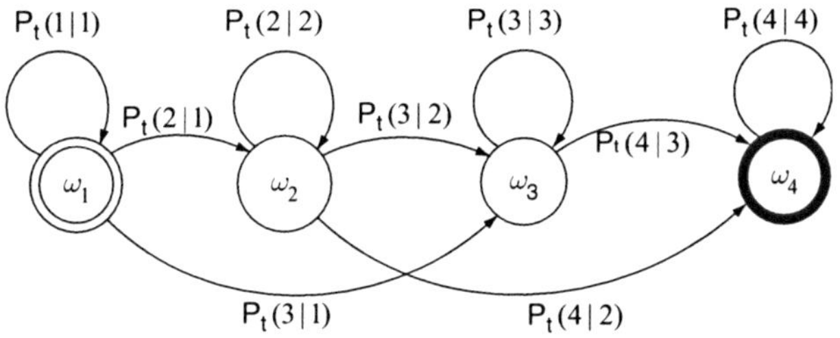

Figure 4.13 A four-state left–right model

with ω_K. Thus, $P_0(k) = \delta(k,1)$ and $P_t(k|K) = \delta(k,K)$. Sometimes, an additional constraint is that large jumps are forbidden. Such a constraint is enforced by letting $P_t(k|\ell) = 0$ for all $k > \ell + \Delta$. Left–right models find applications in processes where the sequence of states must obey some ordering over time. An example is the stroke of a tennis player. For instance, the service of the player follows a sequence like: 'take position behind the base line', 'bring racket over the shoulder behind the back', 'bring up the ball with the left arm', etc.

In a hidden Markov model the state variable $x(i)$ is observable only through its measurements $z(i)$. Now, suppose that a sequence $Z(i) = \{z(0), z(1), \ldots, z(i)\}$ of measurements has been observed. Some applications require the numerical evaluation of the probability $P(Z(i))$ of particular sequence. An example is the recognition of a stroke of a tennis player. We can model each type of stroke by an HMM that is specific for that type, thus having as many HMMs as there are types of strokes. In order to recognize the stroke, we calculate for each type of stroke the probability $P(Z(i)|\text{type of stroke})$ and select the one with maximum probability.

For a given HMM, and a fixed sequence $Z(i)$ of acquired measurements, $P(Z(i))$ can be calculated by using the joint probability of having the measurements $Z(i)$ together with a specific sequence of state variables, i.e. $X(i) = \{x(0), x(1), \ldots, x(i)\}$. First we calculate the joint probability $P(X(i), Z(i))$:

$$\begin{aligned} P(X(i), Z(i)) &= P(Z(i)|X(i))P(X(i)) \\ &= \left(\prod_{j=0}^{i} P_z(z(j)|x(j))\right)\left(P_0(x(0))\prod_{j=1}^{i} P_t(x(j)|x(j-1))\right) \\ &= P_0(x(0))P_z(z(0)|x(0))\prod_{j=1}^{i}(P_z(z(j)|x(j))P_t(x(j)|x(j-1))) \end{aligned} \tag{4.57}$$

Here, use has been made of the assumption that each measurement $z(i)$ only depends on $x(i)$. $P(Z(i))$ follows from summation over all possible state sequences:

$$P(Z(i)) = \sum_{\text{all } X(i)} P((X(i), Z(i)) \tag{4.58}$$

Since there are K^i different state sequences, the direct implementation of (4.57) and (4.58) requires on the order of $(i+1)K^{i+1}$ operations. Even for modest values of i, the number of operations is already impractical.

A more economical approach is to calculate $P(Z(i))$ by means of a recursion. For that, consider the probability $P(Z(i), x(i))$. This probability can be calculated from the previous time step $i - 1$ using the following expression:

$$\begin{aligned} P(Z(i), x(i)) &= \sum_{x(i-1)=1}^{K} P(Z(i), x(i), x(i-1)) \\ &= \sum_{x(i-1)=1}^{K} P(z(i), x(i)|Z(i-1), x(i-1))P(Z(i-1), x(i-1)) \\ &= \sum_{x(i-1)=1}^{K} P(z(i), x(i)|x(i-1))P(Z(i-1), x(i-1)) \\ &= P(z(i)|x(i)) \sum_{x(i-1)=1}^{K} P_t(x(i)|x(i-1))P(Z(i-1), x(i-1)) \end{aligned} \tag{4.59}$$

The recursion must be initiated with $P(z(0), x(0)) = P_0(x(0))P_z(z(0)|x(0))$. The probability $P(Z(i))$ can be retrieved from $P(Z(i), x(i))$ by:

$$P(Z(i)) = \sum_{x(i)=1}^{K} P(Z(i), x(i)) \tag{4.60}$$

The so-called *forward algorithm*[6] uses the array $F(i, x(i)) = P(Z(i), x(i))$ to implement the recursion:

---

[6] The adjective 'forward' refers to the fact that the algorithm proceeds forwards in time. Section 4.3.3 introduces the backward algorithm.

Algorithm 4.1: The forward algorithm

1. Initialization:

$$F(0, x(0) = P_0(x(0))P_z(z(0)|x(0)) \quad \text{for} \quad x(0) = 1, \ldots, K$$

2. Recursion:
   for $i = 1, 2, 3, 4, \ldots$
   - for $x(i) = 1, \ldots, K$

$$F(i, x(i)) = P_z(z(i)|x(i)) \sum_{x(i-1)=1}^{K} F(i-1, x(i-1))P_t(x(i)|x(i-1))$$

   - $P(Z(i)) = \sum_{x(i)=1}^{K} F(i, x(i))$

In each recursion step, the sum consists of K terms, and the number of possible values of $x(i)$ is also K. Therefore, such a step requires on the order of K^2 calculations. The computational complexity for i time steps is on the order of $(i+1)K^2$.

4.3.2 Online state estimation

We now focus our attention on the situation of having a single HMM, where the sequence of measurements is processed online so as to obtain real-time estimates $\hat{x}(i|i)$ of the states. This problem completely fits within the framework of Section 4.1. As such, the solution provided by (4.8) and (4.9) is valid, albeit that the integrals must be replaced by summations.

However, in line with the previous section, an alternative solution will be presented that is equivalent to the one of Section 4.1. The alternative solution is obtained by deduction of the posterior probability:

$$P(x(i)|Z(i)) = \frac{P(Z(i), x(i))}{P(Z(i))} \tag{4.61}$$

In view of the fact that $Z(i)$ are the acquired measurements (and as such known and fixed) the maximization of $P(x(i)|Z(i))$ is equivalent to the maximization of $P(Z(i), x(i))$. Therefore, the MAP estimate is found as:

$$\hat{x}_{MAP}(i|i) = \arg\max_k \{P(Z(i), k))\} \tag{4.62}$$

The probability $P(Z(i), x(i))$ follows from the forward algorithm.

Example 4.8 Online license plate detection in videos
This example demonstrates the ability of HMMs to find the license plate of a vehicle in a video. Figure 4.14 is a typical example of one frame of such a video. The task is to find all the pixels that correspond to the license plate. Such a task is the first step in a license plate recognition system.

A major characteristic of video is that a frame is scanned line-by-line, and that each video line is acquired from left to right. The real-time processing of each line individually is preferable because the throughput requirement of the application is demanding. Therefore, each line is individually modelled as an HMM. The hidden state of a pixel is determined by whether the pixel corresponds to a license plate or not.

The measurements are embedded in the video line. See Figure 4.15. However, the video signal needs to be processed in order to map it onto a finite measurement space. Simply quantizing the signal to a finite number of levels does not suffice because the amplitudes of the signal alone are not very informative. The main characteristic of a license plate in a video line is a typical pattern of dark-bright and bright-dark transitions due to the dark characters against a bright background, or vice versa. The image acquisition is such that the camera–object distance is about constant for all vehicles. Therefore, the statistical properties of the succession of transitions are typical for the imaged license plate regardless of the type of vehicle.

Figure 4.14 License plate detection

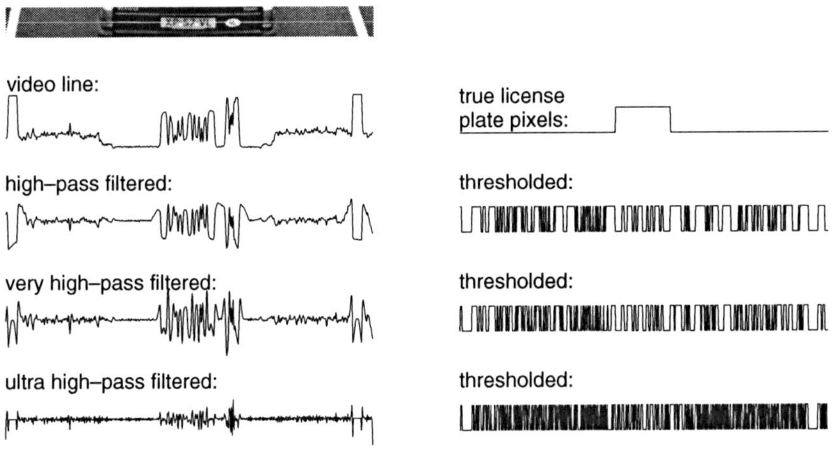

Figure 4.15 Definitions of the measurements associated with a video line

One possibility to decode the succession of transitions is to apply a filter bank and to threshold the output of each filter, thus yielding a set of binary signals. In Figure 4.15, three high-pass filters have been applied with three different cut-off frequencies. Using high-pass filters has the advantage that the thresholds can be zero. As such, the results do not depend on the contrast and brightness of the image. The three binary signals define a measurement signal $z(i)$ consisting of $N = 8$ symbols. Figure 4.16 shows these symbols for one video line. Here, the symbols are encoded as integers from 1 up to 8.

Due to the spatial context of the three binary signals we cannot model the measurements as memoryless symbols. The trick to avoid

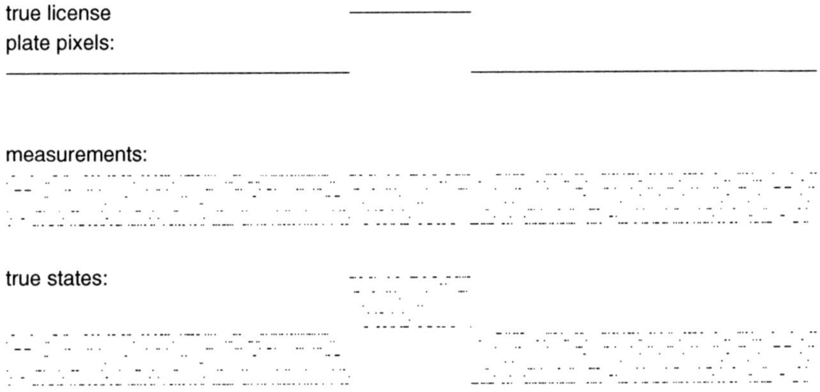

Figure 4.16 States and measurements of a video line

this problem is to embed the measurement $z(i)$ in the state variable $x(i)$. This can be done by encoding the state variable as integers from 1 up to 16. If i is not a license plate pixel, we define the state as $x(i) = z(i)$. If i is a license plate pixel, we define $x(i) = z(i) + 8$. With that, $K = 16$. Figure 4.16 shows these states for one video line.

The embedding of the measurements in the state variables is a form of *state augmentation*. Originally, the number of states was 2, but after this particular state augmentation, the number becomes 16. The advantage of the augmentation is that the dependence, which does exist between any pair $z(i), z(j)$ of measurements, is now properly modelled by means of the transition probability of the states. Yet, the model still meets all the requirements of an HMM. However, due to our definition of the state, the relation between state and measurement becomes deterministic. The observation probability degenerates into:

$$P_z(n|k) = \begin{cases} 1 & \text{if n} = k \text{ and } k \leq 8 \\ 1 & \text{if n} = k - 8 \text{ and } k > 8 \\ 0 & \text{elsewhere} \end{cases}$$

In order to define the HMM, the probabilities $P_0(k)$ and $P_t(k|\ell)$ must be specified. We used a supervised learning procedure to estimate $P_0(k)$ and $P_t(k|\ell)$. For that purpose, 30 images of 30 different vehicles, similar to the one in Figure 4.14, were used. For each image, the license plate area was manually indexed. Histogramming was used to estimate the probabilities.

Application of the online estimation to the video line shown in Figures 4.15 and 4.16 yields results like those shown in Figure 4.17. The figure shows the posterior probability for having a license plate. According to our definition of the state, the posterior probability of having a license plate pixel is $P(x(i) > 8|Z(i))$. Since by definition online estimation is causal, the rise and decay of this probability shows a delay. Consequently, the estimated position of the license plate is biased towards the right. Figure 4.18 shows the detected license plate pixels.

4.3.3 Offline state estimation

In non-real-time applications the sequence of measurements can be buffered before the state estimation takes place. The advantage is that not only 'past and present' measurements can be used, but also 'future'

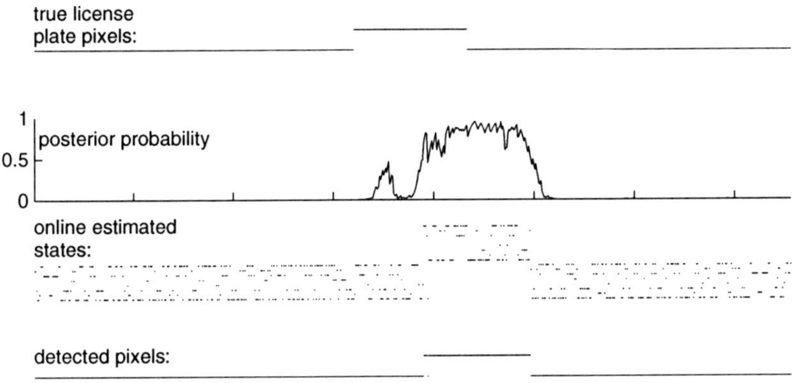

Figure 4.17 Online state estimation

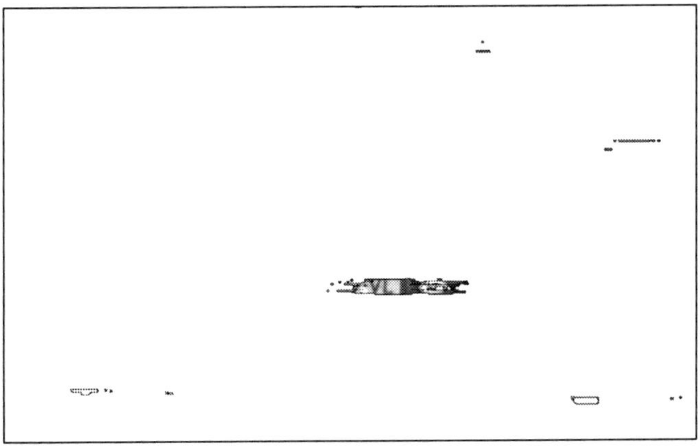

Figure 4.18 Detected license plate pixels using online estimation

measurements. Exactly these measurements can prevent the delay that inherently occurs in online estimation.

The problem is formulated as follows. Given a sequence $Z(I) = \{z(0), \ldots, z(I)\}$ of $I + 1$ measurements of a given HMM, determine the optimal estimate of the sequence $x(0), \ldots, x(I)$ of the underlying states.

Up to now, the adjective 'optimal' meant that we determined the individual posterior probability $P(x(i)|\text{measurements})$ for each time point individually, and that some cost function was applied to determine the estimate with the minimal risk. For instance, the adoption of a uniform cost function for each state leads to an estimate that maximizes the individual posterior probability. Such an estimate minimizes the probability of having an erroneous decision for such a state.

Minimizing the error probabilities of all individually estimated states does not imply that the sequence of states is estimated with minimal error probability. It might even occur that a sequence of 'individually best' estimated states contains a forbidden transition, i.e. a transition for which $P_t(x(i)|x(i-1)) = 0$. In order to circumvent this, we need a criterion that involves all states jointly.

This section discusses two offline estimation methods. One is an 'individually best' solution. The other is an 'overall best' solution.

Individually most likely states

Here, the strategy is to determine the posterior probability $P(x(i)|Z(I))$, and then to determine the MAP estimate: $\hat{x}(i|I) = \arg\max P(x(i)|Z(I))$. As said before, this method minimizes the error probabilities of the individual states. As such, it maximizes the expected number of correctly estimated states.

Section 4.3.1 discussed the forward algorithm, a recursive algorithm for the calculation of the probability $P(x(i), Z(i))$. We now introduce the *backward algorithm* which calculates the probability $P(z(i+1), \ldots, z(I)|x(i))$. During each recursion step of the algorithm, the probability $P(z(j), \ldots, z(I)|x(j-1))$ is derived from $P(z(j+1), \ldots, Z(I)|x(j))$. The recursion proceeds as follows:

$$
\begin{aligned}
&P(z(j), \ldots, z(I)|x(j-1)) \\
&= \sum_{x(j)=1}^{K} P_t(x(j)|x(j-1))P_z(z(j)|x(j))P(z(j+1), \ldots, Z(I)|x(j))
\end{aligned}
\tag{4.63}
$$

The algorithm starts with $j = I$, and proceeds backwards in time, i.e. $I, I-1, I-2, \ldots$ until finally $j = i+1$. In the first step, the expression $P(z(I+1)|x(I))$ appears. Since that probability does not exist (because $z(I+1)$ is not available), it should be replaced by 1 to have the proper initialization.

The availability of the forward and backward probabilities suffices for the calculation of the posterior probability:

$$
\begin{aligned}
P(x(i)|Z(I)) &= \frac{P(x(i), Z(I))}{P(Z(I))} \\
&= \frac{P(z(i+1), \ldots, z(I)|x(i), Z(i))P(x(i), Z(i))}{P(Z(I))} \\
&= \frac{P(z(i+1), \ldots, z(I)|x(i))P(x(i), Z(i))}{P(Z(I))}
\end{aligned}
\tag{4.64}
$$

As said before, the individually most likely state is the one which maximizes $P(x(i)|Z(I))$. The denominator of (4.64) is not relevant for this maximization since it does not depend on $x(i)$.

The complete *forward–backward algorithm* is as follows:

Algorithm 4.2: The forward–backward algorithm

1. Perform the forward algorithm as given in Section 4.3.1, resulting in the array $F(i,k)$ with $i = 0, \cdots, I$ and $k = 1, \cdots, K$
2. Backward algorithm:

   - Initialization:
     $B(I,k) = 1 \quad \text{for} \quad k = 1, \cdots, K$

   - Recursion:

     for $i = I-1, I-2, \cdots, 0 \quad \text{and} \quad x(i) = 1, \cdots, K$

$$F(i,x(i)) = \sum_{x(i+1)=1}^{K} P_t(x(i)|x(i+1))P_z(z(i+1)|x(i+1)) \\ F(i+1, x(i+1))$$

3. MAP estimation of the states:

$$\hat{x}_{MAP}(i|I) = \underset{k=1,\ldots,K}{\arg\max}\{B(i,k)F(i,k)\}$$

The forward–backward algorithm has a computational complexity that is on the order of $(I+1)K^2$. The algorithm is therefore feasible.

The most likely state sequence

A criterion that involves the whole sequence of states is the overall uniform cost function. The function is zero when the whole sequence is estimated without any error. It is unit if one or more states are estimated erroneously. Application of this cost function within a Bayesian framework leads to a solution that maximizes the overall posterior probability:

$$\hat{x}(0), \ldots, \hat{x}(I) = \underset{x(0),\ldots,x(I)}{\arg\max}\{P(x(0), \ldots x(I)|Z(I))\} \tag{4.65}$$

The computation of this most likely state sequence is done efficiently by means of a recursion that proceeds forwards in time. The goal of this recursion is to keep track of the following subsequences:

$$\hat{x}(0),\ldots,\hat{x}(i-1) = \underset{x(0),\ldots,x(i-1)}{\arg\max}\{P(x(0),\ldots x(i-1),x(i)|Z(i))\} \tag{4.66}$$

For each value of $x(i)$, this formulation defines a particular partial sequence. Such a sequence is the most likely partial sequence from time zero and ending at a particular value $x(i)$ at time i given the measurements $z(0),\ldots,z(i)$. Since $x(i)$ can have K different values, there are K partial sequences for each value of i. Instead of using (4.66), we can equivalently use

$$\hat{x}(0),\ldots,\hat{x}(i-1) = \underset{x(0),\ldots,x(i-1)}{\arg\max}\{P(x(0),\ldots x(i-1),x(i),Z(i))\} \tag{4.67}$$

because $P(X(i)|Z(i)) = P(X(i),Z(i))P(Z(i))$ and $Z(i)$ is fixed.

In each recursion step the maximal probability of the path ending in $x(i)$ given $Z(i)$ is transformed into the maximal probability of the path ending in $x(i+1)$ given $Z(i+1)$. For that purpose, we use the following equality:

$$\begin{aligned}
&P(x(0),\ldots,x(i),x(i+1),Z(i+1))\\
&\quad = P(x(i+1),z(i+1)|x(0),\ldots,x(i),Z(i))P(x(0),\ldots,x(i),Z(i))\\
&\quad = P_z(z(i+1)|x(i+1))P_t(x(i+1)|x(i))P(x(0),\ldots,x(i),Z(i))
\end{aligned} \tag{4.68}$$

Here, the Markov condition has been used together with the assumption that the measurements are memoryless.

The maximization of the probability proceeds as follows:

$$\begin{aligned}
&\max_{x(0),\cdots,x(i)}\{P(x(0),\cdots,x(i),x(i+1),Z(i+1))\}\\
&\quad = \max_{x(0),\cdots,x(i)}\{P_z(z(i+1)|x(i+1))P_t(x(i+1)|x(i))P(x(0),\cdots,x(i),Z(i))\}\\
&\quad = P_z(z(i+1)|x(i+1))\max_{x(i)}\Big\{P_t(x(i+1)|x(i)).\\
&\qquad\qquad \max_{x(0),\ldots,x(i-1)}\{P(x(0),\cdots,x(i-1),x(i),Z(i))\}\Big\}
\end{aligned} \tag{4.69}$$

The value of $x(i)$ that maximizes $P(x(0),\ldots,x(i),x(i+1),Z(i+1))$ is a function of $x(i+1)$:

$$\hat{x}(i|x(i+1)) = \arg\max_{x(i)}\left\{ P_t(x(i+1)|x(i)) \max_{x(0),\cdots,x(i-1)}\{P(x(0),\cdots,.x(i-1),x(i),Z(i))\}\right\} \tag{4.70}$$

The so-called *Viterbi algorithm* uses the recursive equation in (4.69) and the corresponding optimal state dependency expressed in (4.70) to find the optimal path. For that, we define the array $Q(i,x(i)) = \max_{x(0),\ldots,x(i-1)}\{P(x(0),\ldots,x(i-1),x(i),Z(i))\}$.

Algorithm 4.3: The Viterbi algorithm

1. Initialization:

   for $x(0) = 1,\cdots,K$
   - $Q(0,x(0)) = P_0(x(0))P_z(z(0)|x(0))$
   - $R(0,x(0)) = 0$

2. Recursion:

   for $i = 2,\cdots,I$ and $x(i) = 1,\cdots,K$
   - $Q(i,x(i)) = \max_{x(i-1)}\{Q(i-1,x(i-1)P_t(x(i)|x(i-1)))\}P_z(z(i)|x(i))$
   - $R(i,x(i)) = \max_{x(i-1)}\{Q(i-1,x(i-1)P_t(x(i)|x(i-1)))\}$

3. Termination:

   - $P = \max_{x(I)}\{Q(I,x(I))\}$
   - $\hat{x}(I|I) = \arg\max_{x(I)}\{Q(I,x(I))\}$

4. Backtracking:

   for $i = I-1, I-2,\cdots,0$
   - $\hat{x}(i|I) = R(i+1,\hat{x}(i+1,I))$

The computational structure of the Viterbi algorithm is comparable to that of the forward algorithm. The computational complexity is also on the order of $(i+1)K^2$.

Example 4.9 Offline license plate detection in videos
Figure 4.19 shows the results of the two offline state estimators applied to the video line shown in Figure 4.15. Figure 4.20 provides the results of the whole image shown in Figure 4.14.

Both methods are able to prevent the delay that is inherent in online estimation. Nevertheless, both methods show some falsely detected license plate pixels on the right side of the plate. These errors are caused by a sticker containing some text. Apparently, the statistical properties of the image of this sticker are similar to the one of a license plate.

A comparison between the individually estimated states and the jointly estimated states shows that the latter are more coherent, and that the former are more fragmented. Clearly, such a fragmentation increases the probability of having erroneous transitions of estimated states. However, usual the resulting erroneous regions are small. The jointly estimated states do not show many of these unwanted transitions, but if they occur, then they are more serious because they result in a larger erroneous region.

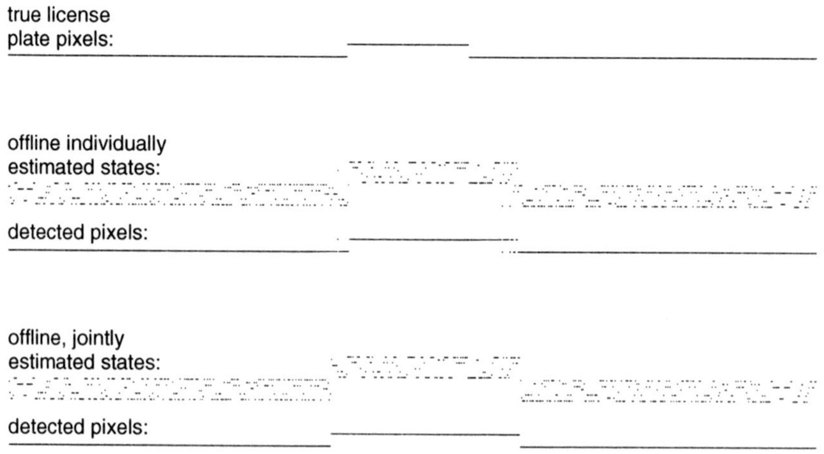

Figure 4.19 Offline state estimation

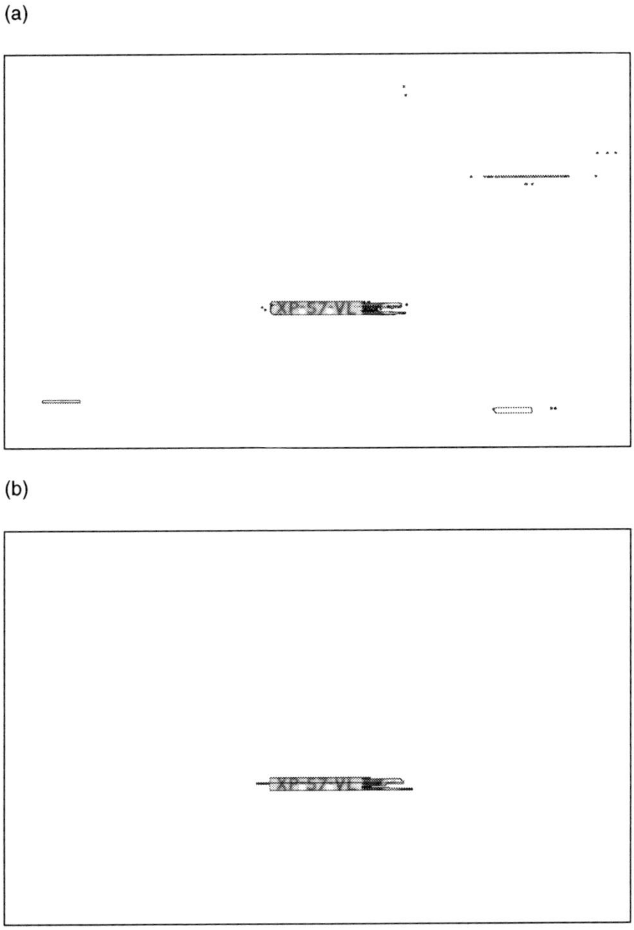

Figure 4.20 Detected license plate pixels using offline estimation. (a) Individually estimated states. (b) Jointly estimated states

MATLAB functions for HMM

The MATLAB functions for the analysis of hidden Markov models are found in the Statistics toolbox. There are five functions:

- `hmmgenerate`: Given $P_t(\cdot|\cdot)$ and $P_z(\cdot|\cdot)$, generate a sequence of states and observations.
- `hmmdecode`: Given $P_t(\cdot|\cdot), P_z(\cdot|\cdot)$ and a sequence of observations, calculate the posterior probabilities of the states.
- `hmmestimate`: Given a sequence of states and observations, estimate $P_t(\cdot|\cdot)$ and $P_z(\cdot|\cdot)$.

`hmmtrain`: Given a sequence of observations, estimate $P_t(\cdot|\cdot)$ and $P_z(\cdot|\cdot)$.

`hmmviterbi`: Given $P_t(\cdot|\cdot), P_z(\cdot|\cdot)$ and a sequence of observations, calculate the most likely state sequence.

The function `hmmtrain()` implements the so-called Baum–Welch algorithm.

4.4 MIXED STATES AND THE PARTICLE FILTER

Sections 4.2 and 4.3 focused on special cases of the general online estimation problem. The topic of Section 4.2 was continuous state estimation, and in particular the linear-Gaussian case or approximations of that. Section 4.3 discussed discrete state estimation. We now return to the general scheme of Figure 4.2. The current section introduces the family of *particle filters* (PF). It is a group of estimators that try to implement the general case. These estimators use random samples to represent the probability densities, just as in the case of Parzen estimation; see Section 5.3.1. As such, particle filters are able to handle nonlinear, non-Gaussian systems, continuous states, discrete states and even combinations. In the sequel we use probability densities (which in the discrete case must be replaced by probability functions).

4.4.1 Importance sampling

A *Monte Carlo* simulation uses a set of random samples generated from a known distribution to estimate the expectation of any function of that distribution. More specifically, let $\mathbf{x}^{(k)}, k = 1, \ldots, K$ be samples drawn from a conditional probability density $p(\mathbf{x}|\mathbf{z})$. Then, the expectation of any function $\mathbf{g}(\mathbf{x})$ can be estimated by:

$$\mathrm{E}[\mathbf{g}(\mathbf{x})|\mathbf{z}] \cong \frac{1}{K}\sum_{k=1}^{K}\mathbf{g}\left(\mathbf{x}^{(k)}\right) \tag{4.71}$$

Under mild conditions, the right-hand side asymptotically approximates the expectation as K increases. For instance, the conditional expectation and covariance matrix are found by substitution of $\mathbf{g}(\mathbf{x}) = \mathbf{x}$ and $\mathbf{g}(\mathbf{x}) = (\mathbf{x} - \hat{\hat{\mathbf{x}}})(\mathbf{x} - \hat{\hat{\mathbf{x}}})^T$, respectively.

In the particle filter, the set $\mathbf{x}^{(k)}$ depends on the time index i. It represents the posterior density $p(\mathbf{x}(i)|\mathbf{Z}(i))$. The samples are called the *particles*. The density can be estimated from the particles by some kernel-based method, for instance, the Parzen estimator to be discussed in Section 5.3.1.

A problem in the particle filter is that we do not know the posterior density beforehand. The solution for that is to get the samples from some other density, say $q(\mathbf{x})$, called the *proposal density*. The various members of the PF family differ (among other things) in their choice of this density. The expectation of $\mathbf{g}(\mathbf{x})$ w.r.t. $p(\mathbf{x}|\mathbf{z})$ becomes:

$$
\begin{aligned}
\mathrm{E}[\mathbf{g}(\mathbf{x})|\mathbf{z}] &= \int \mathbf{g}(\mathbf{x})p(\mathbf{x}|\mathbf{z})d\mathbf{x} \\
&= \int \mathbf{g}(\mathbf{x})\frac{p(\mathbf{x}|\mathbf{z})}{q(\mathbf{x})}q(\mathbf{x})d\mathbf{x} \\
&= \int \mathbf{g}(\mathbf{x})\frac{p(\mathbf{z}|\mathbf{x})p(\mathbf{x})}{p(\mathbf{z})q(\mathbf{x})}q(\mathbf{x})d\mathbf{x} \\
&= \int \mathbf{g}(\mathbf{x})\frac{w(\mathbf{x})}{p(\mathbf{z})}q(\mathbf{x})d\mathbf{x} \quad \text{with: } w(\mathbf{x}) = \frac{p(\mathbf{z}|\mathbf{x})p(\mathbf{x})}{q(\mathbf{x})} \\
&= \frac{1}{p(\mathbf{z})}\int \mathbf{g}(\mathbf{x})w(\mathbf{x})q(\mathbf{x})d\mathbf{x}
\end{aligned}
\tag{4.72}
$$

The factor $1/p(\mathbf{z})$ is a normalizing constant. It can be eliminated as follows:

$$
\begin{aligned}
p(\mathbf{z}) &= \int p(\mathbf{z}|\mathbf{x})p(\mathbf{x})d\mathbf{x} \\
&= \int p(\mathbf{z}|\mathbf{x})p(\mathbf{x})\frac{q(\mathbf{x})}{q(\mathbf{x})}d\mathbf{x} \\
&= \int w(\mathbf{x})q(\mathbf{x})d\mathbf{x}
\end{aligned}
\tag{4.73}
$$

Using (4.72) and (4.73) we can estimate $\mathrm{E}[\mathbf{g}(\mathbf{x})|\mathbf{z}]$ by means of a set of samples drawn from $q(\mathbf{x})$:

$$
\mathrm{E}[\mathbf{g}(\mathbf{x})|\mathbf{z}] \cong \frac{\sum_{k=1}^{K} w(\mathbf{x}^{(k)})\mathbf{g}(\mathbf{x}^{(k)})}{\sum_{k=1}^{K} w(\mathbf{x}^{(k)})}
\tag{4.74}
$$

Being the ratio of two estimates, $\mathrm{E}[\mathbf{g}(\mathbf{x})|\mathbf{z}]$ is a biased estimate. However, under mild conditions, $\mathrm{E}[\mathbf{g}(\mathbf{x})|\mathbf{z}]$ is asymptotically unbiased and consistent as K increases. One of the requirements is that $q(\mathbf{x})$ overlaps the support of $p(\mathbf{x})$.

Usually, the shorter notation for the unnormalized importance weights $w^{(k)} = w(\mathbf{x}^{(k)})$ is used. The so-called *normalized importance weights* are $w^{(k)}_{norm} = w^{(k)}/\sum w^{(k)}$. With that, expression (4.74) simplifies to:

$$\mathrm{E}[\mathbf{g}(\mathbf{x})|\mathbf{z}] \cong \sum_{k=1}^{K} w^{(k)}_{norm}\mathbf{g}\left(\mathbf{x}^{(k)}\right) \tag{4.75}$$

4.4.2 Resampling by selection

Importance sampling provides us with samples $\mathbf{x}^{(k)}$ and weights $w^{(k)}_{norm}$. Taken together, they represent the density $p(\mathbf{x}|\mathbf{z})$. However, we can transform this representation to a new set of samples with equal weights. The procedure to do that is *selection*. The purpose is to delete samples with low weights, and to retain multiple copies of samples with high weights. The number of samples does not change by this; K is kept constant. The various members from the PF family may differ in the way they select the samples. However, an often used method is to draw the samples with replacement according to a multinomial distribution with probabilities $w^{(k)}_{norm}$.

Such a procedure is easily accomplished by calculation of the cumulative weights:

$$w^{(k)}_{cum} = \sum_{j=1}^{k} w^{(j)}_{norm} \tag{4.76}$$

We generate K random numbers $r^{(k)}$ with $k = 1, \ldots, K$. These numbers must be uniformly distributed between 0 and 1. Then, the k-th sample $\mathbf{x}^{(k)}_{selected}$ in the new set is a copy of the j-th sample $\mathbf{x}^{(j)}$ where j is the smallest integer for which $w^{(j)}_{cum} \geq r^{(k)}$.

Figure 4.21 is an illustration. The figure shows a density $p(x)$ and a proposal density $q(x)$. Samples $x^{(k)}$ from $q(x)$ can represent $p(x)$ if they are provided with weights $w^{(k)}_{norm} \propto p(x^{(k)})/q(x^{(k)})$. These weights are visualized in Figure 4.21(d) by the radii of the circles. Resampling by selection gives an unweighted representation of $p(x)$. In Figure 4.21(e), multiple copies of one sample are depicted as a pile. The height of the pile stands for the multiplicity of the copy.

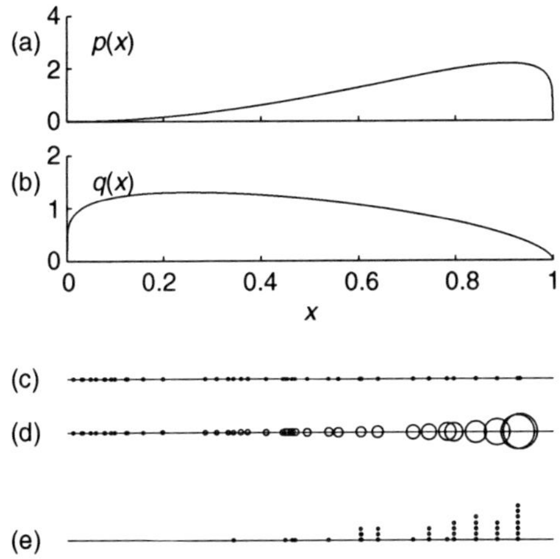

Figure 4.21 Representation of a probability density. (a) A density $p(x)$. (b) The proposal density $q(x)$. (c) 40 samples of $q(x)$. (d) Importance sampling of $p(x)$ using the 40 samples from $q(x)$. (e) Selected samples from (d) as an equally weighted sample representation of $p(x)$

4.4.3 The condensation algorithm

One of the simplest applications of importance sampling combined with resampling by selection is in the so-called *condensation algorithm* ('*cond*itional *dens*ity optimiz*ation*'). The algorithm follows the general scheme of Figure 4.2. The prediction density $p(\mathbf{x}(i)|\mathbf{Z}(i-1))$ is used as the proposal density $q(\mathbf{x})$. So, at time i, we assume that a set $\mathbf{x}^{(k)}$ is available which is an unweighted representation of $p(\mathbf{x}(i)|\mathbf{Z}(i-1))$. We use importance sampling to find the posterior density $p(\mathbf{x}(i)|\mathbf{Z}(i))$. For that purpose we make the following substitutions in (4.72):

$$
\begin{aligned}
p(\mathbf{x}) &\to p(\mathbf{x}(i)|\mathbf{Z}(i-1))\\
p(\mathbf{x}|\mathbf{z}) &\to p(\mathbf{x}(i)|\mathbf{z}(i), \mathbf{Z}(i-1)) = p(\mathbf{x}(i)|\mathbf{Z}(i))\\
q(\mathbf{x}) &\to p(\mathbf{x}(i)|\mathbf{Z}(i-1))\\
p(\mathbf{z}|\mathbf{x}) &\to p(\mathbf{z}(i)|\mathbf{x}(i), \mathbf{Z}(i-1)) = p(\mathbf{z}(i)|\mathbf{x}(i))
\end{aligned}
$$

The weights $w_{norm}^{(k)}$ that define the representation of $p(\mathbf{x}(i)|\mathbf{z}(i))$ is obtained from:

$$w^{(k)} = p(\mathbf{z}(i)|\mathbf{x}^{(k)}) \tag{4.77}$$

Next, resampling by selection provides an unweighted representation $\mathbf{x}_{selected}^{(k)}$. The last step is the prediction. Using $\mathbf{x}_{selected}^{(k)}$ as a representation for $p(\mathbf{x}(i)|\mathbf{Z}(i))$, the representation of $p(\mathbf{x}(i+1)|\mathbf{Z}(i))$ is found by generating one new sample $\mathbf{x}^{(k)}$ for each sample $\mathbf{x}^{(k)}_{selected}$ using $p(\mathbf{x}(i+1)|\mathbf{x}^{(k)}_{selected})$ as the density to draw from. The algorithm is as follows:

Algorithm 4.4: The condensation algorithm

1. Initialization
   - Set $i = 0$
   - Draw K samples $\mathbf{x}^{(k)}, k = 1, \ldots, K$, from the prior probability density $p(\mathbf{x}(0))$
2. Update using importance sampling:
   - Set the importance weights equal to: $w^{(k)} = p(\mathbf{z}(i)|\mathbf{x}^{(k)})$
   - Calculate the normalized importance weights: $w^{(k)}_{norm} = w^{(k)} / \sum w^{(k)}$
3. Resample by selection:
   - Calculate the cumulative weights $w^{(k)}_{cum} = \sum_{j=1}^{k} w^{(j)}_{norm}$
   - for $k = 1, \ldots, K$:
     - Generate a random number r uniformly distributed in $[0, 1]$
     - Find the smallest j such that $w^{(j)}_{cum} \geq r^{(k)}$
     - Set $\mathbf{x}^{(k)}_{selected} = \mathbf{x}^{(j)}$
4. Predict:
   - Set $i = i + 1$
   - for $k = 1, \ldots, K$:
     - Draw sample $\mathbf{x}^{(k)}$, from the density $p(\mathbf{x}(i)|\mathbf{x}(i-1) = \mathbf{x}^{(k)}_{selected})$
5. Go to 2

After step 2, the posterior density is available in terms of the samples $\mathbf{x}^{(k)}$ and the weights $w^{(k)}_{norm}$. The MMSE estimate and the associated error covariance matrix can be obtained from (4.75). For insance, the MMSE is obtained by substitution of $\mathbf{g}(\mathbf{x}) = \mathbf{x}$. Since we have a representation of the posterior density, estimates associated with other criteria can be obtained as well.

The calculation of the importance weights in step 2 involves the conditional density $p(\mathbf{z}(i)|\mathbf{x}^{(k)})$. In the case of the nonlinear measurement functions of the type $\mathbf{z}(i) = \mathbf{h}(\mathbf{x}(i)) + \mathbf{v}(i)$, it all boils down to calculating the density of the measurement noise for $\mathbf{v}(i) = \mathbf{z}(i) - \mathbf{h}(\mathbf{x}^{(k)})$. For instance, if $\mathbf{v}(i)$ is zero mean, Gaussian with covariance matrix $\mathbf{C}_\mathbf{v}$, the weights are calculated as:

$$w^{(k)} = constant \times \exp\left(-\frac{1}{2}(\mathbf{z}(i) - \mathbf{h}(\mathbf{x}^{(k)}))^T \mathbf{C}_\mathbf{v}^{-1} (\mathbf{z}(i) - \mathbf{h}(\mathbf{x}^{(k)}))\right)$$

The actual value of *constant* is irrelevant because of the normalization.

The drawing of new samples in the prediction step involves the state equation. If $\mathbf{x}(i+1) = \mathbf{f}(\mathbf{x}(i), \mathbf{u}(i)) + \mathbf{w}(i)$, then the drawing is governed by the density of $\mathbf{w}(i)$.

The advantages of the particle filtering are obvious. Nonlinearities of both the state equation and the measurement function are handled smoothly without the necessity to calculate Jacobian matrices. However, the method works well only if enough particles are used. Especially for large dimensions of the state vector the required number becomes large. If the number is not sufficient, then the particles are not able to represent the density $p(\mathbf{x}(i)|\mathbf{Z}(i-1))$. Particularly, if for some values of $\mathbf{x}(i)$ the likelihood $p(\mathbf{z}(i)|\mathbf{x}(i))$ is very large, while on these locations $p(\mathbf{x}(i)|\mathbf{Z}(i-1))$ is small, the particle filtering may not converge. It occurs frequently then that all weights become zero except one which becomes unit. The particle filter is said to be *degenerated*.

Example 4.10 Particle filtering applied to volume density estimation
The problem of estimating the volume density of a substance mixed with a liquid is introduced in Example 4.1. The model, expressed in equation (4.3), is nonlinear and non-Gaussian. The state vector consists of two continuous variables (volume and density), and one discrete variable (the state of the on/off controller). The measurement system, expressed in equation (4.40), is nonlinear with additive Gaussian noise. Example 4.6 has shown that the EKF is able to estimate the density, but only by using an approximate model of the process in which the discrete state is removed. The price to pay for such a rough approximation is that the estimation error of the density and (particularly) the volume has a large magnitude.

The particle filter does not need such a rough approximation because it can handle the discrete state variable. In addition, the particle filter can cope with discontinuities. These discontinuities appear here because of the discrete on/off control, but also because the input flow of the substance occurs in chunks at some random points in time.

The particle filter implemented in this example uses the process model given in (4.3), and the measurement model of (4.40). The parameters used are tabulated in Example 4.5. Other parameters are: $V_{low} = 3990$ (litre) and $V_{high} = 4010$ (litre). The random points of the substance are modelled as a Poisson process with mean time between two points $\lambda = 100\Delta = 100$ (s). The chunks have an uniform distribution between 7 and 13 (litre). Results of the particle filter using 10000 particles are shown in Figure 4.22. The figure shows an example of a cloud of particles. Clearly, such a cloud is not represented by a Gaussian distribution. In fact, the distribution is

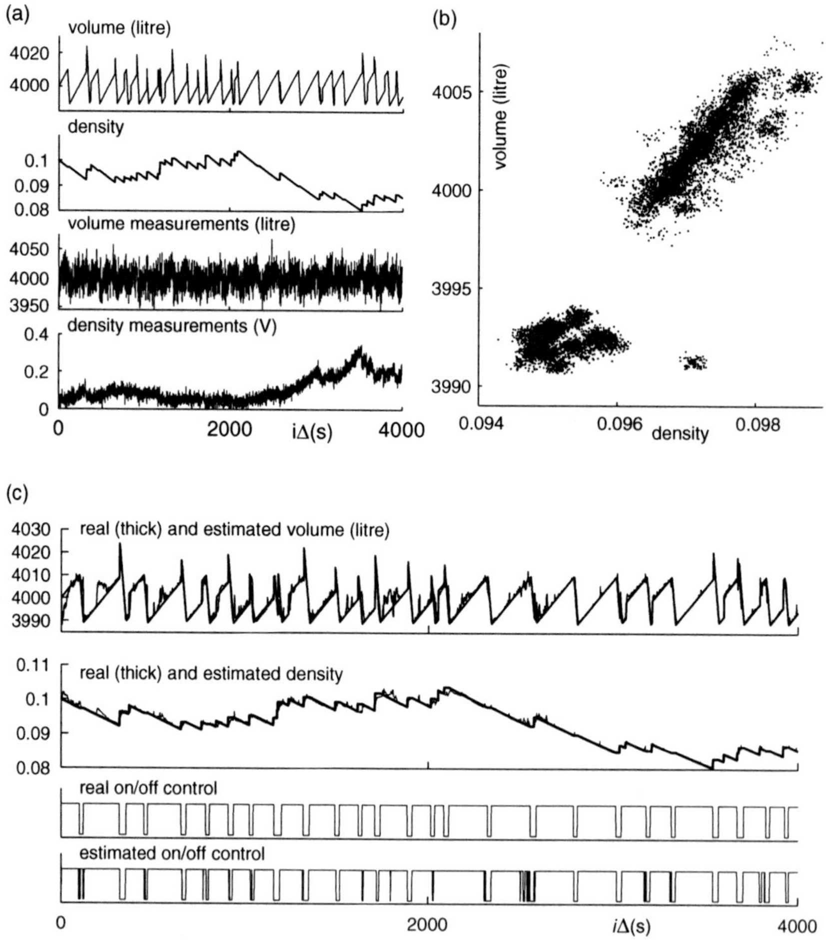

Figure 4.22 Application of particle filtering to the density estimation problem. (a) Real states and measurements. (b) The particles obtained at $i = 511$. (c) Results

multi-modal due to the uncertainty of the moment at which the on/off control switches its state.

In contrast with the Kalman filter, the particle filter is able to estimate the fluctuations of the volume. In addition, the estimation of the density is much more accurate. The price to pay is the computational cost.

MATLAB functions for particle filtering

Many MATLAB users have already implemented particle filters, but no formal toolbox yet exists. Section 9.3 contains a listing of MATLAB code that implements the condensation algorithm. Details of the implementation are also given.

4.5 SELECTED BIBLIOGRAPHY

Some introductory books on Kalman filtering and its applications are Anderson and Moore (1979), Bar-Shalom and Li (1993), Gelb *et al.* (1974), Grewal and Andrews (2001). Hidden Markov models are described in Rabiner (1989). Tutorials on particle filtering are found in Arulampalam *et al.* (2002) and Merwe *et al.* (2000). These tutorials also describe some shortcomings of the particle filter, and possible remedies. Seminal papers for Kalman filtering, particle filtering and unscented Kalman filtering are Kalman (1960), Gordon *et al.* (1993) and Julier and Uhlmann (1997), respectively. Linear systems with random inputs, among which the AR models, are studied in Box and Jenkins (1976). The topic of statistical linearization is treated in Gelb *et al.* (1974). The condensation algorithm is due to Isard and Blake (1996). The Baum-Welch algorithm is described in Rabiner (1986).

Anderson, B.D. and Moore, J.B., *Optimal Filtering*, Prentice Hall, Englewood Cliffs, NJ, 1979.

Arulampalam, M.S., Maskell, S., Gordon, N. and Clapp, T., A tutorial on particle filters for online nonlinear/non-Gaussian Bayesian tracking, *IEEE Transactions on Signal Processing*, 50(2), 174–88, February 2002.

Bar-Shalom, Y. and Li, X.R., *Estimation and Tracking – Principles, Techniques, and Software*, Artech House, Boston, 1993.

Box, G.E.P. and Jenkins, G.M., *Time Series Analysis: Forecasting and Control*, Holden-Day, San Francisco, 1976.

Gelb, A., Kasper, J.F., Nash, R.A., Price, C.F. and Sutherland, A.A., *Applied Optimal Estimation*, MIT Press, Cambridge, MA, 1974.

Gordon, N.J., Salmond, D.J. and Smith, A.F.M., Novel approach to nonlinear/nonGaussian Bayesian state estimation, *IEE Proceedings-F*, 140(2), 107–13, 1993.

Grewal, M.S. and Andrews, A.P., *Kalman Filtering – Theory and Practice Using MATLAB*, Second edition, Wiley, New York, 2001.

Isard, M. and Blake, A. Contour tracking by stochastic propagation of conditional density, *European Conference on Computer Vision*, pp. 343–56, Cambridge, UK, 1996.

Julier, S.J. and Uhlmann, J.K., A new extension of the Kalman filter to nonlinear systems, *Proc. of AeroSense: the 11th International Symposium on Aerospace/Defence Sensing, Simulation and Controls*, Vol. Multi Sensor, Tracking and Resource Management II, Orlando, Florida, 1997.

Kalman, R.E., A new approach to linear filtering and prediction problems, *ASME Journal of Basic Engineering*, 82, 34–45, 1960.

Merwe, R. van der, Doucet, A., Freitas N. de and Wan, E. The unscented particle filter, Technical report CUED/F-INFENG/TR 380, Cambridge University Engineering Department, 2000.

Rabiner, L.R., A tutorial on hidden Markov models and selected applications in speech recognition, *Proceedings of the IEEE*, 77(2), 257–85, February 1989.

4.6 EXERCISES

1. Consider the following model for a random constant:

$$\begin{aligned} x(i+1) &= x(i) \\ z(i) &= x(i) + v(i) \quad v(i) \text{ is white noise with variance } \sigma_v^2 \end{aligned}$$

   The prior knowledge is $\mathrm{E}[x(0)] = 0$ and $\sigma_{x(0)}^2 = \infty$. Give the expression for the solution of the discrete Lyapunov equation. (0).

2. For the random constant model given in exercise 1, give expressions for the innovation matrix $\mathbf{S}(i)$, the Kalman gain matrix $\mathbf{K}(i)$, the error covariance matrix $\mathbf{C}(i|i)$ and the prediction matrix $\mathbf{C}(i|i-1)$ for the first few time steps. That is, for $i = 0, 1, 2$ and 3. Explain the results. (0).
3. In exercise 2, can you find the expressions for arbitrary i. Can you also prove that these expressions are correct? Hint: use induction. Explain the results. (*).
4. Consider the following time-invariant scalar linear-Gaussian system

$$\begin{aligned} x(i+1) &= \alpha x(i) + w(i) \quad w(i) \text{ is white noise with variance } \sigma_w^2 \\ z(i) &= x(i) + v(i) \quad v(i) \text{ is white noise with variance } \sigma_v^2 \end{aligned}$$

   The prior knowledge is $\mathrm{E}x0 = 0$ and $\sigma_{x0}^2 = \infty$. What is the condition for the existence of the solution of the discrete Lyapunov equation? If this condition is met, give an expression for that solution. (0).

5. For the system described in exercise 4, give the steady state solution. That is, give expressions for $\mathbf{S}(i), \mathbf{K}(i), \mathbf{C}(i|i)$ and $\mathbf{C}(i|i-1)$ if $i \to \infty$. (0).
6. For the system given in exercise 4, give expressions for $\mathbf{S}(i), \mathbf{K}(i), \mathbf{C}(i|i)$ and $\mathbf{C}(i|i-1)$ for the first few time steps, that is, $i = 0, 1, 2$ and 3. Explain the results. (0).

7. In exercise 4, can you find the expressions for $\mathbf{S}(i), \mathbf{K}(i), \mathbf{C}(i|i)$ and $\mathbf{C}(i|i-1)$ for arbitrary i? (*).
8. Autoregressive models and MATLAB's Control System Toolbox: consider the following second order autoregressive (AR) model:

$$\begin{aligned} x(i+1) &= \frac{1}{2}x(i) + \frac{1}{4}x(i-1) + w(i) \\ z(i) &= x(i) + v(i) \end{aligned}$$

Using the functions `tf()` and `ss()` from the Control Toolbox, convert this AR model into an equivalent state space model, that is, $\mathbf{x}(i+1) = \mathbf{F}\mathbf{x}(i) + \mathbf{G}w(i)$ and $z(i) = \mathbf{H}\mathbf{x}(i) + v(i)$. (Use `help tf` and `help ss` to find out how these functions should be used.) Assuming that $w(i)$ and $v(i)$ are white noise sequences with variances $\sigma_w^2 = 1$ and $\sigma_v^2 = 1$, use the function `dlyap()` to find the solution of the discrete Lyapunov equation, and the function `kalman()` (or `dlqe()`) to find the solution for the steady state Kalman gain and corresponding error covariance matrices. Hint: the output variable of the command `ss()` is a 'struct' whose fields are printed by typing `struct(ss)`. (*).
9. Moving average models: repeat exercise 8, but now considering the so-called first order moving average (MA) model:

$$\begin{aligned} x(i+1) &= \frac{1}{2}(w(i) + w(i-1)) \quad \sigma_w^2 = 1 \\ z(i) &= x(i) + v(i) \qquad \sigma_v^2 = 1 \end{aligned} \qquad (*)$$

10. Autoregressive, moving average models: repeat exercise 8, but now considering the so-called ARMA(2, 1) model:

$$\begin{aligned} x(i+1) = \frac{1}{2}x(i) + \frac{1}{2}x(i-1) + \frac{1}{2}(w(i) + w(i-1)) \quad & \sigma_w^2 = 1 \\ z(i) = x(i) + v(i) \qquad & \sigma_v^2 = 1 \end{aligned} \qquad (*)$$

11. Simulate the processes mentioned in exercise 1, 8, 9 and 10, using MATLAB, and apply the Kalman filters.

5 Supervised Learning

One method for the development of a classifier or an estimator is the so-called *model-based* approach. Here, the required availability of the conditional probability densities and the prior probabilities are obtained by means of general knowledge of the physical process and the sensory system in terms of mathematical models. The development of the estimators for the backscattering coefficient, discussed in Chapter 3, follows such an approach.

In many other applications, modelling the process is very difficult if not impossible. For instance, in the mechanical parts application, discussed in Chapter 2, the visual appearance of the objects depends on many factors that are difficult to model. The alternative to the model-based approach is the *learning from examples* paradigm. Here, it is assumed that in a given application a population of objects is available. From this population, some objects are selected. These selected objects are called the *samples*. Each sample is presented to the sensory system which returns the measurement vector associated with that sample. The purpose of learning (or *training*) is to use these measurement vectors of the samples to build a classifier or an estimator.

The problem of learning has two versions: *supervised* and *unsupervised*, that is, with or without knowing the true class/parameter of the sample. See Figure 5.1. This chapter addresses the first version. Chapter 7 deals with unsupervised learning.

Classification, Parameter Estimation and State Estimation: An Engineering Approach using MATLAB
F. van der Heijden, R.P.W. Duin, D. de Ridder and D.M.J. Tax
 ISBN: 0-470-09013-8

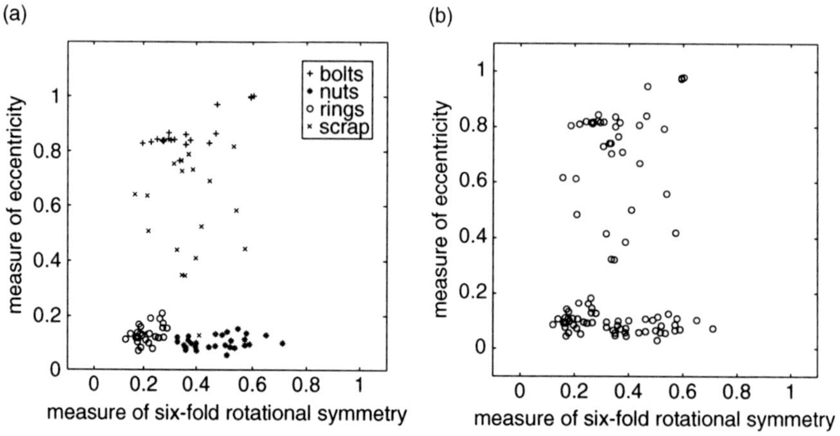

Figure 5.1 Training sets. (a) Labelled. (b) Unlabelled

The chapter starts with a section on the representation of training sets. In Sections 5.2 and 5.3 two approaches to supervised learning are discussed: parametric and nonparametric learning. Section 5.4 addresses the problem of how to evaluate a classifier empirically. The discussion here is restricted to classification problems only. However, many techniques that are useful for classification problems are also useful for estimation problems. Especially Section 5.2 (parametric learning) is useful for estimation problems too.

5.1 TRAINING SETS

The set of samples is usually called the *training set* (or: *learning data* or *design set*). The selection of samples should occur randomly from the population. In almost all cases it is assumed that the samples are i.i.d., independent and identically distributed. This means that all samples are selected from the same population of objects (in the simplest case, with equal probability). Furthermore, the probability of one member of the population being selected is not allowed to depend on the selection of other members of the population.

Figure 5.1 shows scatter diagrams of the mechanical parts application of Chapter 2. In Figure 5.1(a) the samples are provided with a label carrying the information of the true class of the corresponding object. There are several methods to find the true class of a sample, e.g. manual inspection, additional measurements, destructive analysis, etc. Often, these methods are expensive and therefore allowed only if the number of samples is not too large.

The number of samples in the training set is denoted by N_S. The samples are enumerated by the symbol $n = 1, \ldots, N_S$. Object n has a measurement vector $\mathbf{z}_n$. The true class of the n-th object is denoted by $\theta_n \in \Omega$. Then, a labelled training set T_S contains samples ($\mathbf{z}_n$, θ_n) each one consisting of a measurement vector and its true class:

$$T_S = \{(\mathbf{z}_n, \theta_n)\} \quad \text{with} \quad n = 1, \ldots, N_S \tag{5.1}$$

Another representation of the data set is obtained if we split the training set according to their true classes:

$$T_S = \{\mathbf{z}_{k,n}\} \quad \text{with} \quad k = 1, \ldots, K \quad \text{and} \quad n = 1, \ldots, N_k \tag{5.2}$$

where N_k is the number of samples with class ω_k and $K = |\Omega|$ is the number of classes. A representation equivalent to (5.2) is to introduce a set T_k for each class.

$$T_k = \{(\mathbf{z}_n, \theta_n) | \theta_n = \omega_k\} \quad \text{with} \quad k = 1, \ldots, K \quad \text{and} \quad n = 1, \ldots, N_k \tag{5.3}$$

It is understood that the numberings of samples used in these three representations do not coincide. Since the representations are equivalent, we have:

$$N_S = \sum_{k=1}^{K} N_k \tag{5.4}$$

In PRTools, data sets are always represented as in (5.1). In order to obtain representations as in (5.3), separate data sets for each of the classes will have to be constructed. In Listing 5.1 these two ways are shown. It is assumed that `dat` is an $N \times d$ matrix containing the measurements, and `lab` an $N \times 1$ matrix containing the class labels.

Listing 5.1
Two methods of representing data sets in PRTools. The first method is used almost exclusively.

```
% Create a standard MATLAB dataset from data and labels.
% Method (5.1):
```

```
dat=[ 0.1 0.9 ; 0.3 0.95 ; 0.2 0.7 ];
lab={ 'class 1', 'class 2', 'class 3' };
z=dataset(dat,lab);
% Method (5.3):
[nlab,lablist]=getnlab(z);     % Extract the numeric labels
[m,k,c]=getsize(z);            % Extract number of classes
for i=1:c
  T{i}=seldat(z,i);
end;
```

5.2 PARAMETRIC LEARNING

The basic assumption in parametric learning is that the only unknown factors are parameters of the probability densities involved. Thus, learning from samples boils down to finding the suitable values of these parameters. The process is analogous to parameter estimation discussed in Chapter 3. The difference is that the parameters in Chapter 3 describe a physical process whereas the parameters discussed here are parameters of the probability densities of the measurements of the objects. Moreover, in parametric learning a set of many measurement vectors is available rather than just a single vector. Despite these two differences, the concepts from Chapter 3 are fully applicable to the current chapter.

Suppose that $\mathbf{z}_n$ are the samples coming from a same class ω_k. These samples are repeated realizations of a single random vector $\mathbf{z}$. An alternative view is to associate the samples with single realizations coming from a set of random vectors with identical probability densities. Thus, a training set T_k consists of N_k mutually independent, random vectors $\mathbf{z}_n$. The joint probability density of these vectors is

$$p(\mathbf{z}_1, \mathbf{z}_2, \ldots, \mathbf{z}_{N_k}|\omega_k, \boldsymbol{\alpha}_k) = \prod_{n=1}^{N_k} p(\mathbf{z}_n|\omega_k, \boldsymbol{\alpha}_k) \qquad (5.5)$$

$\boldsymbol{\alpha}_k$ is the unknown parameter vector of the conditional probability density $p(\mathbf{z}|\omega_k, \boldsymbol{\alpha}_k)$. Since in parametric learning we assume that the form of $p(\mathbf{z}|\omega_k, \boldsymbol{\alpha}_k)$ is known (only the parameter vector $\boldsymbol{\alpha}_k$ is unknown), the complete machinery of Bayesian estimation (minimum risk, MMSE estimation, MAP estimation, ML estimation) becomes available to find estimators for the parameter vector $\boldsymbol{\alpha}_k$: see Section 3.1. Known concepts to evaluate these estimators (bias and variance) also apply. The next subsections discuss some special cases for the probability densities.

5.2.1 Gaussian distribution, mean unknown

Let us assume that under class ω_k the measurement vector $\mathbf{z}$ is a Gaussian random vector with known covariance matrix $\mathbf{C}_k$ and unknown expectation vector $\boldsymbol{\mu}_k$. No prior knowledge is available concerning this unknown vector. The purpose is to find an estimator for $\boldsymbol{\mu}_k$.

Since no prior knowledge about $\boldsymbol{\mu}_k$ is assumed, a maximum likelihood estimator seems appropriate (Section 3.1.4). Substitution of (5.5) in (3.22) gives the following general expression of a maximum likelihood estimator:

$$\begin{aligned}\hat{\boldsymbol{\mu}}_k &= \underset{\boldsymbol{\mu}}{\operatorname{argmax}}\left\{\prod_{n=1}^{N_k} p(\mathbf{z}_n|\omega_k,\boldsymbol{\mu})\right\}\\ &= \underset{\boldsymbol{\mu}}{\operatorname{argmax}}\left\{\sum_{n=1}^{N_k} \ln(p(\mathbf{z}_n|\omega_k,\boldsymbol{\mu}))\right\}\end{aligned} \tag{5.6}$$

The logarithms introduced in the last line transform the product into a summation. This is only a technical matter which facilitates the maximization.

Knowing that $\mathbf{z}$ is Gaussian, the likelihood of $\boldsymbol{\mu}_k$ from a single observation $\mathbf{z}_n$ is:

$$p(\mathbf{z}_n|\omega_k,\boldsymbol{\mu}_k) = \frac{1}{\sqrt{(2\pi)^N|\mathbf{C}_k|}}\exp\left(-\frac{1}{2}(\mathbf{z}_n-\boldsymbol{\mu}_k)^T\mathbf{C}_k^{-1}(\mathbf{z}_n-\boldsymbol{\mu}_k)\right) \tag{5.7}$$

Upon substitution of (5.7) in (5.6), rearrangement of terms and elimination of irrelevant terms, we have:

$$\begin{aligned}\hat{\boldsymbol{\mu}}_k &= \underset{\boldsymbol{\mu}}{\operatorname{argmin}}\left\{\sum_{n=1}^{N_k}(\mathbf{z}_n-\boldsymbol{\mu})^T\mathbf{C}_k^{-1}(\mathbf{z}_n-\boldsymbol{\mu})\right\}\\ &= \underset{\boldsymbol{\mu}}{\operatorname{argmin}}\left\{\sum_{n=1}^{N_k}\mathbf{z}_n^T\mathbf{C}_k^{-1}\mathbf{z}_n + \sum_{n=1}^{N_k}\boldsymbol{\mu}^T\mathbf{C}_k^{-1}\boldsymbol{\mu} - 2\sum_{n=1}^{N_k}\mathbf{z}_n^t\mathbf{C}_k^{-1}\boldsymbol{\mu}\right\}\end{aligned} \tag{5.8}$$

Differentiating the expression between braces with respect to $\boldsymbol{\mu}$ (Appendix B.4) and equating the result to zero yields the *average* or *sample mean* calculated over the training set:

$$\hat{\boldsymbol{\mu}}_k = \frac{1}{N_k}\sum_{n=1}^{N_k}\mathbf{z}_n \tag{5.9}$$

Being a sum of Gaussian random variables, this estimate has a Gaussian distribution too. The expectation of the estimate is:

$$\mathrm{E}[\hat{\boldsymbol{\mu}}_k] = \frac{1}{N_k}\sum_{n=1}^{N_k}\mathrm{E}[\mathbf{z}_n] = \frac{1}{N_k}\sum_{n=1}^{N_k}\boldsymbol{\mu}_k = \boldsymbol{\mu}_k \tag{5.10}$$

where $\boldsymbol{\mu}_k$ is the true expectation of $\mathbf{z}$. Hence, the estimation is unbiased. The covariance matrix of the estimation error is found as:

$$\mathbf{C}_{\hat{\mu}_k} = \mathrm{E}\left[(\hat{\boldsymbol{\mu}}_k - \boldsymbol{\mu}_k)(\hat{\boldsymbol{\mu}}_k - \boldsymbol{\mu}_k)^T\right] = \frac{1}{N_k}\mathbf{C}_k \tag{5.11}$$

The proof is left as an exercise for the reader.

5.2.2 Gaussian distribution, covariance matrix unknown

Next, we consider the case where under class ω_k the measurement vector $\mathbf{z}$ is a Gaussian random vector with unknown covariance matrix $\mathbf{C}_k$. For the moment we assume that the expectation vector $\boldsymbol{\mu}_k$ is known. No prior knowledge is available. The purpose is to find an estimator for $\mathbf{C}_k$.

The maximum likelihood estimate follows from (5.5) and (5.7):

$$\hat{\mathbf{C}}_k = \underset{\mathbf{C}}{\mathrm{argmax}}\left\{\sum_{n=1}^{N_k}\ln(p(\mathbf{z}_n|\omega_k, \mathbf{C}))\right\} = \frac{1}{N_k}\sum_{n=1}^{N_k}(\mathbf{z}_n - \boldsymbol{\mu}_k)(\mathbf{z}_n - \boldsymbol{\mu}_k)^T \tag{5.12}$$

The last step in (5.12) is non-trivial. The proof is rather technical and will be omitted. However, the result is plausible since the estimate is the average of the N_k matrices $(\mathbf{z}_n - \boldsymbol{\mu}_k)(\mathbf{z}_n - \boldsymbol{\mu}_k)^T$ whereas the true covariance matrix is the expectation of $(\mathbf{z} - \boldsymbol{\mu}_k)(\mathbf{z} - \boldsymbol{\mu}_k)^T$.

The probability distribution of the random variables in $\hat{\mathbf{C}}_k$ is a *Wishart distribution*. The estimator is unbiased. The variances of the elements of $\hat{\mathbf{C}}_k$ are:

$$\mathrm{Var}[\hat{\mathbf{C}}_{k_{i,j}}] = \frac{1}{N_k}\left(\mathbf{C}_{k_{i,i}}\mathbf{C}_{k_{j,j}} + \mathbf{C}^2_{k_{i,j}}\right) \tag{5.13}$$

5.2.3 Gaussian distribution, mean and covariance matrix both unknown

If both the expectation vector and the covariance matrix are unknown, the estimation problem becomes more complicated because then we have to estimate the expectation vector and covariance matrix simultaneously. It can be deduced that the following estimators for C_k and $\boldsymbol{\mu}_k$ are unbiased:

$$\hat{\boldsymbol{\mu}}_k = \frac{1}{N_k}\sum_{n=1}^{N_k} \mathbf{z}_n$$
$$\hat{C}_k = \frac{1}{N_k - 1}\sum_{n=1}^{N_k} (\mathbf{z}_n - \hat{\boldsymbol{\mu}}_k)(\mathbf{z}_n - \hat{\boldsymbol{\mu}}_k)^T \tag{5.14}$$

$\hat{C}_k$ is called the *sample covariance*. Comparing (5.12) with (5.14) we note two differences. In the latter expression the unknown expectation has been replaced with the sample mean. Furthermore, the divisor N_k has been replaced with N_k-1. Apparently, the lack of knowledge of $\boldsymbol{\mu}_k$ in (5.14) makes it necessary to sacrifice one degree of freedom in the averaging operator. For large N_K, the difference between (5.12) and (5.14) vanishes.

In classification problems, often the inverse C_k^{-1} is needed, for instance, in quantities like: $\mathbf{z}^T C_k^{-1}\boldsymbol{\mu}$, $\mathbf{z}^T C_k^{-1}\mathbf{z}$, etc. Often, $\hat{C}_k^{-1}$ is used as an estimate of C_k^{-1}. To determine the number of samples required such that $\hat{C}_k^{-1}$ becomes an accurate estimate of C_k^{-1}, the variance of $\hat{C}_k$, given in (5.13), is not very helpful. To see this it is instructive to rewrite the inverse as (see Appendix B.5 and C.3.2):

$$C_k^{-1} = V_k \Lambda_k^{-1} V_k^T \tag{5.15}$$

where Λ_k is a diagonal matrix containing the eigenvalues of C_k. Clearly, the behaviour of C_k^{-1} is strongly affected by small eigenvalues in Λ_k. In fact, the number of nonzero eigenvalues in the estimate $\hat{C}_k$ given in (5.14) cannot exceed $N_k - 1$. If $N_k - 1$ is smaller than the dimension N of the measurement vector, the estimate $\hat{C}_k$ is not invertible. Therefore, we must require that N_k is (much) larger than N. As a rule of thumb, the number of samples must be chosen such that at least $N_k > 5N$.

In order to reduce the sensitivity to statistical errors, we might also want to *regularize* the inverse operation. Suppose that $\hat{\Lambda}$ is the diagonal matrix containing the eigenvalues of the estimated covariance matrix $\hat{\mathbf{C}}$ (we conveniently drop the index k for a moment). $\hat{\mathbf{V}}$ is the matrix containing the eigenvectors. Then, we can define a regularized inverse operation as follows:

$$\hat{\mathbf{C}}^{-1}_{regularized} = \hat{\mathbf{V}}\left((1-\gamma)\hat{\Lambda} + \gamma\frac{trace(\hat{\Lambda})}{N}\mathbf{I}\right)^{-1}\hat{\mathbf{V}}^T \quad 0 \le \gamma \le 1 \tag{5.16}$$

where $trace(\hat{\Lambda})/N$ is the average of the eigenvalues of $\hat{\mathbf{C}}$. γ is a regularization parameter. The effect is that the influence of the smallest eigenvalues is tamed. A simpler implementation of (5.16) is (see exercise 2):

$$\hat{\mathbf{C}}^{-1}_{regularized} = \left((1-\gamma)\hat{\mathbf{C}} + \gamma\frac{trace(\hat{\mathbf{C}})}{N}\mathbf{I}\right)^{-1} \tag{5.17}$$

Another method to regularize a covariance matrix estimate is by suppressing all off-diagonal elements to some extent. This is achieved by multiplying these elements by a factor that is selected between 0 and 1.

Example 5.1 Classification of mechanical parts, Gaussian assumption

We now return to the example in Chapter 2 where mechanical parts like nuts and bolts, etc. must be classified in order to sort them. See Figure 2.2. In Listing 5.2 the PRTools procedure for training and visualizing the classifiers is given. Two classifiers are trained: a linear classifier (`ldc`) and a quadratic classifier (`qdc`). The trained classifiers are stored in `w_l` and `w_q`, respectively. Using the `plotc` function, the decision boundaries can be plotted. In principle this visualization is only possible in 2D. For data with more than two measurements, the classifiers cannot be visualized.

Listing 5.2

PRTools code for training and plotting linear and quadratic discriminants under assumption of normal distributions of the conditional probability densities.

```
load nutsbolts;         % Load the mechanical parts dataset
w_l = ldc(z,0,0.7);     % Train a linear classifier on z
```

```
w_q=qdc(z,0,0.5);          % Train a quadratic classifier on z
figure; scatterd(z);       % Show scatter diagram of z
plotc(w_l);                % Plot the first classifier
plotc(w_q,':');            % Plot the second classifier
[0.4 0.2]*w_l*labeld       % Classify a new object with z=[0.4 0.2]
```

Figure 5.2 shows the decision boundaries obtained from the data shown in Figure 5.1(a) assuming Gaussian distributions for each class. The discriminant in Figure 5.2(a) assumes that the covariance matrices for different classes are the same. This assumption yields a Mahalanobis distance classifier. The effect of the regularization is that the classifier tends to approach the Euclidean distance classifier. Figure 5.2(b) assumes unequal covariance matrices. The effect of the regularization here is that the decision boundaries tend to approach circle segments.

5.2.4 Estimation of the prior probabilities

The prior probability of a class is denoted by $P(\omega_k)$. There are exactly K classes. Having a labelled training set with N_S samples (randomly selected from a population), the number N_k of samples with class ω_k has a so-called *multinomial distribution*. If $K = 2$ the distribution is *binomial*. See Appendix C.1.3.

The multinomial distribution is fully defined by K parameters. In addition to the $K - 1$ parameters that are necessary to define the prior

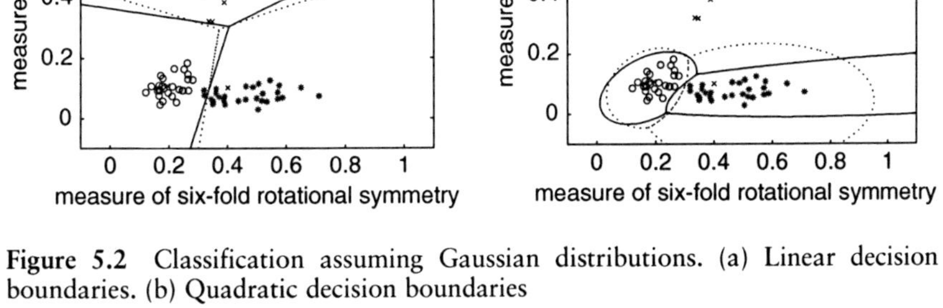

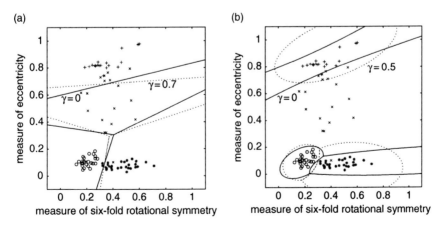

Figure 5.2 Classification assuming Gaussian distributions. (a) Linear decision boundaries. (b) Quadratic decision boundaries

probabilities, we need to specify one extra parameter, N_S, which is the number of samples.

Intuitively, the following estimator is appropriate:

$$\hat{P}(\omega_k) = \frac{N_k}{N_S} \tag{5.18}$$

The expectation of N_k equals $N_S P(\omega_k)$. Therefore, $\hat{P}(\omega_k)$ is an unbiased estimate of $P(\omega_k)$. The variance of a multinomial distributed variable is $N_S P(\omega_k)(1 - P(\omega_k))$. Consequently, the variance of the estimate is:

$$\mathrm{Var}[\hat{P}(\omega_k)] = \frac{P(\omega_k)(1 - P(\omega_k))}{N_S} \tag{5.19}$$

This shows that the estimator is consistent. That is if $N_S \to \infty$, then $\mathrm{Var}[\hat{P}(\omega_k)] \to 0$. The required number of samples follows from the constraint that $\sqrt{\mathrm{Var}[\hat{P}(\omega_k)]} << P(\omega_k)$. For instance, if for some class we anticipate that $P(\omega_k) = 0.01$, and the permitted relative error is 20%, i.e. $\sqrt{\mathrm{Var}[\hat{P}(\omega_k)]} = 0.2P(\omega_k)$, then N_S must be about 2500 in order to obtain the required precision.

5.2.5 Binary measurements

Another example of a multinomial distribution occurs when the measurement vector $\mathbf{z}$ can only take a finite number of states. For instance, if the sensory system is such that each element in the measurement vector is binary, i.e. either '1' or '0', then the number of states the vector can take is at most 2^N. Such a binary vector can be replaced with an equivalent scalar z that only takes integer values from 1 up to 2^N. The conditional probability density $p(\mathbf{z}|\omega_k)$ turns into a probability function $P(z|\omega_k)$. Let $N_k(z)$ be the number of samples in the training set with measurement z and class ω_k. $N_k(z)$ has a multinomial distribution.

At first sight, one would think that estimating $P(z|\omega_k)$ is the same type of problem as estimating the prior probabilities such as discussed in the previous section:

$$\hat{P}(z|\omega_k) = \frac{N_k(z)}{N_k} \tag{5.20}$$

$$\mathrm{Var}[\hat{P}(z|\omega_k)] = \frac{P(z|\omega_k)(1 - P(z|\omega_k))}{N_k} \tag{5.21}$$

For small N and a large training set, this estimator indeed suffices. However, if N is too large, the estimator fails. A small example demonstrates this. Suppose the dimension of the vector is $N = 10$. Then the total number of states is $2^{10} \approx 10^3$. Therefore, some states will have a probability of less than 10^{-3}. The uncertainty of the estimated probabilities must be a fraction of that, say 10^{-4}. The number of samples, N_k, needed to guarantee such a precision is on the order of 10^5 or more. Needless to say that in many applications 10^5 samples is much too expensive. Moreover, with even a slight increase of N the required number of samples becomes much larger.

One way to avoid a large variance is to incorporate more prior knowledge. For instance, without the availability of a training set, it is known beforehand that all parameters are bounded by $0 \leq P(z|\omega_k) \leq 1$. If nothing further is known, we could first 'guess' that all states are equally likely: $P(z|\omega_k) = 2^{-N}$. Based on this guess, the estimator takes the form:

$$\hat{P}(z|\omega_k) = \frac{N_k(z) + 1}{N_k + 2^N} \tag{5.22}$$

The variance of the estimate is:

$$\mathrm{Var}[\hat{P}(z|\omega_k)] = \frac{N_k P(z|\omega_k)(1 - P(z|\omega_k))}{(N_k + 2^N)^2} \tag{5.23}$$

Comparing (5.22) and (5.23) with (5.20) and (5.21) the conclusion is that the variance of the estimate is reduced at the cost of a small bias. See also exercise 4.

5.3 NONPARAMETRIC LEARNING

Nonparametric methods are learning methods for which prior knowledge about the functional form of the conditional probability distributions is not available or is not used explicitly. The name may suggest that no parameters are involved. However, in fact these methods often require *more* parameters than parametric methods. The difference is that in nonparametric methods the parameters are not the parameters of the conditional distributions.

At first sight, nonparametric learning seems to be more difficult than parametric learning because nonparametric methods exploit less knowledge. For some nonparametric methods this is indeed the case. In principle, these types of nonparametric methods can handle arbitrary types of conditional distributions. Their generality is high. The downside of this advantage is that large to very large training sets are needed to compensate the lacking knowledge about the densities.

Other nonparametric learning methods cannot handle arbitrary types of conditional distributions. The classifiers being trained are constrained to some preset computational structure of their decision function. By this, the corresponding decision boundaries are also constrained. An example is the linear classifier already mentioned in Section 2.1.2. Here, the decision boundaries are linear (hyper)planes. The advantage of incorporating constraints is that fewer samples in the training set are needed. The stronger the constraints are, the fewer samples are needed. However, good classifiers can only be obtained if the constraints that are used match the type of the underlying problem-specific distributions. Hence, in constraining the computational structure of the classifier implicit knowledge of the distribution is needed.

5.3.1 Parzen estimation and histogramming

The objective of *Parzen estimation* and histogramming is to obtain estimates of the conditional probability densities. This is done without much prior knowledge of these densities. As before, the estimation is based on a labelled training set T_S. We use the representation according to (5.3), i.e. we split the training set into K subsets T_k, each having N_k samples all belonging to class ω_k. The goal is to estimate the conditional density $p(\mathbf{z}|\omega_k)$ for arbitrary $\mathbf{z}$.

A simple way to reach the goal is to partition the measurement space into a finite number of disjoint regions R_i, called *bins*, and to count the number of samples that falls in each of these bins. The estimated probability density within a bin is proportional to that count. This technique is called *histogramming*. Suppose that $N_{k,i}$ is the number of samples with class ω_k that fall within the i-th bin. Then the probability density within the i-th bin is estimated as:

$$\hat{p}(\mathbf{z}|\omega_k) = \frac{N_{k,i}}{\text{Volume}(R_i) \times N_k} \quad \text{with} \quad \mathbf{z} \in R_i \tag{5.24}$$

For each class, the number $N_{k,i}$ has a multinomial distribution with parameters

$$P_{k,i} = \int_{\mathbf{z} \in R_i} p(\mathbf{z}|\omega_k) d\mathbf{z} \quad \text{with} \quad i = 1, \ldots, N_{bin}$$

where N_{bin} is the number of bins. The statistical properties of $\hat{p}(\mathbf{z}|\omega_k)$ follows from arguments that are identical to those used in Section 5.2.5. In fact, if we quantize the measurement vector to, for instance, the nearest centre of gravity of the bins, we end up in a situation similar to the one of Section 5.2.5. The conclusion is that histogramming works fine if the number of samples within each bin is sufficiently large. With a given size of the training set, the size of the bins must be large enough to assure a minimum number of samples per bin. Hence, with a small training set, or a large dimension of the measurement space, the resolution of the estimation will be very poor.

Parzen estimation can be considered as a refinement of histogramming. The first step in the development of the estimator is to consider only one sample from the training set. Suppose that $\mathbf{z}_j \in T_k$. Then, we are certain that at this position in the measurement space the density is nonzero, i.e. $p(\mathbf{z}_j|\omega_k) \neq 0$. Under the assumption that $p(\mathbf{z}|\omega_k)$ is continuous over the entire measurement space it follows that in a small neighbourhood of $\mathbf{z}_j$ the density is likely to be nonzero too. However, the further we move away from $\mathbf{z}_j$, the less we can say about $p(\mathbf{z}|\omega_k)$. The basic idea behind Parzen estimation is that the knowledge gained by the observation of $\mathbf{z}_j$ is represented by a function positioned at $\mathbf{z}_j$ and with an influence restricted to a small vicinity of $\mathbf{z}_j$. Such a function is called the *kernel* of the estimator. It represents the contribution of $\mathbf{z}_j$ to the estimate. Summing together the contributions of all vectors in the training set yields the final estimate.

Let $\rho(\mathbf{z}, \mathbf{z}_j)$ be a distance measure (Appendix A.2) defined in the measurement space. The knowledge gained by the observation $\mathbf{z}_j \in T_k$ is represented by the kernel $h(\rho(\mathbf{z}, \mathbf{z}_j))$ where $h(\cdot)$ is a function $\mathbb{R}^+ \to \mathbb{R}^+$ such that $h(\rho(\mathbf{z}, \mathbf{z}_j))$ has its maximum at $\mathbf{z} = \mathbf{z}_j$, i.e. at $\rho(\mathbf{z}, \mathbf{z}_j) = 0$. Furthermore, $h(\rho(\cdot, \cdot))$ must be monotonically decreasing as $\rho(\cdot, \cdot)$ increases, and $h(\rho(\cdot,\cdot))$ must be normalized to one, i.e. $\int h(\rho(\mathbf{z},\mathbf{z}_j))d\mathbf{z} = 1$ where the integration extends over the entire measurement space.

The contribution of a single observation $\mathbf{z}_j$ is $h(\rho(\mathbf{z},\mathbf{z}_j))$. The contributions of all observations are summed to yield the final Parzen estimate:

$$\hat{p}(\mathbf{z}|\omega_k) = \frac{1}{N_k}\sum_{\mathbf{z}_j \in T_k} h\big(\rho(\mathbf{z},\mathbf{z}_j)\big) \tag{5.25}$$

The kernel $h(\rho(\cdot,\cdot))$ can be regarded as an interpolation function that interpolates between the samples of the training set.

Figure 5.3 gives an example of Parzen estimation in a one-dimensional measurement space. The plot is generated by the code in Listing 5.3. The true distribution is zero for negative z and has a peak value near $z = 1$ after which it slowly decays to zero. Fifty samples are available (shown at the bottom of the figure). The interpolation function chosen is a Gaussian function with width σ_h. The distance measure is Euclidean. Figure 5.3(a) and Figure 5.3(b) show the estimations using $\sigma_h = 1$ and $\sigma_h = 0.2$, respectively. These graphs illustrate a phenomenon related to the choice of the interpolation function. If the interpolation function is peaked, the influence of a sample is very local, and the variance of the estimator is large. But if the interpolation is smooth, the variance

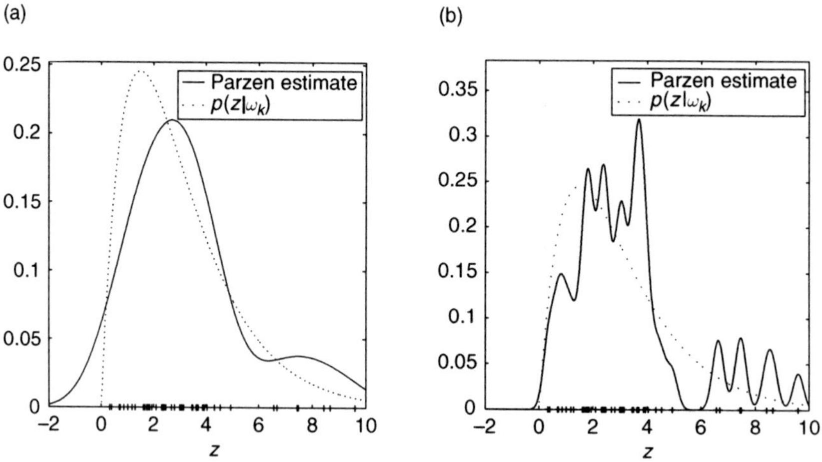

Figure 5.3 Parzen estimation of a density function using 50 samples. (a) $\sigma_h = 1$. (b) $\sigma_h = 0.2$

decreases, but at the same time the estimate becomes a smoothed version of the true density. That is, the estimate becomes biased. By changing the width of the interpolation function one can balance between the bias and the variance. Of course, both types of errors can be reduced by enlargement of the training set.

Listing 5.3

The following PRTools listing generates plots similar to those in Figure 5.3. It generates 50 samples from a $\Gamma(a, b)$ distribution with $a = 2$ and $b = 1.5$, and then estimates the density using the Parzen method with two different kernel widths.

```
n=50; a=2; b=1.5;
x=(-2:0.01:10)'; y=gampdf(x,a,b);           % Generate function
z=dataset(gamrnd(a,b,n,1),genlab(n));       % Generate dataset
w=parzenm(z,1);                             % Parzen, sigma=1
figure; scatterd(z); axis([-2 10 0 0.3]);
plotm(w,1); hold on; plot(x,y,':');
w=parzenm(z,0.2);                           % Parzen, sigma=0.2
figure; scatterd(z); axis([-2 10 0 0.3]);
plotm(w,1); hold on; plot(x,y,':');
```

In the N-dimensional case, various interpolation functions are useful. A popular one is the Gaussian function:

$$\begin{aligned} \rho(\mathbf{z}, \mathbf{z}_j) &= \sqrt{(\mathbf{z} - \mathbf{z}_j)^T \mathbf{C}^{-1} (\mathbf{z} - \mathbf{z}_j)} \\ h(\rho) &= \frac{1}{\sigma_h^N \sqrt{(2\pi)^N |\mathbf{C}|}} \exp\left(-\frac{\rho^2}{2\sigma_h^2} \right) \end{aligned} \tag{5.26}$$

The matrix C must be symmetric and positive definite (Appendix B.5). The metric is according to the Mahalanobis distance. The constant σ_h controls the size of the influence zone of $h(\cdot)$. It can be chosen smaller as the number of samples in the training set increases. If the training set is very large, the actual choice of C is less important. If the set is not very large, a suitable choice is the covariance matrix $\hat{\mathbf{C}}_k$ determined according to (5.14).

The following algorithm implements a classification based on Parzen estimation:

Algorithm 5.1 Parzen classification
Input: a labelled training set T_S, an unlabelled test set T.

1. *Determination of* σ_h: maximize the log-likelihood of the training set T_S by varying σ_h using leave-one-out estimation (see Section 5.4). In other words, select σ_h such that

$$\sum_{k=1}^{K}\sum_{j=1}^{N_k}\ln\left(\hat{p}(\mathbf{z}_{k,j}|\omega_k)\right)$$

   is maximized. Here, $\mathbf{z}_{k,j}$ is the j-th sample from the k-th class, which is left out during the estimation of $\hat{p}(\mathbf{z}_{k,j}|\omega_k)$.
2. *Density estimation*: compute for each sample $\mathbf{z}$ in the test set the density for each class:

$$\hat{p}(\mathbf{z}|\omega_k) = \frac{1}{N_k}\sum_{\mathbf{z}_j \in T_k}\frac{1}{\sigma_h^N\sqrt{(2\pi)^N}}\exp\left(-\frac{\|\mathbf{z}-\mathbf{z}_j\|^2}{2\sigma_h^2}\right)$$

3. *Classification*: assign the samples in T to the class with the maximal posterior probability:

$$\hat{\omega} = \omega_k \quad \text{with} \quad k = \underset{i=1,\cdots,K}{\operatorname{argmax}}\left\{\hat{p}(\mathbf{z}|\omega_i)\hat{P}(\omega_i)\right\}$$

Output: the labels $\hat{\omega}$ of T.

Example 5.2 Classification of mechanical parts, Parzen estimation
We return to Example 2.2 in Chapter 2, where mechanical parts like nuts and bolts, etc. must be classified in order to sort them. Application of Algorithm 5.1 with Gaussians as the kernels and estimated covariance matrices as the weighting matrices yields $\sigma_h = 0.0485$ as the optimal sigma. Figure 5.4(a) presents the estimated overall density. The corresponding decision boundaries are shown in Figure 5.4(b). To show that the choice of σ_h significantly influences the decision boundaries, in Figure 5.4(c) a similar density plot is shown, for which σ_h was set to 0.0175. The density estimate is more peaked, and the decision boundaries (Figure 5.4(d)) are less smooth.

Figures 5.4(a–d) were generated using MATLAB code similar to that given in Listing 5.3.

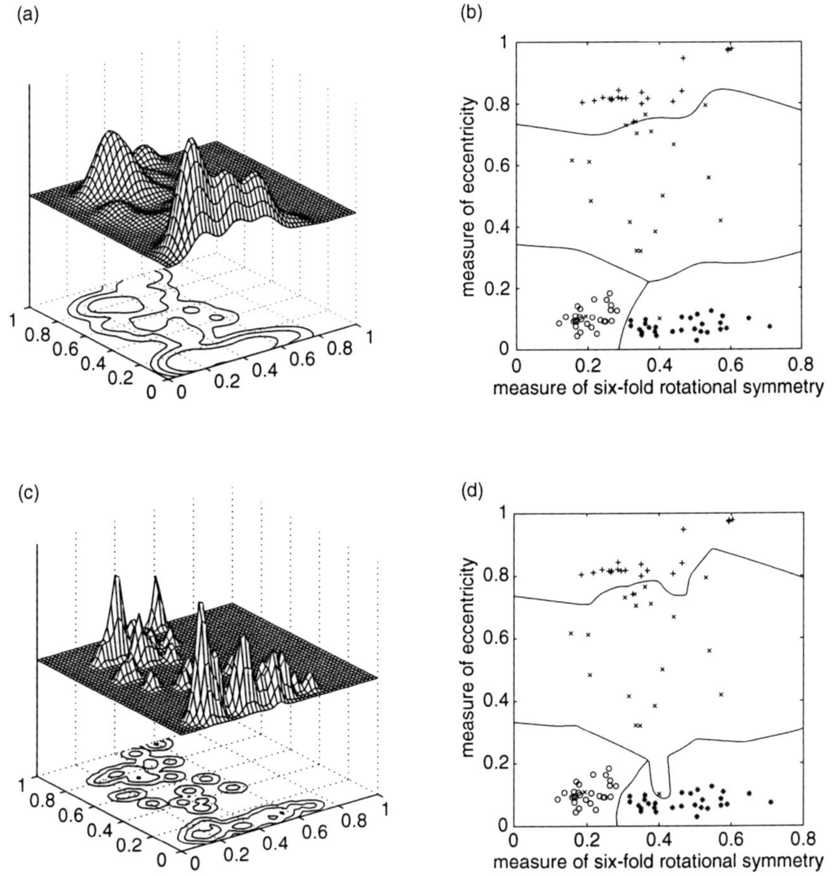

Figure 5.4 Probability densities of the measurements shown in Figure 5.1. (a) The 3D plot of the Parzen estimate of the unconditional density together with a 2D contour plot of this density on the ground plane. The parameter σ_h was set to 0.0485. (b) The resulting decision boundaries. (c) Same as (a), but with σ_h set to 0.0175. (d) Same as (b), for the density estimate shown in (c)

5.3.2 Nearest neighbour classification

In Parzen estimation, each sample in the training set contributes in a like manner to the estimate. The estimation process is space-invariant. Consequently, the trade-off which exists between resolution and variance is a global one. A refinement would be to have an estimator with high resolution in regions where the training set is dense, and with low resolution in other regions. The advantage is that the balance between resolution and variance can be adjusted locally.

Nearest neighbour estimation is a method that implements such a refinement. The method is based on the following observation. Let $R(\mathbf{z}) \subset \mathbb{R}^N$ be a hypersphere with volume V. The centre of $R(\mathbf{z})$ is $\mathbf{z}$. If the number of samples in the training set T_k is N_k, then the probability of having exactly n samples within $R(\mathbf{z})$ has a binomial distribution with expectation:

$$\mathrm{E}[n] = N_k \int_{\mathbf{y} \in R(\mathbf{z})} p(\mathbf{y}|\omega_k)d\mathbf{y} \approx N_k V p(\mathbf{z}|\omega_k) \tag{5.27}$$

Suppose that the radius of the sphere around $\mathbf{z}$ is selected such that this sphere contains exactly κ samples. It is obvious that this radius depends on the position $\mathbf{z}$ in the measurement space. Therefore, the volume will depend on $\mathbf{z}$. We have to write $V(\mathbf{z})$ instead of V. With that, an estimate of the density is:

$$\hat{p}(\mathbf{z}|\omega_k) = \frac{\kappa}{N_k V(\mathbf{z})} \tag{5.28}$$

The expression shows that in regions where $p(\mathbf{z}|\omega_k)$ is large, the volume is expected to be small. This is similar to having a small interpolation zone. If, on the other hand, $p(\mathbf{z}|\omega_k)$ is small, the sphere needs to grow in order to collect the required κ samples.

The parameter κ controls the balance between the bias and variance. This is like the parameter σ_h in Parzen estimation. The choice of κ should be such that:

$$\begin{aligned} \kappa &\to \infty \quad \text{as} \quad N_k \to \infty \quad \text{in order to obtain a low variance} \\ \kappa/N_k &\to 0 \quad \text{as} \quad N_k \to \infty \quad \text{in order to obtain a low bias} \end{aligned} \tag{5.29}$$

A suitable choice is to make κ proportional to $\sqrt{N_k}$.

Nearest neighbour estimation is of practical interest because it paves the way to a classification technique that directly uses the training set, i.e. without explicitly estimating probability densities. The development of this technique is as follows. We consider the entire training set and use the representation T_S as in (5.1). The total number of samples is N_S. Estimates of the prior probabilities follow from (5.18): $\hat{P}(\omega_k) = N_k/N_S$.

As before, let $R(\mathbf{z}) \subset \mathbb{R}^N$ be a hypersphere with volume $V(\mathbf{z})$. In order to classify a vector $\mathbf{z}$ we select the radius of the sphere around $\mathbf{z}$ such that this sphere contains exactly κ samples taken from T_S. These samples are called the *κ-nearest neighbours*[1] of **z**. Let κ_k denote the number of samples found with class ω_k. An estimate of the conditional density is (see (5.28)):

$$\hat{p}(\mathbf{z}|\omega_k) \approx \frac{\kappa_k}{N_k V(\mathbf{z})} \tag{5.30}$$

Combination of (5.18) and (5.30) in the Bayes classification with uniform cost function (2.12) produces the following suboptimal classification:

$$\begin{aligned}\hat{\omega}(\mathbf{z}) = \omega_k \quad \text{with} \quad k &= \operatorname*{argmax}_{i=1,\ldots,K}\left\{\hat{p}(\mathbf{z}|\omega_i)\hat{P}(\omega_i)\right\} = \operatorname*{argmax}_{i=1,\ldots,K}\left\{\frac{\kappa_i}{N_i V(\mathbf{z})}\frac{N_i}{N_S}\right\} \\ &= \operatorname*{argmax}_{i=1,\ldots,K}\{\kappa_i\}\end{aligned} \tag{5.31}$$

The interpretation of this classification is simple. The class assigned to a vector **z** is the class with the maximum number of votes coming from κ samples nearest to **z**. In literature, this classification method is known as *k-nearest neighbour rule classification* (k-NNR, but in our nomenclature κ-NNR). The special case in which $\kappa = 1$ is simply referred to as *nearest neighbour rule classification* (NNR or 1-NNR).

Example 5.3 Classification of mechanical parts, NNR classification
PRTools can be used to perform κ-nearest neighbour classification. Listing 5.4 shows how a κ-nearest neighbour classifier is trained on the mechanical parts data set of Example 2.2. If κ is not specified, it will be found by minimizing the leave-one-out error. For this data set, the optimal κ is 7. The resulting decision boundaries are shown in Figure 5.5(a). If we set κ to 1, the classifier will classify all samples in the training set correctly (see Figure 5.5(b)), but its performance on a test set will be worse.

[1] The literature about nearest neighbour classification often uses the symbol k to denote the number of samples in a volume. However, in order to avoid confusion with symbols like ω_k, T_k, etc. we prefer to use κ.

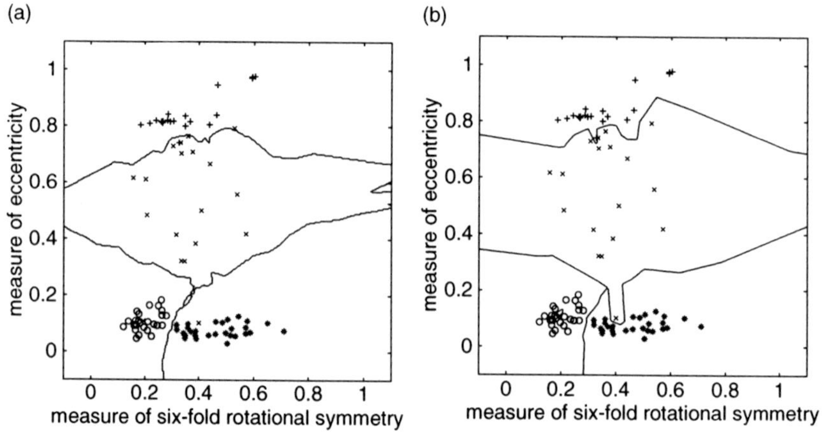

Figure 5.5 Application of κ-NNR classification. (a) $\kappa = 7$. (b) $\kappa = 1$

Listing 5.4
PRTools code for finding and plotting an optimal κ-nearest neighbour classifier and a one-nearest neighbour classifier.

```
load nutsbolts;              % Load the dataset
[w,k] = knnc(z);             % Train a k-NNR
disp(k);                     % Show the optimal k found
figure; scatterd(z)          % Plot the dataset
plotc(w);                    % Plot the decision boundaries
w = knnc(z,1);               % Train a 1-NNR
figure; scatterd(z);         % Plot the dataset
plotc(w);                    % Plot the decision boundaries
```

The analysis of the performance of κ-nearest neighbour classification is difficult. This holds true especially if the number of samples in the training set is finite. In the limiting case, when the number of samples grows to infinity, some bounds on the error rate can be given. Let the minimum error rate, i.e. the error rate of a Bayes classifier with uniform cost function, be denoted by $E_{\min}$. See Section 2.1.1. Since $E_{\min}$ is the minimum error rate among all classifiers, the error rate of a κ-NNR, denoted E_κ, is bounded by:

$$E_{\min} \leq E_\kappa \tag{5.32}$$

It can be shown that for the 1-NNR the following upper bound holds:

$$E_1 \leq E_{\min}\left(2 - \frac{K}{K-1}E_{\min}\right) \leq 2E_{\min} \tag{5.33}$$

Apparently, replacing the true probability densities with estimations based on the first nearest neighbour gives an error rate that is at most twice the minimum. Thus, at least half of the classification information in a dense training set is contained in the first nearest neighbour.

In the two-class problem ($K = 2$) the following bound can be proven:

$$\begin{aligned} E_\kappa &\leq E_{\min} + \frac{E_1}{\sqrt{0.5(\kappa-1)\pi}} && \text{if } \kappa \text{ is odd} \\ E_\kappa &= E_{\kappa-1} && \text{if } \kappa \text{ is even} \end{aligned} \tag{5.34}$$

The importance of (5.34) is that it shows that the performance of the κ-NNR approximates the optimum as κ increases. This asymptotic optimality holds true only if the training set is dense ($N_S \to \infty$). Nevertheless, even in the small sized training set given in Figure 5.5 it can be seen that the 7-NNR is superior to the 1-NNR. The topology of the compartments in Figure 5.5(b) is very specific for the given training set. This is in contrast with the topology of the 7-NNR in Figure 5.5(a). The 7-NNR *generalizes* better than the 1-NNR.

Unfortunately, κ-NNR classifiers also have some serious disadvantages:

- A distance measure is needed in order to decide which sample in the training set is nearest. Usually the Euclidean distance measure is chosen, but this choice needs not to be optimal. A systematic method to determine the optimal measure is hard to find.
- The optimality is reached only when $\kappa \to \infty$. But since at the same time it is required that $\kappa/N_S \to 0$, the demand on the size of the training set is very high. If the size is not large enough, κ-NNR classification may be far from optimal.
- If the training set is large, the computational complexity of κ-NNR classification becomes a serious burden.

Many attempts have been made to remedy the last drawback. One method is to design fast algorithms with suitably chosen data structures (e.g. hierarchical data structures combined with a pre-ordered training set).

Another method is to preprocess the training set so as to speed up the search for nearest neighbours. There are two principles on which this reduction can be based: *editing* and *condensing*. The first principle is to edit the training set such that the (expensive) κ-NNR can be replaced with the 1-NNR. The strategy is to remove those samples from the training set that when used in the 1-NNR would cause erroneous results. An algorithm that accomplishes this is the so-called *multi-edit algorithm*. The following algorithm is from Devijver and Kittler.

Algorithm 5.2 Multi-edit
Input: a labelled training set T_S.

1. *Diffusion*: Partition the training set T_S randomly into L disjunct subsets: $T'_1, T'_2, \ldots, T'_L$ with $T_S = \cup_l T'_l$, and $L \geq 3$.
2. *Classification*: classify the samples in $T\prime_\ell$ using 1-NNR classification with $T'_{(\ell+1) \bmod L}$ as training set.
3. *Editing*: discard all the samples that were misclassified at step 2.
4. *Confusion*: pool all the remaining samples to constitute a new training set T_S.
5. *Termination*: if the last I iterations produced no editing, then exit with the final training set, else go to step 1.

Output: a subset of T_S.

The subsets created in step 1 are regarded as independent random selections. A minimum of three subsets is required in order to avoid a two-way interaction between two subsets. Because in the first step the subsets are randomized, it cannot be guaranteed that, if during one iteration no changes in the training set occurred, changes in further iterations are ruled out. Therefore, the algorithm does not stop immediately after an iteration with no changes has occurred.

The effect of the algorithm is that ambiguous samples in the training set are removed. This eliminates the need to use the κ-NNR. The 1-NNR can be used instead.

The second principle, called condensing, aims to remove samples that do not affect the classification in any way. This is helpful to reduce the computational cost. The algorithm – also from Devijver and Kittler – is used to eliminate all samples in the training set that are irrelevant.

Algorithm 5.3 Condensing
Input: a labeled training set T_S.

1. *Initiation*: set up two new training sets T_{STORE} and $T_{GRABBAG}$; place the first sample of T_S in T_{STORE}, all other samples in $T_{GRABBAG}$.

2. *Condensing*: use 1-NNR classification with the current T_{STORE} to classify a sample in $T_{GRABBAG}$; if classified correctly, the sample is retained in $T_{GRABBAG}$, otherwise it is moved from $T_{GRABBAG}$ to T_{STORE}; repeat this operation for all other samples in $T_{GRABBAG}$.
3. *Termination*: if one complete pass is made through step 2 with no transfer from $T_{GRABBAG}$ to T_{STORE}, or if $T_{GRABBAG}$ is empty, then terminate; else go to step 2.

Output: a subset of T_S.

The effect of this algorithm is that in regions where the training set is overcrowded with samples of the same class most of these samples will be removed. The remaining set will, hopefully, contain samples close to the Bayes decision boundaries.

Example 5.4 Classification of mechanical parts, editing and condensation

An example of a multi-edited training set is given in Figure 5.6(a). The decision boundaries of the 1-NNR classifier are also shown. It can be seen that the topology of the resulting decision function is in accordance with the one of the 7-NNR given in Figure 5.5(a). Hence, multi-editing improves the generalization property.

Figure 5.6(b) shows that condensing can be successful when applied to a multi-edited training set. The decision boundaries in Figure 5.6(b)

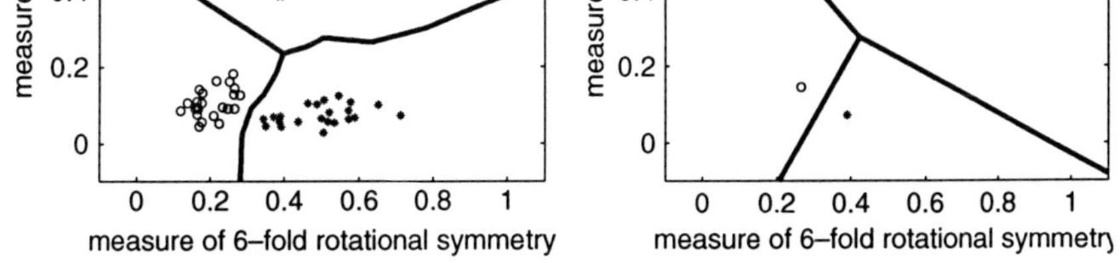

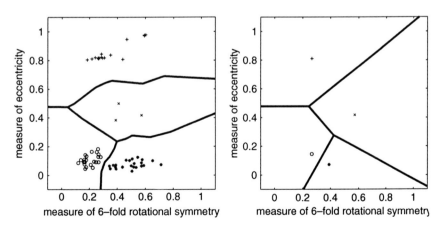

Figure 5.6 Application of editing and condensing. (a) Edited training set. (b) Edited and condensed training set

are close to those in Figure 5.6(a), especially in the more important areas of the measurement space.

The basic PRTools code used to generate Figure 5.6 is given in Listing 5.5.

Listing 5.5

PRTools code for finding and plotting one-nearest neighbour classifiers on both an edited and a condensed data set. The function `edicon` takes a distance matrix as input. In PRTools, calculating a distance matrix is implemented as a mapping `proxm`, so `z * proxm(z)` is the distance matrix between all samples in `z`. See Section 7.2.

```
load nutsbolts;                          % Load the dataset z
J=edicon(z*proxm(z),3,5,[]);             % Edit z
w=knnc(z(J,:),1);                        % Train a 1-NNR
figure; scatterd(z(J,:)); plotc(w);
J=edicon(z*proxm(z),3,5,10);             % Edit and condense z
w=knnc(z(J,:),1);                        % Train a 1-NNR
figure; scatterd(z(J,:)); plotc(w);
```

If a non-edited training set is fed into the condensing algorithm, it may result in erroneous decision boundaries, especially in areas of the measurement space where the training set is ambiguous.

5.3.3 Linear discriminant functions

Discriminant functions are functions $g_k(\mathbf{z})$, $k = 1, \ldots, K$ that are used in a decision function as follows:

$$\hat{\omega}(\mathbf{z}) = \omega_n \quad \text{with: } n = \underset{k=1,\ldots,K}{\text{argmax}}\{g_k(\mathbf{z})\} \tag{5.35}$$

Clearly, if $g_k(\mathbf{z})$ are the posterior probabilities $P(\omega_k|\mathbf{z})$, the decision function becomes a Bayes decision function with a uniform cost function. Since the posterior probabilities are not known, the strategy is to replace the probabilities with some predefined functions $g_k(\mathbf{z})$ whose parameters should be learned from a labelled training set.

An assumption often made is that the samples in the training set can be classified correctly with linear decision boundaries. In that case, the discriminant functions take the form of:

$$g_k(\mathbf{z}) = \mathbf{w}_k^T \mathbf{z} + w_k \tag{5.36}$$

Functions of this type are called *linear discriminant functions*. In fact, these functions implement a linear machine. See also Section 2.1.2.

The notation can be simplified by the introduction of an augmented measurement vector $\mathbf{y}$, defined as:

$$\mathbf{y} = \begin{bmatrix} \mathbf{z} \\ 1 \end{bmatrix} \tag{5.37}$$

With that, the discriminant functions become:

$$g_k(\mathbf{y}) = \mathbf{w}_k^T \mathbf{y} \tag{5.38}$$

where the scalar w_k in (5.36) has been embedded in the vector $\mathbf{w}_k$ by augmenting the latter with the extra element w_k.

The augmentation can also be used for a generalization that allows for nonlinear machines. For instance, a quadratic machine is obtained with:

$$\mathbf{y}(\mathbf{z}) = \begin{bmatrix} \mathbf{z} & 1 & z_0^2 & z_1^2 & \cdots & z_{N-1}^2 & z_0 z_1 & z_0 z_2 & \cdots & z_{N-1} z_N \end{bmatrix}^T \tag{5.39}$$

The corresponding functions $g_k(\mathbf{y}) = \mathbf{w}_k^T \mathbf{y}(\mathbf{z})$ are called *generalized linear discriminant functions*.

Discriminant functions depend on a set of parameters. In (5.38) these parameters are the vectors $\mathbf{w}_k$. In essence, the learning process boils down to a search for parameters such that with these parameters the decision function in (5.35) correctly classifies all samples in the training set.

The basic approach to find the parameters is to define a performance measure that depends on both the training set and the set of parameters. Adjustment of the parameters such that the performance measure is maximized gives the optimal decision function; see Figure 5.7.

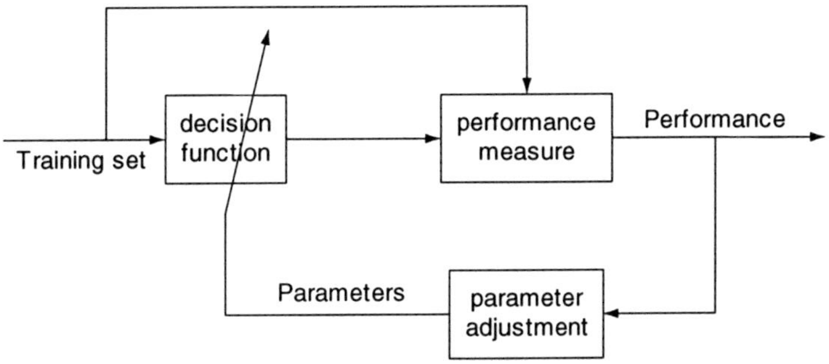

Figure 5.7 Training by means of performance optimization

Strategies to adjust the parameters may be further categorized into 'iterative' and 'non-iterative'. Non-iterative schemes are found when the performance measure allows for an analytic solution of the optimization. For instance, suppose that the set of parameters is denoted by $\mathbf{w}$ and that the performance measure is a continuous function $J(\mathbf{w})$ of $\mathbf{w}$. The optimal solution is one which maximizes $J(\mathbf{w})$. Hence, the solution must satisfy $\frac{\partial J(\mathbf{w})}{\partial \mathbf{w}} = \mathbf{0}$.

In iterative strategies the procedure to find a solution is numerical. Samples from the training set are fed into the decision function. The classes found are compared with the true classes. The result controls the adjustment of the parameters. The adjustment is in a direction which improves the performance. By repeating this procedure it is hoped that the parameters iterate towards the optimal solution.

The most popular search strategy is the *gradient ascent* method (also called *steepest ascent*)[1]. Suppose that the performance measure $J(\mathbf{w})$ is a continuous function of the parameters contained in $\mathbf{w}$. Furthermore, suppose that $\nabla J(\mathbf{w}) = \frac{\partial J(\mathbf{w})}{\partial \mathbf{w}}$ is the gradient vector. Then the gradient ascent method updates the parameters according to:

$$\mathbf{w}(i+1) = \mathbf{w}(i) + \eta(i)\nabla J(\mathbf{w}(i)) \tag{5.40}$$

where $\mathbf{w}(i)$ is the parameter obtained in the i-th iteration. $\eta(i)$ is the so-called *learning rate*. If $\eta(i)$ is selected too small, the process converges very slowly, but if it is too large, the process may overshoot the maximum, or oscillate near the maximum. Hence, a compromise must be found.

Different choices of the performance measures and different search strategies lead to a multitude of different learning methods. This section confines itself to two-class problems. From the many iterative, gradient-based methods we only discuss 'perceptron learning' and the 'least squared error learning'. Perhaps the practical significance of these two methods is not very large, but they are introductory to the more involved techniques of succeeding sections.

Perceptron learning

In a two-class problem, the decision function expressed in (5.35) is equivalent to a test $g_1(\mathbf{y}) - g_2(\mathbf{y}) > 0$. If the test fails, it is decided for ω_2, otherwise for ω_1. The test can be accomplished equally well with a single linear function:

---

[1] Equivalently, we define $J(\mathbf{w})$ as an error measure. A *gradient descent* method should be applied to minimize it.

$$g(\mathbf{y}) = \mathbf{w}^T\mathbf{y} \tag{5.41}$$

defined as $g(\mathbf{y}) = g_1(\mathbf{y}) - g_2(\mathbf{y})$. The so-called *perceptron*, graphically represented in Figure 5.8, is a computational structure that implements $g(\mathbf{y})$. The two possible classes are encoded in the output as '1' and '−1'.

A simple performance measure of a classifier is obtained by applying the training set to the classifier, and to count the samples that are erroneously classified. Obviously, such a performance measure – actually an error measure – should be minimized. The disadvantage of this measure is that it is not a continuous function of y. Therefore, the gradient is not well defined.

The performance measure of the perceptron is based on the following observation. Suppose that a sample $\mathbf{y}_n$ is misclassified. Thus, if the true class of the sample is ω_1, then $g(\mathbf{y}_n) = \mathbf{w}^T\mathbf{y}_n$ is negative, and if the true class is ω_2, then $g(\mathbf{y}_n) = \mathbf{w}^T\mathbf{y}_n$ is positive. In the former case we would like to correct $\mathbf{w}^T\mathbf{y}_n$ with a positive constant, in the latter case with a negative constant. We define $Y_1(\mathbf{w})$ as the set containing all ω_1 samples in the training set that are misclassified, and $Y_2(\mathbf{w})$ as the set of all misclassified ω_2 samples. Then:

$$J_{perceptron}(\mathbf{w}) = -\sum_{\mathbf{y}\in Y_1} \mathbf{w}^T\mathbf{y} + \sum_{\mathbf{y}\in Y_2} \mathbf{w}^T\mathbf{y} \tag{5.42}$$

This measure is continuous in w and its gradient is:

$$\nabla J_{perceptron}(\mathbf{w}) = -\sum_{\mathbf{y}\in Y_1} \mathbf{y} + \sum_{\mathbf{y}\in Y_2} \mathbf{y} \tag{5.43}$$

Application of the gradient descent, see (5.40), gives the following learning rule:

$$\mathbf{w}(i+1) = \mathbf{w}(i) - \eta\left(-\sum_{\mathbf{y}\in Y_1} \mathbf{y} + \sum_{\mathbf{y}\in Y_2} \mathbf{y}\right) \tag{5.44}$$

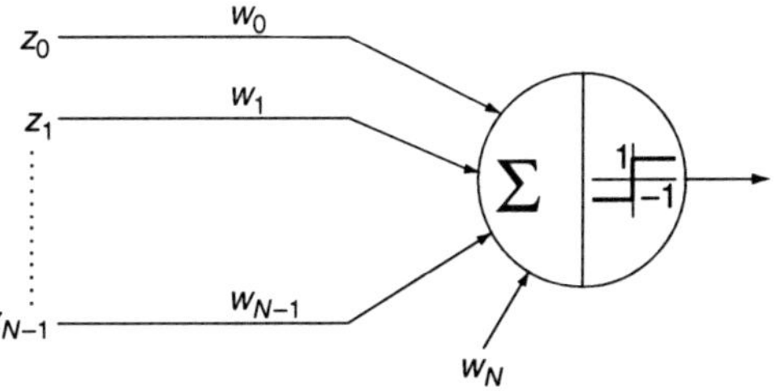

Figure 5.8 The perceptron

where i is the iteration count. The iteration procedure stops when $\mathbf{w}(i+1) = \mathbf{w}(i)$, i.e. when all samples in the training set are classified correctly. If such a solution exists, that is if the training set is *linearly separable*, the perceptron learning rule will find it.

Instead of processing the full training set in one update step (so-called *batch* processing) we can also cycle through the training set and update the weight vector whenever a misclassified sample has been encountered (*single sample* processing). If $\mathbf{y}_n$ is a misclassified sample, then the learning rule becomes:

$$\mathbf{w}(i+1) = \mathbf{w}(i) + \eta c_n \mathbf{y}_n \tag{5.45}$$

The variable c_n is $+1$ if $\mathbf{y}_n$ is a misclassified ω_1 sample. If $\mathbf{y}_n$ is a misclassified ω_2 sample, then $c_n = -1$.

The *error correction* procedure of (5.45) can be extended to the multi-class problem as follows. Let $g_k(\mathbf{y}) = \mathbf{w}_k^T \mathbf{y}$ as before. We cycle through the training set and update the weight vectors $\mathbf{w}_k$ and $\mathbf{w}_j$ whenever a ω_k sample is classified as class ω_j:

$$\begin{aligned} \mathbf{w}_k &\to \mathbf{w}_k + \eta \mathbf{y}_n \\ \mathbf{w}_j &\to \mathbf{w}_j - \eta \mathbf{y}_n \end{aligned} \tag{5.46}$$

The procedure will converge in a finite number of iterations provided that the training set is linearly separable. Perceptron training is illustrated in Example 5.5.

Least squared error learning

A disadvantage of the perceptron learning rule is that it only works well in separable cases. If the training set is not separable, the iterative procedure often tends to fluctuate around some value. The procedure is terminated at some arbitrary point, but it becomes questionable then whether the corresponding solution is useful.

Non-separable cases can be handled if we change the performance measure such that its maximization boils down to solving a set of linear equations. Such a situation is created if we introduce a set of so-called *target vectors*. A target vector $\mathbf{t}_n$ is a K-dimensional vector associated with the (augmented) sample $\mathbf{y}_n$. Its value reflects the desired response of the discriminant function to $\mathbf{y}_n$. The simplest one is *place coding*:

$$t_{n,k} = \begin{cases} 1 & \text{if } \theta_n = \omega_k \\ 0 & \text{otherwise} \end{cases} \tag{5.47}$$

(We recall that θ_n is the class label of sample $\mathbf{y}_n$.) This target function aims at a classification with minimum error rate.

We now can apply a least squares criterion to find the weight vectors:

$$J_{LS} = \sum_{n=1}^{N_S} \sum_{k=1}^{K} \left(\mathbf{w}_k^T \mathbf{y}_n - t_{n,k}\right)^2 \tag{5.48}$$

The values of $\mathbf{w}_k$ that minimize J_{LS} are the weight vectors of the least squared error criterion.

The solution can be found by rephrasing the problem in a different notation. Let $\mathbf{Y} = [\,\mathbf{y}_1 \ldots \mathbf{y}_{N_S}\,]^T$ be a $N_S \times (N+1)$ matrix, $\mathbf{W} = [\,\mathbf{w}_1 \ldots \mathbf{w}_K\,]$ a $(N+1) \times K$ matrix, and $\mathbf{T} = [\,\mathbf{t}_1 \ldots \mathbf{t}_{N_S}\,]^T$ a $N_S \times K$ matrix. Then:

$$J_{LS} = \|\mathbf{YW} - \mathbf{T}\|^2 \tag{5.49}$$

where $\|\cdot\|^2$ is the Euclidean matrix norm, i.e. the sum of squared elements. The value of $\mathbf{W}$ that minimizes J_{LS} is the LS solution to the problem:

$$\mathbf{W}_{LS} = \left(\mathbf{Y}^T\mathbf{Y}\right)^{-1}\mathbf{Y}^T\mathbf{T} \tag{5.50}$$

Of course, the solution is only valid if $(\mathbf{Y}^T\mathbf{Y})^{-1}$ exists. The matrix $(\mathbf{Y}^T\mathbf{Y})^{-1}\mathbf{Y}^T$ is the pseudo inverse of $\mathbf{Y}$. See (3.25).

An interesting target function is:

$$t_{n,k} = C(\omega_k|\theta_n) \tag{5.51}$$

Here, $\mathbf{t}_n$ embeds the cost that is involved if the assigned class is ω_k whereas the true class is θ_n. This target function aims at a classification with minimal risk and the discriminant function $g_k(\mathbf{y})$ attempts to approximate the risk $\sum_{i=1}^{K} C(\omega_k|\omega_i)P(\omega_i|\mathbf{y})$ by linear LS fitting. The decision function in (5.35) should now involve a minimization rather than a maximization.

Example 5.5 illustrates how the least squared error classifier can be found in PRTools.

Example 5.5 Classification of mechanical parts, perceptron and least squared error classifier

Decision boundaries for the mechanical parts example are shown in Figure 5.9(a) (perceptron) and Figure 5.9(b) (least squared error

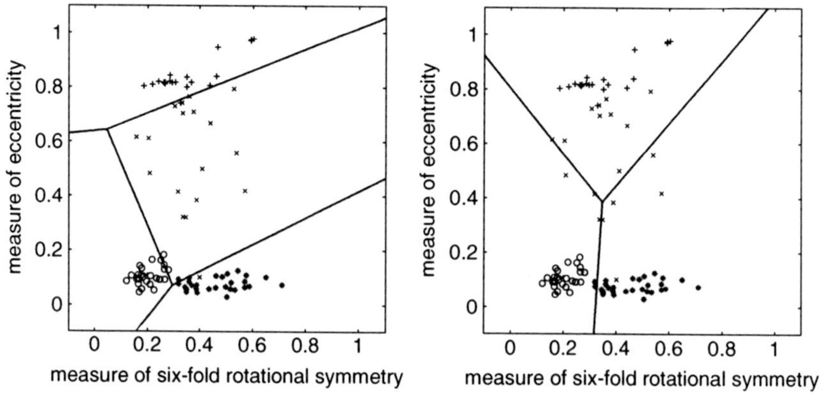

Figure 5.9 Application of two linear classifiers. (a) Linear perceptron. (b) Least squared error classifier

classifier). These plots were generated by the code shown in Listing 5.6. In PRTools, the linear perceptron classifier is implemented as `perlc`; the least squared error classifier is called `fisherc`. For `perlc` to find a good perceptron, the learning rate η had to be set to 0.01. Training was stopped after 1000 iterations. Interestingly, the least squared error classifier is not able to separate the data successfully, because the 'scrap' class is not linearly separable from the other classes.

Listing 5.6
PRTools code for finding and plotting a linear perceptron and least squared error classifier on the mechanical parts data set.

```
load nutsbolts;                       % Load the dataset
w=perlc(z,1000,0.01);                 % Train a linear perceptron
figure; scatterd(z); plotc(w);
w=fisherc(z);                         % Train a LS error classifier
figure; scatterd(z); plotc(w);
```

5.3.4 The support vector classifier

The basic support vector classifier is very similar to the perceptron. Both are linear classifiers, assuming separable data. In perceptron learning, the iterative procedure is stopped when all samples in the training set are classified correctly. For linearly separable data, this means that the found perceptron is one solution arbitrarily selected from an (in principle) infinite set of solutions. In contrast, the support vector classifier chooses one particular solution: the classifier which separates the classes with maximal *margin*. The margin is defined as the width of the largest 'tube' not containing samples that can be drawn around the decision boundary; see Figure 5.10. It can be proven that this particular solution has the highest generalization ability.

Mathematically, this can be expressed as follows. Assume we have training samples $\mathbf{z}_n, n = 1,.., N_S$ (not augmented with an extra element) and for each sample a label $c_n \in \{1, -1\}$, indicating to which of the two classes the sample belongs. Then a linear classifier $g(\mathbf{z}) = \mathbf{w}^T\mathbf{z} + b$ is sought, such that:

$$\begin{array}{lll} \mathbf{w}^T\mathbf{z}_n + b \geq 1 & \text{if} \quad c_n = +1 & \\ & & \text{for all } n \\ \mathbf{w}^T\mathbf{z}_n + b \leq -1 & \text{if} \quad c_n = -1 & \end{array} \tag{5.52}$$

These two constraints can be rewritten into one inequality:

$$c_n(\mathbf{w}^T\mathbf{z}_n + b) \geq 1 \tag{5.53}$$

The gradient vector of $g(\mathbf{z})$ is $\mathbf{w}$. Therefore, the square of the margin is inversely proportional to $\|\mathbf{w}\|^2 = \mathbf{w}^T\mathbf{w}$. To maximize the margin, we have to minimize $\|\mathbf{w}\|^2$. Using *Lagrange multipliers*, we can incorporate the constraints (5.53) into the minimization:

$$L = \frac{1}{2}\|\mathbf{w}\|^2 + \sum_{n=1}^{N_S} \alpha_n \left(c_n\left[\mathbf{w}^T\mathbf{z}_n + b\right] - 1\right), \quad \alpha_n \geq 0 \tag{5.54}$$

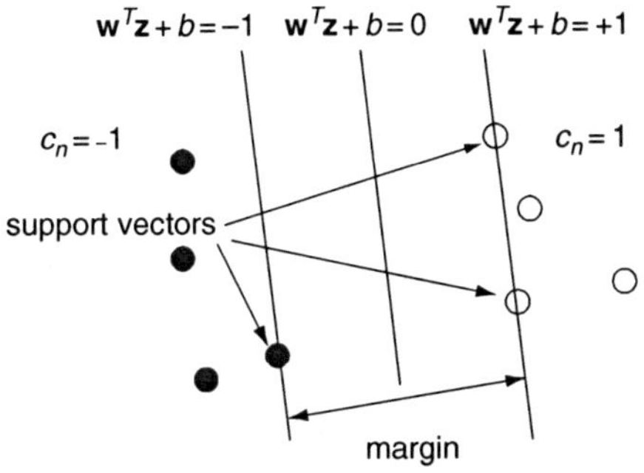

Figure 5.10 The linear support vector classifier

L should be minimized with respect to $\mathbf{w}$ and b, and maximized with respect to the Lagrange multipliers α_n. Setting the partial derivates of L w.r.t. $\mathbf{w}$ and b to zero results in the constraints:

$$\begin{aligned} \mathbf{w} &= \sum_{n=1}^{N_S} \alpha_n c_n \mathbf{z}_n \\ \sum_{n=1}^{N_S} c_n \alpha_n &= 0 \end{aligned} \tag{5.55}$$

Resubstituting this into (5.54) gives the so-called *dual form*:

$$L = \sum_{n=1}^{N_S} \alpha_n - \frac{1}{2} \sum_{n=1}^{N_S} \sum_{m=1}^{N_S} c_n c_m \alpha_n \alpha_m \mathbf{z}_n^T \mathbf{z}_m, \quad \alpha_n \geq 0 \tag{5.56}$$

L should be maximized with respect to the α_n. This is a quadratic optimization problem, for which standard software packages are available. After optimization, the α_n are used in (5.55) to find $\mathbf{w}$. In typical problems, the solution is sparse, meaning that many of the α_n become 0. Samples $\mathbf{z}_n$ for which $\alpha_n = 0$ are not required in the computation of $\mathbf{w}$. The remaining samples $\mathbf{z}_n$ (for which $\alpha_n > 0$) are called *support vectors*.

This formulation of the support vector classifier is of limited use: it only covers a linear classifier for separable data. To construct nonlinear boundaries, discriminant functions, introduced in (5.39), can be applied. The data is transformed from the measurement space to a new feature space. This can be done efficiently in this case because in formulation (5.56) all samples are coupled to other samples by an inner product. For instance, when all polynomial terms up to degree 2 are used (as in (5.39)), we can write:

$$\mathbf{y}(\mathbf{z}_n)^T \mathbf{y}(\mathbf{z}_m) = (\mathbf{z}_n^T \mathbf{z}_m + 1)^2 = K(\mathbf{z}_n, \mathbf{z}_m) \tag{5.57}$$

This can be generalized further: instead of $(\mathbf{z}_n^T \mathbf{z}_m + 1)^2$ any integer degree $(\mathbf{z}_n^T \mathbf{z}_m + 1)^d$ with $d > 1$ can be used. Due to the fact that only the inner products between the samples are considered, the very expensive explicit expansion is avoided. The resulting decision boundary is a d-th degree polynomial in the measurement space. The classifier $\mathbf{w}$ cannot easily be expressed explicitly (as in (5.55)). However, we are only interested in the classification result. And this is in terms of the inner product between the object $\mathbf{z}$ to be classified and the classifier (compare also with (5.36)):

$$g(\mathbf{z}) = \mathbf{w}^T\mathbf{y}(\mathbf{z}) = \sum_{n=1}^{N_S} K(\mathbf{z}, \mathbf{z}_n) \tag{5.58}$$

Replacing the inner product by a more general kernel function is called the *kernel trick*. Besides polynomial kernels, other kernels have been proposed. The Gaussian kernel with $\sigma^2\mathbf{I}$ as weighting matrix (the *radial basis function* kernel, RBF kernel) is frequently used in practice:

$$K(\mathbf{z}_n, \mathbf{z}_m) = \exp\left(-\frac{\|\mathbf{z}_n - \mathbf{z}_m\|^2}{\sigma^2}\right) \tag{5.59}$$

For very small values of σ, this kernel gives very detailed boundaries, while for high values very smooth boundaries are obtained.

In order to cope with overlapping classes, the support vector classifier can be extended to have some samples erroneously classified. For that, the hard constraints (5.53) are replaced by soft constraints:

$$\begin{aligned} \mathbf{w}^T\mathbf{z}_n + b &\geq 1 - \xi_n \quad \text{if} \quad c_n = 1 \\ \mathbf{w}^T\mathbf{z}_n + b &\leq -1 + \xi_n \quad \text{if} \quad c_n = -1 \end{aligned} \tag{5.60}$$

Here so-called slack variables $\xi_n \geq 0$ are introduced. These should be minimized in combination with the $\mathbf{w}^2$. The optimization problem is thus changed into:

$$\begin{aligned} L = \frac{1}{2}\mathbf{w}^2 + C\sum_{n=1}^{N_S}\xi_n + \sum_{n=1}^{N_S}\alpha_n(c_n\left[\mathbf{w}^T\mathbf{z}_n + b\right] - 1 + \xi_n) \\ + \sum_{n=1}^{N_S}\gamma_n\xi_n, \quad \alpha_n, \gamma_n \geq 0 \end{aligned} \tag{5.61}$$

The second term expresses our desire to have the slack variables as small as possible. C is a trade-off parameter that determines the balance between having a large overall margin at the cost of more erroneously classified samples, or having a small margin with less erroneously classified samples. The last term holds the Lagrange multipliers that are needed to assure that $\xi_n \geq 0$.

The dual formulation of this problem is the same as (5.56). Its derivation is left as an exercise for the reader; see exercise 6. The only difference is that an extra upper bound on α_n is introduced: $\alpha_n \leq C$.

It basically means that the influence of a single object on the description of the classifier is limited. This upper bound avoids that noisy objects with a very large weight completely determine the weight vector and thus the classifier. The parameter C has a large influence on the final solution, in particular when the classification problem contains overlapping class distributions. It should be set carefully. Unfortunately, it is not clear beforehand what a suitable value for C will be. It depends on both the data and the type of kernel function which is used. No generally applicable number can be given. The only option in a practical application is to run cross-validation (Section 5.4) to optimize C.

The support vector classifier has many advantages. A unique global optimum for its parameters can be found using standard optimization software. Nonlinear boundaries can be used without much extra computational effort. Moreover, its performance is very competitive with other methods. A drawback is that the problem complexity is not of the order of the dimension of the samples, but of the order of the number of samples. For large sample sizes ($N_S > 1000$) general quadratic programming software will often fail and special-purpose optimizers using problem-specific speedups have to be used to solve the optimization.

A second drawback is that, like the perceptron, the classifier is basically a two-class classifier. The simplest solution for obtaining a classifier with more than two classes is to train K classifiers to distinguish one class from the rest (similar to the place coding mentioned above). The classifier with the highest output $\mathbf{w}_k^T\mathbf{z} + b$ then determines the class label. Although the solution is simple to implement, and works reasonable well, it can lead to problems because the output value of the support vector classifier is only determined by the margin between the classes it is trained for, and is not optimized to be used for a confidence estimation. Other methods train K classifiers simultaneously, incorporating the one-class-against-the-rest labelling directly into the constraints of the optimization. This gives again a quadratic optimization problem, but the number of constraints increases significantly which complicates the optimization.

Example 5.6 Classification of mechanical parts, support vector classifiers

Decision boundaries found by support vector classifiers for the mechanical parts example are shown in Figure 5.11. These plots were generated by the code shown in Listing 5.7. In Figure 5.11(a), the kernel used was a polynomial one with degree $d = 2$ (a quadratic kernel); in Figure 5.11(b), it was a Gaussian kernel with a width

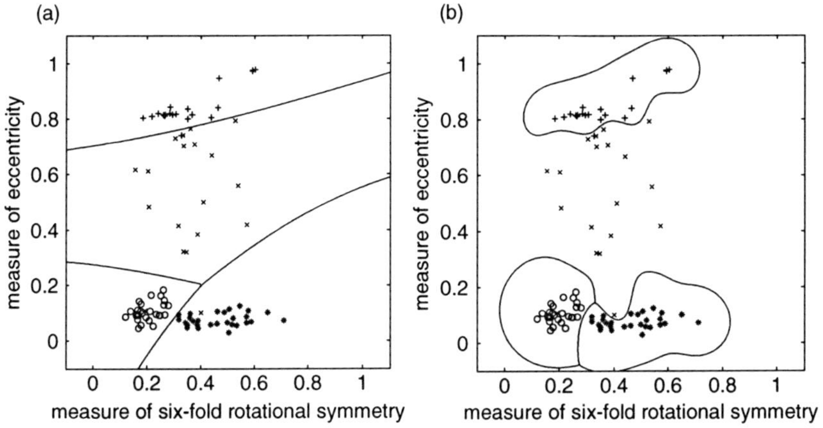

Figure 5.11 Application of two support vector classifiers. (a) Polynomial kernel, $d = 2$, $C = 100$. (b) Gaussian kernel, $\sigma = 0.1$, $C = 100$

$\sigma = 0.1$. In both cases, the trade-off parameter C was set to 100; if it was set smaller, especially the support vector classifier with the polynomial kernel did not find good results. Note in Figure 5.11(b) how the decision boundary is built up by Gaussians around the support vectors, and so forms a closed boundary around the classes.

Listing 5.7
PRTools code for finding and plotting two different support vector classifiers.

```
load nutsbolts;                    % Load the dataset
w=svc(z,'p',2,100);                % Train a quadratic kernel svc
figure; scatterd(z); plotc(w);
w=svc(z,'r',0.1,100);              % Train a Gaussian kernel svc
figure; scatterd(z); plotc(w);
```

5.3.5 The feed-forward neural network

A neural network extends the perceptron in another way: it combines the output of several perceptrons by another perceptron. A single perceptron is called a *neuron* in neural network terminology. Like a perceptron, a neuron computes the weighted sum of the inputs. However, instead of a sign function, a more general *transfer* function is applied.

A transfer function used often is the sigmoid function, a continuous version of the sign function:

$$g(\mathbf{y}) = f(\mathbf{w}^T\mathbf{y}) = \frac{1}{1 + \exp(-\mathbf{w}^T\mathbf{y})} \tag{5.62}$$

where $\mathbf{y}$ is the vector $\mathbf{z}$ augmented with a constant value 1. The vector $\mathbf{w}$ is called the weight vector, and the specific weight corresponding to the constant value 1 in $\mathbf{z}$ is called the *bias weight*.

In principle, several layers of different numbers of neurons can be constructed. For an example, see Figure 5.12. Neurons which are not directly connected to the input or output are called *hidden neurons*. The hidden neurons are organized in (hidden) layers. If all neurons in the network compute their output based only on the output of neurons in previous layers, the network is called a feed-forward neural network. In a feed-forward neural network, no loops are allowed (neurons cannot get their input from next layers).

Assume that we have only one hidden layer with H hidden neurons. The output of the total neural network is:

$$g_k(\mathbf{y}) = f\left(\sum_{h=1}^{H} v_{k,h} f(\mathbf{w}_h^T\mathbf{y}) + v_{k,H+1}\right) \tag{5.63}$$

Here, $\mathbf{w}_h$ is the weight vector of the inputs to hidden neuron h, and $\mathbf{v}_k$ is the weight vector of the inputs to output neuron k. Analogous to least

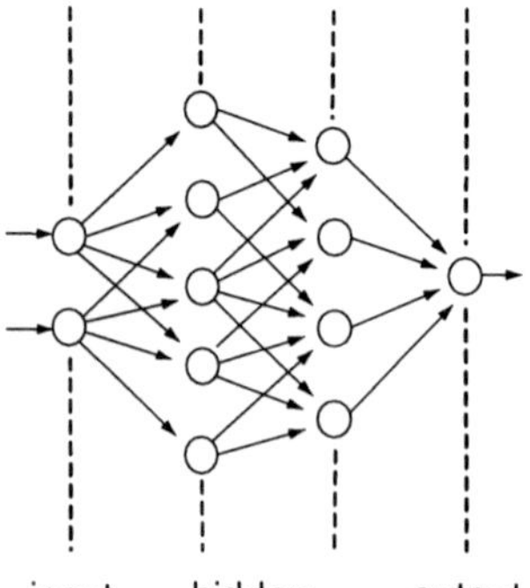

Figure 5.12 A two-layer feed-forward neural network with two input dimensions and one output (for presentation purposes, not all connections have been drawn)

squared error fitting, we define a sum of squared errors between the output of the neural network and the *target vector*:

$$J_{SE} = \frac{1}{2}\sum_{n=1}^{N_S}\sum_{k=1}^{K}(g_k(\mathbf{y}_n) - t_{n,k})^2 \tag{5.64}$$

The target vector is usually created by place coding: $t_{n,k} = 1$ if the label of sample $\mathbf{y}_n$ is ω_k, otherwise it is 0. However, as the sigmoid function lies in the range $<0, 1>$, the values 0 and 1 are hard to reach, and as a result the weights will grow very large. To prevent this, often targets are chosen that are easier to reach, e.g. 0.8 and 0.2.

Because all neurons have continuous transfer functions, it is possible to compute the derivative of this error J_{SE} with respect to the weights. The weights can then be updated using gradient descent. Using the chain rule, the updates of $v_{k,h}$ are easy to compute:

$$\Delta v_{k,h} = -\eta\frac{\partial J_{SE}}{\partial v_{k,h}} = \sum_{n=1}^{N_S}(g_k(\mathbf{y}_n) - t_{n,k})\dot{f}\left(\sum_{h=1}^{H} v_{k,h} f(\mathbf{w}_h^T\mathbf{y}) + v_{k,H+1}\right) f(\mathbf{w}_h^T\mathbf{y}) \tag{5.65}$$

The derivation of the gradient with respect to $w_{h,i}$ is more complicated:

$$\begin{aligned}\Delta w_{h,i} &= -\eta\frac{\partial J_{SE}}{\partial w_{h,i}}\\ &= \sum_{k=1}^{K}\sum_{n=1}^{N_S}(g_k(\mathbf{y}_n) - t_{n,k}) v_{k,h}\dot{f}(\mathbf{w}_h^T\mathbf{y}) y_i \dot{f}\left(\sum_{h=1}^{H} v_{k,h} f(\mathbf{w}_h^T\mathbf{y}) + v_{k,H+1}\right)\end{aligned} \tag{5.66}$$

For the computation of equation (5.66) many elements of equation (5.65) can be reused. This also holds when the network contains more than one hidden layer. When the updates for $v_{k,h}$ are computed first, and those for $w_{h,i}$ are computed from that, we effectively distribute the error between the output and the target value over all weights in the network. We back-propagate the error. The procedure is called *back-propagation* training.

The number of hidden neurons and hidden layers in a neural network controls how nonlinear the decision boundary can be. Unfortunately, it is hard to predict which number of hidden neurons is suited for the task at hand. In practice, we often train a number of neural networks of varying complexity and compare their performance on an independent validation set. The danger of finding a too nonlinear decision boundary is illustrated in Example 5.7.

Example 5.7 Classification of mechanical parts, neural networks
Figure 5.13 shows the decision boundaries found by training two neural networks. The first network, whose decision boundaries is shown in Figure 5.13(a), contains one hidden layer of five units. This gives a reasonably smooth decision boundary. For the decision function shown in Figure 5.13(c), a network was used with two hidden layers of 100 units each. This network has clearly found a highly nonlinear solution, which does not generalize as well as the first network. For example, the '∗' region (nuts) contains one outlying 'x' sample (scrap). The decision boundary bends heavily to include the single 'x' sample within the scrap region. Although such a crumpled curve decreases the squared error, it is undesirable because the outlying 'x' is not likely to occur again at that same location in other realizations.

Note also the spurious region in the right bottom of the plot, in which samples are classified as 'bolt' (denoted by + in the scatterplot). Here too, the network generalizes poorly as it has not seen any examples in this region.

Figures 5.13(b) and (d) show the learn curves that were derived during training the network. One epoch is a training period in which the algorithm has cycled through all training samples. The figures show 'error', which is the fraction of the training samples that are erroneously classified, and 'mse', which is $2J_{SE}/(KN_S)$. The larger the network is, the more epochs are needed before the minimum will be reached. However, sometimes it is better to stop training before actually reaching the minimum because the generalization ability can degenerate in the vicinity of the minimum.

The figures were generated by the code shown in Listing 5.8.

Listing 5.8
PRTools code for training and plotting two neural network classifiers.

```
load nutsbolts;                          % Load the dataset
[w,R] = bpxnc(z,5,500);                  % Train a small
                                           network
figure; scatterd(z); plotc(w);           % Plot the
                                           classifier
```

```
figure; plotyy(R(:,1),R(:,2),R(:,1),R(:,4));    % Plot the learn
                                                  curves
[w,R] =bpxnc(z,[100 100],1000);                 % Train a larger
                                                  network
figure; scatterd(z); plotc(w);                  % Plot the
                                                  classifier
figure; plotyy(R(:,1),R(:,2),R(:,1),R(:,4));    % Plot the learn
                                                  curves
```

5.4 EMPIRICAL EVALUATION

In the preceding sections various methods for training a classifier have been discussed. These methods have led to different types of classifiers and different types of learning rules. However, none of these methods can claim overall superiority above the other because their applicability and effectiveness is largely determined by the specific nature of the problem at hand. Therefore, rather than relying on just one method that has been selected at the beginning of the design process, the designer often examines various methods and selects the one that appears most suitable. For that purpose, each classifier has to be evaluated.

Another reason for performance evaluation stems from the fact that many classifiers have their own parameters that need to be tuned. The optimization of a design criterion using only training data holds the risk of *overfitting* the design, leading to an inadequate ability to generalize. The behaviour of the classifier becomes too specific for the training data at hand, and is less appropriate for future measurement vectors coming from the same application. Particularly, if there are many parameters relative to the size of the training set and the dimension of the measurement vector, the risk of overfitting becomes large (see also Figure 5.13 and Chapter 6). Performance evaluation based on a *validation set* (*test set, evaluation set*), independent from the training set, can be used as a stopping criterion for the parameter tuning process.

A third motivation for performance evaluation is that we would like to have reliable specifications of the design anyhow.

There are many criteria for the performance of a classifier. The probability of misclassification, i.e. the *error rate*, is the most popular one. The analytical expression for the error rate as given in (2.16) is not very useful because, in practice, the conditional probability densities are unknown. However, we can easily obtain an estimate of the error rate by subjecting the classifier to a validation set. The estimated error rate is the fraction of misclassified samples with respect to the size of the validation set.

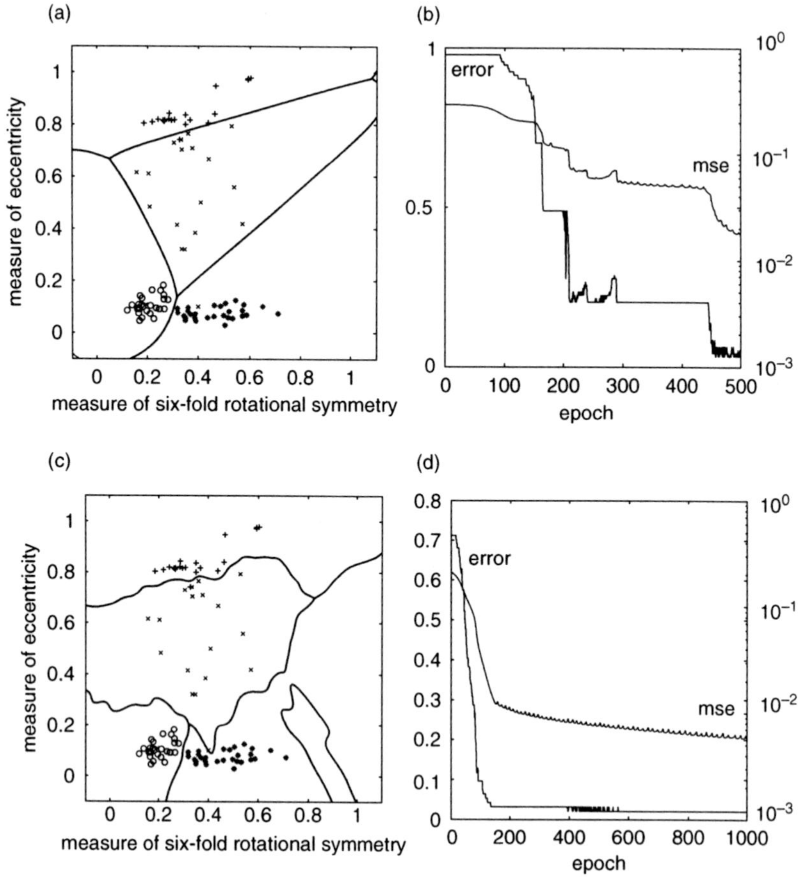

Figure 5.13 Application of two neural networks. (a) One hidden layer of five units (b) Learn curves of (a). (c) Two hidden layers of 100 units each. (d) Learn curves of (c)

Where the classification uses the reject option we need to consider two measures: the error rate and the *reject rate*. These two measures are not fully independent because lowering the reject rate will increase the error rate. A plot of the error rate versus the reject rate visualizes this dependency.

The error rate itself is somewhat crude as a performance measure. In fact, it merely suffices in the case of a uniform cost function. A more profound insight of the behaviour of the classifier is revealed by the so-called *confusion matrix*. The i, j-th element of this matrix is the count of ω_i samples in the validation set to which class ω_j is assigned. The corresponding PRTools function is `confmat()`.

Another design criterion might be the *computational complexity* of a classifier. From an engineering point of view both the processor capability and the storage capability are subjected to constraints. At the same time, the application may limit the computational time needed for the classification. Often, a trade-off must be found between computational complexity and performance. In order to do so, we have to assess the number of operations per classification and the storage requirements. These aspects scale with the number of elements N in the measurement vector, the number of classes K, and for some classifiers the size N_S of the training set. For instance, the number of operations of a quadratic machine is of the order of KN^2, and so is its memory requirement. For binary measurements, the storage requirement is of the order of $K2^N$ (which becomes prohibitive even for moderate N).

Often, the acquisition of labelled samples is laborious and costly. Therefore, it is tempting to use the training set also for evaluation purposes. However, the strategy of 'testing from training data' disguises the phenomenon of overfitting. The estimated error rate is likely to be strongly overoptimistic. To prevent this, the validation set should be independent of the training set.

A straightforward way to accomplish this is the so-called *holdout* method. The available samples are randomly partitioned into a training set and a validation set. Because the validation set is used to estimate only one parameter, i.e. the error rate, and the training set is used to tune all other parameters of the classifier (which may be quite numerous), the training set must be larger than the validation set. As a rule of thumb, the validation set contains about 20% of the data, and the training set the remaining 80%.

Suppose that a validation set consisting of, say, N_{Test} labelled samples is available. Application of the 'classifier-under-test' to this validation set results in n_{error} misclassified samples. If the true error rate of the classifier is denoted by E, then the estimated error rate is:

$$\hat{E} = \frac{n_{error}}{N_{Test}} \tag{5.67}$$

The random variable n_{error} has a binomial distribution with parameters E and N_{Test}. Therefore, the expectation of $\hat{E}$ and its standard deviation is found as:

$$\begin{aligned} \mathrm{E}\left[\hat{E}\right] &= \frac{\mathrm{E}[n_{error}]}{N_{Test}} = \frac{EN_{Test}}{N_{Test}} = E \\ \sigma_{\hat{E}} &= \frac{\sqrt{\mathrm{Var}[n_{error}]}}{N_{Test}} = \frac{\sqrt{N_{Test}(1-E)E}}{N_{Test}} = \sqrt{\frac{(1-E)E}{N_{Test}}} \end{aligned} \tag{5.68}$$

Hence, $\hat{E}$ is an unbiased, consistent estimate of E.

Usually, the required number of samples N_{Test} is defined such that $\hat{E}$ is within some specified margin around the true E with a prescribed probability. It can be calculated from the posterior probability density $p(E|n_{error}, N_{Test})$. See exercise 5. However, we take our ease by simply requiring that the relative uncertainty $\sigma_{\hat{E}}/E$ is equal to some fixed fraction γ. Substitution in (5.68) and solving for N_{Test} we obtain:

$$N_{Test} = \frac{1-E}{\gamma^2 E} \tag{5.69}$$

Figure 5.14 shows the required number of samples for different values of E such that the relative uncertainty is 10%. The figure shows that with $E = 0.01$ the number of samples must be about 10 000.

The holdout method is not economic in handling the available data because only part of that data is used for training. A classifier trained with a reduced training set is expected to be inferior to a classifier trained with all available data. Particularly if the acquisition of labelled data is expensive, we prefer methods that use the data as much as possible for training and yet give unbiased estimates for the error rate. Examples of methods in this category are the *cross-validation* method and the *leave-one-out* method. These methods are computationally expensive, because they require the design of many classifiers.

The cross-validation method randomly partitions the available data into L equally sized subsets $T(\ell)$, $\ell = 1, \ldots, L$. First, the subset $T(1)$ is

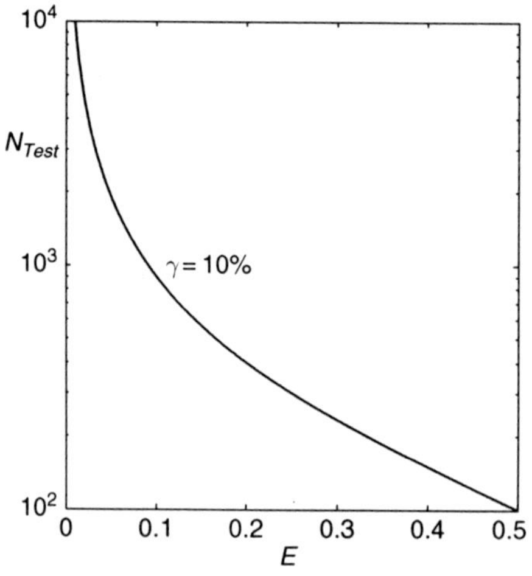

Figure 5.14 Required number of test samples

withheld and a classifier is trained using all the data from the remaining $L-1$ subsets. This classifier is tested using $T(1)$ as validation data, yielding an estimate $\hat{E}(1)$. This procedure is repeated for all other subsets, thus resulting in L estimates $\hat{E}(\ell)$. The last step is the training of the final classifier using all available data. Its error rate is estimated as the average over $\hat{E}(\ell)$. This estimate is a little pessimistic especially if L is small.

The leave-one-out method is essentially the same as the cross-validation method except that now L is as large as possible, i.e. equal to N_S (the number of available samples). The bias of the leave-one-out method is negligible, especially if the training set is large. However, it requires the training of N_S+1 classifiers and it is computationally very expensive if N_S is large.

5.5 REFERENCES

Bishop, C.M., *Neural Networks for Pattern Recognition*, Oxford University Press, Oxford, UK, 1995.

Duda, R.O., Hart, P.E. and Stork, D.G., *Pattern Classification*, Wiley, London, UK, 2001.

Devijver, P.A. and Kittler, J., *Pattern Recognition, a Statistical Approach*, Prentice-Hall, London, UK, 1982.

Haykin, S., *Neural Networks – a Comprehensive Foundation*, 2nd ed., Prentice-Hall, Upper Saddle River, NJ, 1998.

Hertz, J., Krogh, A. and Palmer, R.G., *Introduction to the Theory of Neural Computation*, Addison-Wesley, Reading, MA, 1991.

Ripley, B.D., *Pattern Recognition and Neural Networks*, Cambridge University Press, Cambridge, UK, 1996.

Vapnik, V.N., *Statistical Learning Theory*, Wiley, New York, 1998.

5.6 EXERCISES

1. Prove that if C is the covariance matrix of a random vector **z**, then

$$\frac{1}{N}\mathrm{C}$$

   is the covariance matrix of the average of N realizations of **z**. (0)
2. Show that (5.16) and (5.17) are equivalent. (*)
3. Prove that, for the two-class case, (5.50) is equivalent to the Fisher linear discriminant (6.52). (**)
4. Investigate the behaviour (bias and variance) of the estimators for the conditional probabilities of binary measurements, i.e. (5.20) and (5.22), at the extreme ends. That is, if $N_k << 2^N$ and $N_k >> 2^N$. (**)

5. Assuming that the prior probability density of the error rate of a classifier is uniform between 0 and $1/K$, give an expression of the posterior density $p(E|n_{error}, N_{Test})$ where N_{Test} is the size of an independent validation set and n_{error} is the number of misclassifications of the classifier. $(*)$
6. Derive the dual formulation of the support vector classifier from the primal formulation. Do this by setting the partial derivatives of L to zero, and substituting the results in the primal function. $(*)$
7. Show that the support vector classifiers with slack variables gives almost the same dual formulation as the one without slack variables (5.56). $(*)$
8. Derive the neural network weight update rules (5.65) and (5.66). $(*)$
9. Neural network weights are often initialized to random values in a small range, e.g. $<-0.01, 0.01>$. As training progresses, the weight values quickly increase. However, the support vector classifier tells us that solutions with small norms of the weight vector have high generalization capability. What would be a simple way to assure that the network does not become too nonlinear? $(*)$
10. Given the answer to exercise 9, what will be the effect of using better optimization techniques (such as second-order algorithms) in neural network training? Validate this experimentally using PRTools `lmnc` function. $(*)$

6

Feature Extraction and Selection

In some cases, the dimension N of a measurement vector $\mathbf{z}$, i.e. the number of sensors, can be very high. In image processing, when raw image data is used directly as the input for a classification, the dimension can easily attain values of 10^4 (a 100×100 image) or more. Many elements of $\mathbf{z}$ can be redundant or even irrelevant with respect to the classification process.

For two reasons, the dimension of the measurement vector cannot be taken arbitrarily large. The first reason is that the computational complexity becomes too large. A linear classification machine requires in the order of KN operations (K is the number of classes; see Chapter 2). A quadratic machine needs about KN^2 operations. For a machine acting on binary measurements the memory requirement is on the order of $K2^N$. This, together with the required throughput (number of classifications per second), the state of the art in computer technology and the available budget define an upper bound to N.

A second reason is that an increase of the dimension ultimately causes a decrease of performance. Figure 6.1 illustrates this. Here, we have a measurement space with the dimension N varying between 1 and 13. There are two classes ($K = 2$) with equal prior probabilities. The (true) minimum error rate $E_{\min}$ is the one which would be obtained if all class densities of the problem were fully known. Clearly, the minimum error

Classification, Parameter Estimation and State Estimation: An Engineering Approach using MATLAB
F. van der Heijden, R.P.W. Duin, D. de Ridder and D.M.J. Tax
 ISBN: 0-470-09013-8

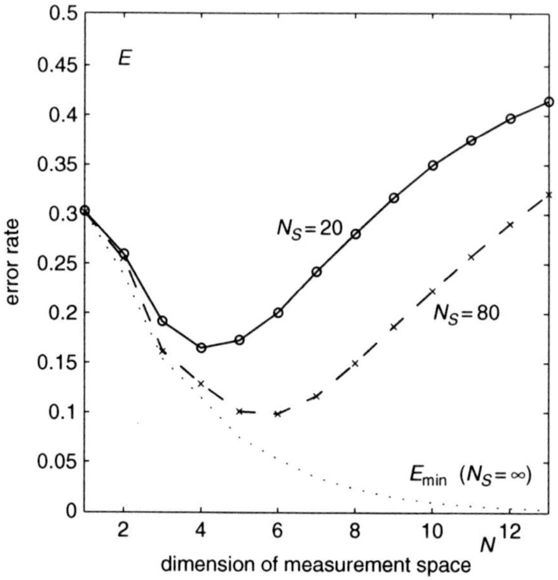

Figure 6.1 Error rates versus dimension of measurement space

rate is a non-increasing function of the number of sensors. Once an element has been added with discriminatory information, the addition of another element cannot destroy this information. Therefore, with growing dimension, class information accumulates.

However, in practice the densities are seldom completely known. Often, the classifiers have to be designed using a (finite) training set instead of using knowledge about the densities. In the example of Figure 6.1 the measurement data is binary. The number of states a vector can take is 2^N. If there are no constraints on the conditional probabilities, then the number of parameters to estimate is in the order of 2^N. The number of samples in the training set must be much larger than this. If not, overfitting occurs and the trained classifier will become too much adapted to the noise in the training data. Figure 6.1 shows that if the size of the training set is $N_S = 20$, the optimal dimension of the measurement vector is about $N = 4$; that is where the error rate E is lowest. Increasing the sample size permits an increase of the dimension. With $N_S = 80$ the optimal dimension is about $N = 6$.

One strategy to prevent overfitting, or at least to reduce its effect, has already been discussed in Chapter 5: incorporating more prior knowledge by restricting the structure of the classifier (for instance, by an appropriate choice of the discriminant function). In the current chapter, an alternative strategy will be discussed: the reduction of the dimension of the measurement vector. An additional advantage of this strategy is that it automatically reduces the computational complexity.

For the reduction of the measurement space, two different approaches exist. One is to discard certain elements of the vector and to select the ones that remain. This type of reduction is *feature selection*. It is discussed in Section 6.2. The other approach is *feature extraction*. Here, the selection of elements takes place in a transformed measurement space. Section 6.3 addresses the problem of how to find suitable transforms. Both methods rely on the availability of optimization criteria. These are discussed in Section 6.1.

6.1 CRITERIA FOR SELECTION AND EXTRACTION

The first step in the design of optimal feature selectors and feature extractors is to define a quantitative criterion that expresses how well such a selector or extractor performs. The second step is to do the actual optimization, i.e. to use that criterion to find the selector/extractor that performs best. Such an optimization can be performed either analytically or numerically.

Within a Bayesian framework 'best' means the one with minimal risk. Often, the cost of misclassification is difficult to assess, or even fully unknown. Therefore, as an optimization criterion the risk is often replaced by the error rate E. Techniques to assess the error rate empirically by means of a validation set are discussed in Section 5.4. However, in this section we need to be able to manipulate the criterion mathematically. Unfortunately, the mathematical structure of the error rate is complex. The current section introduces some alternative, approximate criteria that are simple enough for a mathematical treatment.

In feature selection and feature extraction, these simple criteria are used as alternative performance measures. Preferably, such performance measures have the following properties:

- The measure increases as the average distance between the expectation vectors of different classes increases. This property is based on the assumption that the class information of a measurement vector is mainly in the differences between the class-dependent expectations.
- The measure decreases with increasing noise scattering. This property is based on the assumption that the noise on a measurement

vector does not contain class information, and that the class distinction is obfuscated by this noise.
- The measure is invariant to reversible linear transforms. Suppose that the measurement space is transformed to a feature space, i.e. $\mathbf{y} = \mathbf{Az}$ with $\mathbf{A}$ an invertible matrix, then the measure expressed in the $\mathbf{y}$ space should be exactly the same as the one expressed in the $\mathbf{z}$ space. This property is based on the fact that both spaces carry the same class information.
- The measure is simple to manipulate mathematically. Preferably, the derivatives of the criteria are obtained easily as it is used as an optimization criterion.

From the various measures known in literature (Devijver and Kittler, 1982), two will be discussed. One of them – the interclass/intraclass distance (Section 6.1.1) – applies to the multi-class case. It is useful if class information is mainly found in the differences between expectation vectors in the measurement space, while at the same time the scattering of the measurement vectors (due to noise) is class-independent. The second measure – the Chernoff distance (Section 6.1.2) – is particularly useful in the two-class case because it can then be used to express bounds on the error rate.

Section 6.1.3 concludes with an overview of some other performance measures.

6.1.1 Inter/intra class distance

The inter/intra distance measure is based on the Euclidean distance between pairs of samples in the training set. We assume that the class-dependent distributions are such that the expectation vectors of the different classes are discriminating. If fluctuations of the measurement vectors around these expectations are due to noise, then these fluctuations will not carry any class information. Therefore, our goal is to arrive at a measure that is a monotonically increasing function of the distance between expectation vectors, and a monotonically decreasing function of the scattering around the expectations.

As in Chapter 5, T_S is a (labelled) training set with N_S samples. The classes ω_k are represented by subsets $T_k \subset T_S$, each class having N_k samples ($\Sigma N_k = N_S$). Measurement vectors in T_S – without reference to their class – are denoted by $\mathbf{z}_n$. Measurement vectors in T_k (i.e. vectors coming from class ω_k) are denoted by $\mathbf{z}_{k,n}$.

The development starts with the following definition of the *average squared distance* of pairs of samples in the training set:

$$\overline{\rho^2} = \frac{1}{N_S^2} \sum_{n=1}^{N_S} \sum_{m=1}^{N_S} (\mathbf{z}_n - \mathbf{z}_m)^T (\mathbf{z}_n - \mathbf{z}_m) \tag{6.1}$$

The summand $(\mathbf{z}_n - \mathbf{z}_m)^T(\mathbf{z}_n - \mathbf{z}_m)$ is the squared Euclidean distance between a pair of samples. The sum involves all pairs of samples in the training set. The divisor N_S^2 accounts for the number of terms.

The distance $\overline{\rho^2}$ is useless as a performance measure because none of the desired properties mentioned above are met. Moreover, $\overline{\rho^2}$ is defined without any reference to the labels of the samples. Thus, it does not give any clue about how well the classes in the training set can be discriminated. To correct this, $\overline{\rho^2}$ must be divided into a part describing the average distance between expectation vectors and a part describing distances due to noise scattering. For that purpose, estimations of the conditional expectations ($\boldsymbol{\mu}_k = \mathrm{E}[\mathbf{z}|\omega_k]$) of the measurement vectors are used, along with an estimate of the unconditional expectation ($\boldsymbol{\mu} = \mathrm{E}[\mathbf{z}]$). The sample mean of class ω_k is:

$$\hat{\boldsymbol{\mu}}_k = \frac{1}{N_k} \sum_{n=1}^{N_k} \mathbf{z}_{k,n} \tag{6.2}$$

The sample mean of the entire training set is:

$$\hat{\boldsymbol{\mu}} = \frac{1}{N_S} \sum_{n=1}^{N_S} \mathbf{z}_n \tag{6.3}$$

With these definitions, it can be shown (see exercise 1) that the average squared distance is:

$$\overline{\rho^2} = \frac{2}{N_S} \sum_{k=1}^{K} \left[\sum_{n=1}^{N_k} \left((\mathbf{z}_{k,n} - \hat{\boldsymbol{\mu}}_k)^T (\mathbf{z}_{k,n} - \hat{\boldsymbol{\mu}}_k) + (\hat{\boldsymbol{\mu}} - \hat{\boldsymbol{\mu}}_k)^T (\hat{\boldsymbol{\mu}} - \hat{\boldsymbol{\mu}}_k) \right) \right] \tag{6.4}$$

The first term represents the average squared distance due to the scattering of samples around their class-dependent expectation. The second term corresponds to the average squared distance between class-dependent expectations and the unconditional expectation.

An alternative way to represent these distances is by means of *scatter matrices*. A scatter matrix gives some information about the dispersion of a population of samples around their mean. For instance, the matrix that describes the scattering of vectors from class ω_k is:

$$\mathbf{S}_k = \frac{1}{N_k}\sum_{n=1}^{N_k}(\mathbf{z}_{k,n} - \hat{\boldsymbol{\mu}}_k)(\mathbf{z}_{k,n} - \hat{\boldsymbol{\mu}}_k)^T \tag{6.5}$$

Comparison with equation (5.14) shows that $\mathbf{S}_k$ is close to an unbiased estimate of the class-dependent covariance matrix. In fact, $\mathbf{S}_k$ is the maximum likelihood estimate of $\mathbf{C}_k$. With that, $\mathbf{S}_k$ does not only supply information about the average distance of the scattering, it also supplies information about the eccentricity and orientation of this scattering. This is analogous to the properties of a covariance matrix.

Averaged over all classes the scatter matrix describing the noise is:

$$\mathbf{S}_w = \frac{1}{N_S}\sum_{k=1}^{K} N_k\mathbf{S}_k = \frac{1}{N_S}\sum_{k=1}^{K}\sum_{n=1}^{N_k}(\mathbf{z}_{k,n} - \hat{\boldsymbol{\mu}}_k)(\mathbf{z}_{k,n} - \hat{\boldsymbol{\mu}}_k)^T \tag{6.6}$$

This matrix is the *within-scatter matrix* as it describes the average scattering within classes. Complementary to this is the *between-scatter matrix* $\mathbf{S}_b$ that describes the scattering of the class-dependent sample means around the overall average:

$$\mathbf{S}_b = \frac{1}{N_S}\sum_{k=1}^{K} N_k(\hat{\boldsymbol{\mu}}_k - \hat{\boldsymbol{\mu}})(\hat{\boldsymbol{\mu}}_k - \hat{\boldsymbol{\mu}})^T \tag{6.7}$$

Figure 6.2 illustrates the concepts of within-scatter matrices and between-scatter matrices. The figure shows a scatter diagram of a training set consisting of four classes. A scatter matrix $\mathbf{S}$ corresponds to an ellipse, $\mathbf{z}\mathbf{S}^{-1}\mathbf{z}^T = 1$, that can be thought of as a contour roughly surrounding the associated population of samples. Of course, strictly speaking the correspondence holds true only if the underlying probability density is Gaussian-like. But even if the densities are not Gaussian, the ellipses give an impression of how the population is scattered. In the scatter diagram in Figure 6.2 the within-scatter $\mathbf{S}_w$ is represented by four similar ellipses positioned at the four conditional sample means. The between-scatter $\mathbf{S}_b$ is depicted by an ellipse centred at the mixture sample mean.

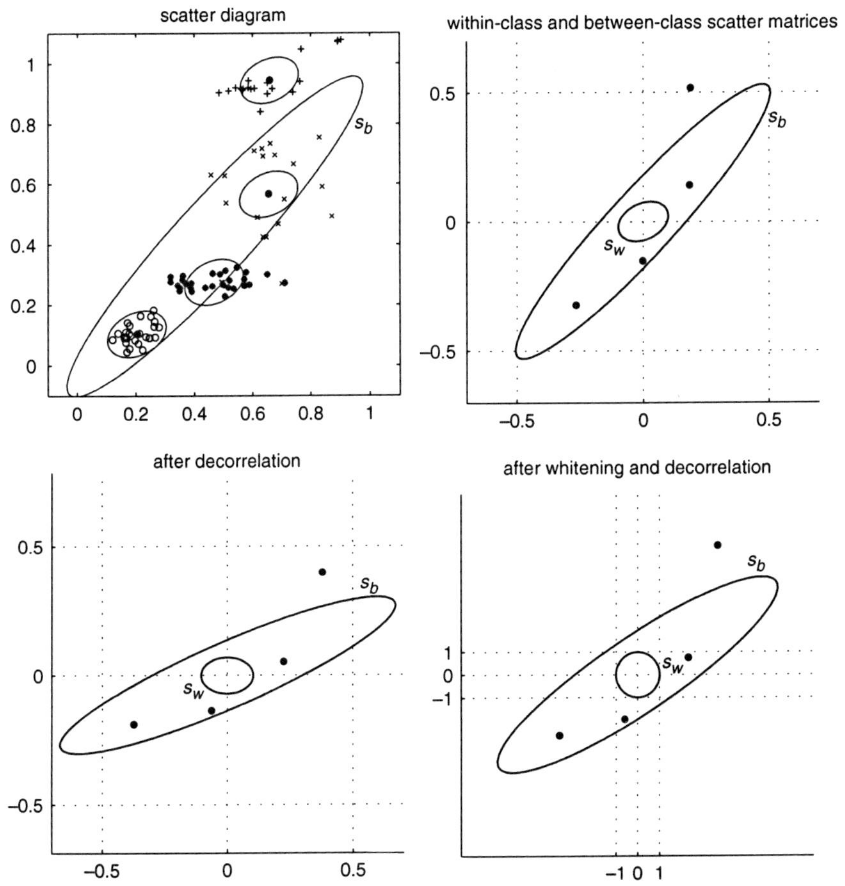

Figure 6.2 The inter/intra class distance

With the definitions in (6.5), (6.6) and (6.7) the average squared distance in (6.4) is proportional to the trace of the matrix $S_w + S_b$; see (b.22):

$$\overline{\rho^2} = 2trace(S_w + S_b) = 2trace(S_w) + 2trace(S_b) \tag{6.8}$$

Indeed, this expression shows that the average distance is composed of a contribution due to differences in expectation and a contribution due to noise. The term $J_{INTRA} = trace(S_w)$ is called the *intraclass distance*. The term $J_{INTER} = trace(S_b)$ is the *interclass distance*. Equation (6.8) also shows that the average distance is not appropriate as a performance measure, since a large value of $\overline{\rho^2}$ does not imply that the classes are well separated, and vice versa.

A performance measure more suited to express the separability of classes is the ratio between interclass and intraclass distance:

$$\frac{J_{INTER}}{J_{INTRA}} = \frac{trace(\mathbf{S}_b)}{trace(\mathbf{S}_w)} \tag{6.9}$$

This measure possesses some of the desired properties of a performance measure. In Figure 6.2, the numerator, $trace(\mathbf{S}_b)$, measures the area of the ellipse associated with $\mathbf{S}_b$. As such, it measures the fluctuations of the conditional expectations around the overall expectation, i.e. the fluctuations of the 'signal'. The denominator, $trace(\mathbf{S}_w)$, measures the area of the ellipse associated with $\mathbf{S}_w$. As such, it measures the fluctuations due to noise. Therefore, $trace(\mathbf{S}_b)/trace(\mathbf{S}_w)$ can be regarded as a 'signal-to-noise ratio'.

Unfortunately, the measure of (6.9) oversees the fact that the ellipse associated with the noise can be quite large, but without having a large intersection with the ellipse associated with the signal. A large $\mathbf{S}_w$ can be quite harmless for the separability of the training set. One way to correct this defect is to transform the measurement space such that the within-scattering becomes *white*, i.e. $\mathbf{S}_w = \mathbf{I}$. For that purpose, we apply a linear operation to all measurement vectors yielding *feature vectors* $\mathbf{y}_n = \mathbf{A}\mathbf{z}_n$. In the transformed space the within- and between-scatter matrices become: $\mathbf{A}\mathbf{S}_w\mathbf{A}^T$ and $\mathbf{A}\mathbf{S}_b\mathbf{A}^T$, respectively. The matrix $\mathbf{A}$ is chosen such that $\mathbf{A}\mathbf{S}_w\mathbf{A}^T = \mathbf{I}$.

The matrix $\mathbf{A}$ can be found by factorization: $\mathbf{S}_w = \mathbf{V}\mathbf{\Lambda}\mathbf{V}^T$ where $\mathbf{\Lambda}$ is a diagonal matrix containing the eigenvalues of $\mathbf{S}_w$, and $\mathbf{V}$ a unitary matrix containing the corresponding eigenvectors; see appendix B.5. With this factorization it follows that $\mathbf{A} = \mathbf{\Lambda}^{-1/2}\mathbf{V}^T$. An illustration of the process is depicted in Figure 6.2. The operation $\mathbf{V}^T$ performs a rotation that aligns the axes of $\mathbf{S}_w$. It decorrelates the noise. The operation $\mathbf{\Lambda}^{-1/2}$ scales the axes. The normalized within-scatter matrix corresponds with a circle with unit radius. In this transformed space, the area of the ellipse associated with the between-scatter is a useable performance measure:

$$\begin{aligned} J_{INTER/INTRA} &= trace(\mathbf{\Lambda}^{-\frac{1}{2}}\mathbf{V}^T\mathbf{S}_b\mathbf{V}\mathbf{\Lambda}^{-\frac{1}{2}}) \\ &= trace(\mathbf{V}\mathbf{\Lambda}^{-1}\mathbf{V}^T\mathbf{S}_b) \\ &= trace(\mathbf{S}_w^{-1}\mathbf{S}_b) \end{aligned} \tag{6.10}$$

This performance measure is called the *inter/intra distance*. It meets all of our requirements stated above.

Example 6.1 Interclass and intraclass distance
Numerical calculations of the example in Figure 6.2 show that before normalization $J_{INTRA} = trace(S_w) = 0.016$ and $J_{INTER} = trace(S_b) = 0.54$. Hence, the ratio between these two is $J_{INTER}/J_{INTRA} = 33.8$. After normalization, $J_{INTRA} = N = 2$, $J_{INTER} = J_{INTER/INTRA} = 59.1$, and $J_{INTER}/J_{INTRA} = trace(S_b)/trace(S_w) = 29.6$. In this example, before normalization, the J_{INTER}/J_{INTRA} measure is too optimistic. The normalization accounts for this phenomenon.

6.1.2 Chernoff–Bhattacharyya distance

The interclass and intraclass distances are based on the Euclidean metric defined in the measurement space. Another possibility is to use a metric based on probability densities. Examples in this category are the *Chernoff distance* (Chernoff, 1952) and the *Bhattacharyya distance* (Bhattacharyya, 1943). These distances are especially useful in the two-class case.

The merit of the Bhattacharyya and Chernoff distance is that an inequality exists with which the distances bound the minimum error rate $E_{\min}$. The inequality is based on the following relationship:

$$\min\{a, b\} \leq \sqrt{ab} \tag{6.11}$$

The inequality holds true for any positive quantities a and b. We will use it in the expression of the minimum error rate. Substitution of (2.15) in (2.16) yields:

$$\begin{aligned} E_{\min} &= \int_{\mathbf{z}} [1 - \max\{P(\omega_1|\mathbf{z}), P(\omega_2|\mathbf{z})\}] p(\mathbf{z}) d\mathbf{z} \\ &= \int_{\mathbf{z}} \min\{p(\mathbf{z}|\omega_1)P(\omega_1), p(\mathbf{z}|\omega_2)P(\omega_2)\} d\mathbf{z} \end{aligned} \tag{6.12}$$

Together with (6.11) we have the following inequality:

$$E_{\min} \leq \sqrt{P(\omega_1)P(\omega_2)} \int_{\mathbf{z}} \sqrt{p(\mathbf{z}|\omega_1)p(\mathbf{z}|\omega_2)} d\mathbf{z} \tag{6.13}$$

The inequality is called the *Bhattacharyya upper bound*. A more compact notation of it is achieved with the so-called Bhattacharyya distance. This performance measure is defined as:

$$J_{BHAT} = -\ln\left[\int_{\mathbf{z}} \sqrt{p(\mathbf{z}|\omega_1)p(\mathbf{z}|\omega_2)}d\mathbf{z}\right] \tag{6.14}$$

With that, the Bhattacharyya upper bound simplifies to:

$$E_{\min} \leq \sqrt{P(\omega_1)P(\omega_2)}\exp(-J_{BHAT}) \tag{6.15}$$

The bound can be made more tight if inequality (6.11) is replaced with the more general inequality $\min\{a, b\} \leq a^s b^{1-s}$. This last inequality holds true for any s, a and b in the interval $[0, 1]$. The inequality leads to the Chernoff distance, defined as:

$$J_C(s) = -\ln\left[\int_{\mathbf{z}} p^s(\mathbf{z}|\omega_1)p^{1-s}(\mathbf{z}|\omega_2)d\mathbf{z}\right] \quad \text{with: } 0 \leq s \leq 1 \tag{6.16}$$

Application of the Chernoff distance in a derivation similar to (6.12) yields:

$$E_{\min} \leq P(\omega_1)^s P(\omega_2)^{1-s}\exp(-J_C(s)) \quad \text{for any} \quad s \in [0, 1] \tag{6.17}$$

The so-called *Chernoff bound* encompasses the Bhattacharyya upper bound. In fact, for $s = 0.5$ the Chernoff distance and the Bhattacharyya distance are equal: $J_{BHAT} = J_C(0.5)$.

There also exists a lower bound based on the Bhattacharyya distance. This bound is expressed as:

$$\frac{1}{2}\left[1 - \sqrt{1 - 4P(\omega_1)P(\omega_2)\exp(-2J_{BHAT})}\right] \leq E_{\min} \tag{6.18}$$

A further simplification occurs when we specify the conditional probability densities. An important application of the Chernoff and Bhattacharyya distance is the Gaussian case. Suppose that these densities have class-dependent expectation vectors $\boldsymbol{\mu}_k$ and covariance matrices C_k, respectively. Then, it can be shown that the Chernoff distance transforms into:

$$J_C(s) = \frac{1}{2}s(1-s)(\mu_2-\mu_1)^T[(1-s)C_1+sC_2]^{-1}(\mu_2-\mu_1) + \frac{1}{2}\ln\left[\frac{|(1-s)C_1+sC_2|}{|C_1|^{1-s}|C_2|^s}\right] \tag{6.19}$$

It can be seen that if the covariance matrices are independent of the classes, e.g. $C_1 = C_2$, the second term vanishes, and the Chernoff and the Bhattacharyya distances become proportional to the Mahalanobis distance *SNR* given in (2.46): $J_{BHAT} = SNR/8$. Figure 6.3(a) shows the corresponding Chernoff and Bhattacharyya upper bounds. In this particular case, the relation between *SNR* and the minimum error rate is easily obtained using expression (2.47). Figure 6.3(a) also shows the Bhattacharyya lower bound.

The dependency of the Chernoff bound on s is depicted in Figure 6.3(b). If $C_1 = C_2$, the Chernoff distance is symmetric in s, and the minimum bound is located at $s = 0.5$ (i.e. the Bhattacharyya upper bound). If the covariance matrices are not equal, the Chernoff distance is not symmetric, and the minimum bound is not guaranteed to be located midway. A numerical optimization procedure can be applied to find the tightest bound.

If in the Gaussian case, the expectation vectors are equal ($\mu_1 = \mu_2$), the first term of (6.19) vanishes, and all class information is represented by the second term. This term corresponds to class information carried by differences in covariance matrices.

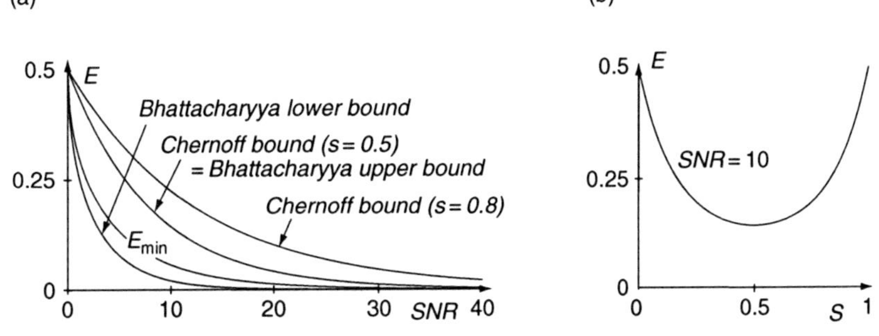

Figure 6.3 Error bounds and the true minimum error for the Gaussian case ($C_1 = C_2$). (a) The minimum error rate with some bounds given by the Chernoff distance. In this example the bound with $s = 0.5$ (Bhattacharyya upper bound) is the most tight. The figure also shows the Bhattacharyya lower bound. (b) The Chernoff bound with dependence on s

6.1.3 Other criteria

The criteria discussed above are certainly not the only ones used in feature selection and extraction. In fact, large families of performance measures exist; see Devijver and Kittler (1982) for an extensive overview. One family is the so-called *probabilistic distance measures*. These measures are based on the observation that large differences between the conditional densities $p(\mathbf{z}|\omega_k)$ result in small error rates. Let us assume a two-class problem with conditional densities $p(\mathbf{z}|\omega_1)$ and $p(\mathbf{z}|\omega_2)$. Then, a probabilistic distance takes the following form:

$$\mathrm{J} = \int_{-\infty}^{\infty} g(p(\mathbf{z}|\omega_1), p(\mathbf{z}|\omega_2))d\mathbf{z} \tag{6.20}$$

The function $g(\cdot,\cdot)$ must be such that J is zero when $p(\mathbf{z}|\omega_1) = p(\mathbf{z}|\omega_2)$, $\forall\mathbf{z}$, and non-negative otherwise. In addition, we require that J attains its maximum whenever the two densities are completely non-overlapping. Obviously, the Bhattacharyya distance (6.14) and the Chernoff distance (6.16) are examples of performance measures based on a probabilistic distance. Other examples are the Matusita measure and the divergence measures:

$$\mathrm{J}_{MATUSITA} = \sqrt{\int \left(\sqrt{p(\mathbf{z}|\omega_1)} - \sqrt{p(\mathbf{z}|\omega_2)}\right)^2 d\mathbf{z}} \tag{6.21}$$

$$\mathrm{J}_{DIVERGENCE} = \int (p(\mathbf{z}|\omega_1) - p(\mathbf{z}|\omega_2)) \ln \frac{p(\mathbf{z}|\omega_1)}{p(\mathbf{z}|\omega_2)} d\mathbf{z} \tag{6.22}$$

These measures are useful for two-class problems. For more classes, a measure is obtained by taking the average of pairs. That is:

$$\mathrm{J} = \sum_{k=1}^{K}\sum_{l=0}^{K} P(\omega_k)P(\omega_l)\mathrm{J}_{k,l} \tag{6.23}$$

where $\mathrm{J}_{k,l}$ is the measure found between the classes ω_k and ω_l.

Another family is the one using the *probabilistic dependence* of the measurement vector $\mathbf{z}$ on the class ω_k. Suppose that the conditional density of $\mathbf{z}$ is not affected by ω_k, i.e. $p(\mathbf{z}|\omega_k) = p(\mathbf{z})$, $\forall\mathbf{z}$, then an observation of $\mathbf{z}$ does not increase our knowledge on ω_k. Therefore, the ability of

$\mathbf{z}$ to discriminate class ω_k from the rest can be expressed in a measure of the probabilistic dependence:

$$\int_{-\infty}^{\infty} g(p(\mathbf{z}|\omega_k), p(\mathbf{z}))d\mathbf{z} \tag{6.24}$$

where the function $g(\cdot,\cdot)$ must have likewise properties as in (6.20). In order to incorporate all classes, a weighted sum of (6.24) is formed to get the final performance measure. As an example, the Chernoff measure now becomes:

$$\mathrm{J}_{C-dep}(s) = -\sum_{k=1}^{K} P(\omega_k) \int p^s(\mathbf{z}|\omega_k)p^{1-s}(\mathbf{z})d\mathbf{z} \tag{6.25}$$

Other dependence measures can be derived from the probabilistic distance measures in likewise manner.

A third family is founded on *information theory* and involves the posterior probabilities $P(\omega_k|\mathbf{z})$. An example is *Shannon's entropy measure*. For a given $\mathbf{z}$, the information of the true class associated with $\mathbf{z}$ is quantified by Shannon by *entropy*:

$$\mathrm{H}(\mathbf{z}) = -\sum_{k=1}^{K} P(\omega_k|\mathbf{z}) \log_2 P(\omega_k|\mathbf{z}) \tag{6.26}$$

Its expectation

$$\mathrm{J}_{SHANNON} = \mathrm{E}[\mathrm{H}(\mathbf{z})] = \int \mathrm{H}(\mathbf{z})p(\mathbf{z})d\mathbf{z} \tag{6.27}$$

is a performance measure suitable for feature selection and extraction.

6.2 FEATURE SELECTION

This section introduces the problem of selecting a subset from the N-dimensional measurement vector such that this subset is most suitable for classification. Such a subset is called a *feature set* and its elements are *features*. The problem is formalized as follows. Let $F(N) = \{z_n|n = 0, \ldots, N-1\}$ be the set with elements from the measurement vector $\mathbf{z}$. Furthermore, let $F_j(D) = \{y_d|d = 0, \ldots, D-1\}$ be a subset of

$F(N)$ consisting of $D < N$ elements taken from $\mathbf{z}$. For each element y_d there exists an element z_n such that $y_d = z_n$. The number of distinguishable subsets for a given D is:

$$q(D) = \binom{N}{D} = \frac{N!}{(N-D)!D!} \tag{6.28}$$

This quantity expresses the number of different combinations that can be made from N elements, each combination containing D elements and with no two combinations containing exactly the same D elements. We will adopt an enumeration of the index j according to: $j = 1, \ldots, q(D)$.

In Section 6.1 the concept of a performance measure has been introduced. These performance measures evaluate the appropriateness of a measurement vector for the classification task. Let $\mathrm{J}(F_j(D))$ be a performance measure related to the subset $F_j(D)$. The particular choice of $\mathrm{J}(.)$ depends on the problem at hand. For instance, in a multi-class problem, the interclass/intraclass distance $\mathrm{J}_{INTER/INTRA}(\cdot)$ could be useful; see (6.10).

The goal of feature selection is to find a subset $\hat{F}(D)$ with dimension D such that this subset outperforms all other subsets with dimension D:

$$\hat{F}(D) = F_i(D) \quad \text{with: } \mathrm{J}(F_i(D)) \geq \mathrm{J}(F_j(D)) \quad \text{for all} \quad j \in \{1, \ldots, q(D)\} \tag{6.29}$$

An exhaustive search for this particular subset would solve the problem of feature selection. However, there is one practical problem. How to accomplish the search process? An exhaustive search requires $q(D)$ evaluations of $\mathrm{J}(F_j(D))$. Even in a simple classification problem, this number is enormous. For instance, let us assume that $N = 20$ and $D = 10$. This gives about 2×10^5 different subsets. A doubling of the dimension to $N = 40$ requires about 10^9 evaluations of the criterion. Needless to say that even in problems with moderate complexity, an exhaustive search is out of the question.

Obviously, a more efficient search strategy is needed. For this, many options exist, but most of them are suboptimal (in the sense that they cannot guarantee that the subset with the best performance will be found). In the remaining part of this section, we will first consider a search strategy called 'branch-and-bound'. This strategy is one of the few that guarantees optimality (under some assumptions). Next, we continue with some of the much faster, but suboptimal strategies.

6.2.1 Branch-and-bound

The search process is accomplished systematically by means of a tree structure. See Figure 6.4. The tree consists of $N - D + 1$ levels. Each level is enumerated by a variable n with n varying from D up to N. A level consists of a number of nodes. A node i at level n corresponds to a subset $F_i(n)$. At the highest level, $n = N$, there is only one node corresponding to the full set $F(N)$. At the lowest level $n = D$ there are $q(D)$ nodes corresponding to the $q(D)$ subsets among which the solution must be found. Levels in between have a number of nodes that is less than or equal to $q(n)$. A node at level n is connected to one node at level $n + 1$ (except the node $F(N)$ at level N). In addition, each node at level n is connected to one or more nodes at level $n - 1$ (except the nodes at level D). In the example of Figure 6.4, $N = 6$ and $D = 2$.

A prerequisite of the *branch-and-bound* strategy is that the performance measure is a monotonically increasing function of the dimension D. The assumption behind it is that if we remove one element from a measurement vector, the performance can only become worse. In practice, this requirement is not always satisfied. If the training set is finite, problems in estimating large numbers of parameters may result in an actual performance increase when fewer measurements are used (see Figure 6.1).

The search process takes place from the highest level ($n = N$) by systematically traversing all levels until finally the lowest level ($n = D$) is reached. Suppose that one branch of the tree has been explored up to the lowest level, and suppose that the best performance measure found so far at level $n = D$ is $\hat{\mathrm{J}}$. Consider a node $F_i(l)$ (at a level $n > D$) which has not been explored as yet. Then, if $\mathrm{J}(F_i(n)) \leq \hat{\mathrm{J}}$, it is

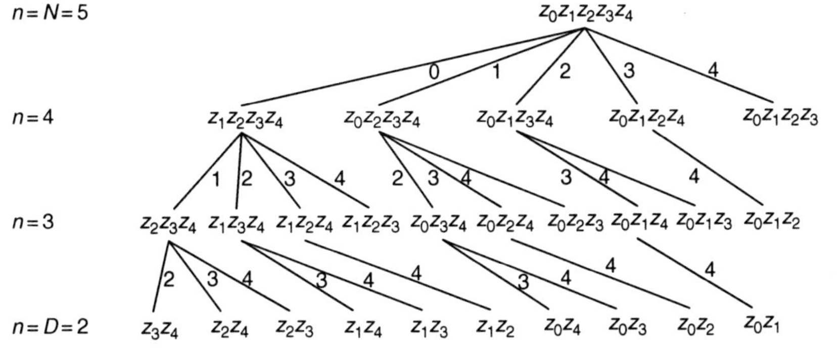

Figure 6.4 A top-down tree structure in behalf of feature selection

unnecessary to consider the nodes below $F_i(n)$. The reason is that the performance measure for these nodes can only become less than $J(F_i(n))$ and thus less than $\hat{J}$.

The following algorithm traverses the tree structure according to a *depth first search* with a backtrack mechanism. In this algorithm the best performance measure found so far at level $n = D$ is stored in a variable $\hat{J}$. Branches whose performance measure is bounded by $\hat{J}$ are skipped. Therefore, relative to exhaustive search the algorithm is much more computationally efficient. The variable S denotes a subset. Initially this subset is empty, and $\hat{J}$ is set to 0. As soon as a combination of D elements has been found that exceeds the current performance, this combination is stored in S and the corresponding measure in $\hat{J}$.

Algorithm 6.1: Branch-and-bound search
Input: a labelled training set on which a performance measure J() is defined.

1. Initiate: $\hat{J} = 0$ and $S = \phi$;
2. Explore-node($F_1(N)$);

Output: The maximum performance measure stored in $\hat{J}$ with the associated subset of $F_1(N)$ stored in S.

Procedure: Explore-node($F_i(n)$)

1. *If* ($J(F_i(n)) \leq \hat{J}$) *then return*;
2. *If* ($n = D$) *then*
   2.1. *If* ($J(F_i(n)) > \hat{J}$) *then*
      2.1.1. $\hat{J} = J(F_i(n))$;
      2.1.2. $S = F_i(n)$;
      2.1.3. *return*;
3. *For all* ($F_j(n-1) \subset F_i(n)$) *do* Explore-node($F_j(n-1)$);
4. *return*;

The algorithm is recursive. The procedure 'Explore-node()' explores the node given in its argument list. If the node is not a leaf, all its children nodes are explored by calling the procedure recursively. The first call is with the full set $F(N)$ as argument.

The algorithm listed above does not specify the exact structure of the tree. The structure follows from the specific implementation of the loop in step 3. This loop also controls the order in which branches are explored. Both aspects influence the computational efficiency of the algorithm. In Figure 6.4, the list of indices of elements that are deleted from a node to form the child nodes follows a specific pattern (see exercise 2). The indices of these elements are shown in the figure. Note that this tree is not minimal because some twigs have their leaves at a level higher than D. Of course, pruning these useless twigs in advance is computationally more efficient.

6.2.2 Suboptimal search

Although the branch-and-bound algorithm can save many calculations relative to an exhaustive search (especially for large values of $q(D)$), it may still require too much computational effort.

Another possible defect of branch-and-bound is the top-down search order. It starts with the full set of measurements and successively deletes elements. However, the assumption that the performance becomes worse as the number of features decreases holds true only in theory; see Figure 6.1. In practice, the finite training set may give rise to an overoptimistic view if the number of measurements is too large. Therefore, a bottom-up search order is preferable. The tree in Figure 6.5 is an example.

Among the many suboptimal methods, we mention a few. A simple method is *sequential forward selection* (SFS). The method starts at the bottom (the root of the tree) with an empty set and proceeds its way to the top (a leaf) without backtracking. At each level of the tree, SFS adds one feature to the current feature set by selecting from the remaining available measurements the element that yields a maximal increase of the performance. A disadvantage of the SFS is that once a feature is

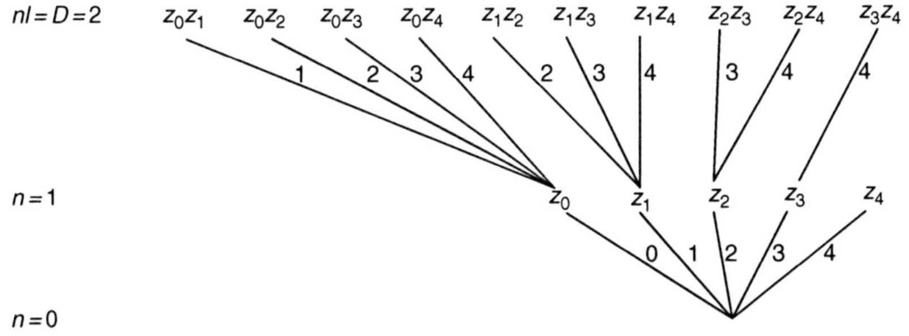

Figure 6.5 A bottom-up tree structure for feature selection

included in the selected sets, it cannot be removed, even if at a higher level in the tree, when more features are added, this feature would be less useful.

The SFS adds one feature at a time to the current set. An improvement is to add more than one, say l, features at a time. Suppose that at a given stage of the selection process we have a set $F_j(n)$ consisting of n features. Then, the next step is to expand this set with l features taken from the remaining $N - n$ measurements. For that we have $(N - n)!/((N - n - l)!\,l!)$ combinations which all must be checked to see which one is most profitable. This strategy is called the *generalized sequential forward selection* (GSFS(l)).

Both the SFS and the GSFS(l) lack a backtracking mechanism. Once a feature has been added, it cannot be undone. Therefore, a further improvement is to add l features at a time, and after that to dispose of some of the features from the set obtained so far. Hence, starting with a set of $F_j(n)$ features we select the combination of l remaining measurements that when added to $F_j(n)$ yields the best performance. From this expanded set, say $F_i(n + l)$, we select a subset of r features and remove it from $F_i(n + l)$ to obtain a set, say $F_k(n + l - r)$. The subset of r features is selected such that $F_k(n + l - r)$ has the best performance. This strategy is called '*Plus l – take away r selection*'.

Example 6.2 Character classification for license plate recognition
In traffic management, automated license plate recognition is useful for a number of tasks, e.g. speed checking and automated toll collection. The functions in a system for license plate recognition include image acquisition, vehicle detection, license plate localization, etc. Once the license plate has been localized in the image, it is partitioned such that each part of the image – a bitmap – holds exactly one character. The classes include all alphanumerical values. Therefore, the number of classes is 36. Figure 6.6(a) provides some examples of bitmaps obtained in this way.

The size of the bitmaps ranges from 15×6 up to 30×15. In order to fix the number of pixels, and also to normalize the position, scale and contrasts of the character within the bitmap, all characters are normalized to 15×11.

Using the raw image data as measurements, the number of measurements per object is $15 \times 11 = 165$. A labelled training set consisting of about 70 000 samples, i.e. about 2000 samples/class, is available. Comparing the number of measurements against the number of samples/class we conclude that overfitting is likely to occur.

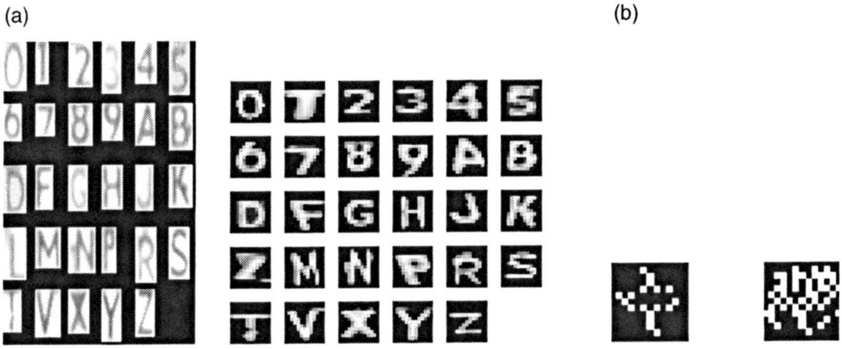

Figure 6.6 Character classification for license plate recognition. (a) Character sets from license plates, before and after normalization. (b) Selected features. The number of features is 18 and 50 respectively

A feature selection procedure, based on the 'plus l-take away r' method ($l = 3$, $r = 2$) and the inter/intra distance (Section 6.1.1) gives feature sets as depicted in Figure 6.6(b). Using a validation set consisting of about 50 000 samples, it was established that 50 features gives the minimal error rate. A number of features above 50 introduces overfitting. The pattern of 18 selected features, as shown in Figure 6.6(b), is one of the intermediate results that were obtained to get the optimal set with 50 features. It indicates which part of a bitmap is most important to recognize the character.

6.2.3 Implementation issues

PRTools offers a large range of feature selection methods. The evaluation criteria are implemented in the function `feateval`, and are basically all inter/intra cluster criteria. Additionally, a κ-nearest neighbour classification error is defined as a criterion. This will give a reliable estimate of the classification complexity of the reduced data set, but can be very computationally intensive. For larger data sets it is therefore recommended to use the simpler inter-intra-cluster measures.

PRTools also offers several search strategies, i.e. the branch-and-bound algorithm, plus-l-takeaway-r, forward selection and backward selection. Feature selection mappings can be found using the function `featselm`. The following listing is an example.

Listing 6.1
PRTools code for performing feature selection.

```
% Create a labeled dataset with 8 features, of which only 2
% are useful, and apply various feature selection methods
z=gendatd(200,8,3,3);

w=featselm(z, 'maha-s', 'forward',2);        % Forward selection
figure; clf; scatterd(z*w);
title(['forward: ' num2str(+w{2})]);
w=featselm(z,'maha-s', 'backward',2);        % Backward selection
figure; clf; scatterd(z*w);
title(['backward: ' num2str(+w{2})]);
w=featselm(z, 'maha-s', 'b&b',2);            % B&B selection
figure; clf; scatterd(z*w);
title(['b&b: ' num2str(+w{2})]);
```

The function `gendatd` creates a data set in which just the first two measurements are informative while all other measurements only contain noise (there the classes completely overlap). The listing shows three possible feature selection methods. All of them are able to retrieve the correct two features. The main difference is in the required computing time: finding two features out of eight is approached most efficiently by the forward selection method, while backward selection is the most inefficient.

6.3 LINEAR FEATURE EXTRACTION

Another approach to reduce the dimension of the measurement vector is to use a transformed space instead of the original measurement space. Suppose that $\mathbf{W}(.)$ is a transformation that maps the measurement space $\mathbb{R}^N$ onto a reduced space $\mathbb{R}^D$, $D \ll N$. Application of the transformation to a measurement vector yields a *feature vector* $\mathbf{y} \in \mathbb{R}^D$:

$$\mathbf{y} = \mathbf{W}(\mathbf{z}) \tag{6.30}$$

Classification is based on the feature vector rather than on the measurement vector; see Figure 6.7.

The advantage of feature extraction above feature selection is that no information from any of the elements of the measurement vector needs to be wasted. Furthermore, in some situations feature extraction is easier than feature selection. A disadvantage of feature extraction is that it requires the determination of some suitable transformation $\mathbf{W}()$. If the

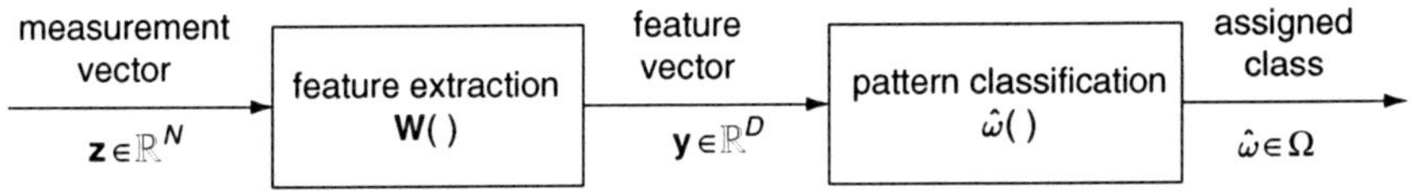

Figure 6.7 Feature extraction

transform chosen is too complex, the ability to generalize from a small data set will be poor. On the other hand, if the transform chosen is too simple, it may constrain the decision boundaries to a form which is inappropriate to discriminate between classes. Another disadvantage is that all measurements will be used, even if some of them are useless. This might be unnecessarily expensive.

This section discusses the design of linear feature extractors. The transformation W() is restricted to the class of linear operations. Such operations can be written as a matrix–vector product:

$$\mathbf{y} = \mathbf{W}\mathbf{z} \tag{6.31}$$

where $\mathbf{W}$ is a $D \times N$ matrix. The reason for the restriction is threefold. First, a linear feature extraction is computationally efficient. Second, in many classification problems – though not all – linear features are appropriate. Third, a restriction to linear operations facilitates the mathematical handling of the problem.

An illustration of the computational efficiency of linear feature extraction is the Gaussian case. If covariance matrices are unequal, the number of calculations is on the order of KN^2; see equation (2.20). Classification based on linear features requires about $DN + KD^2$ calculations. If D is very small compared with N, the extraction saves a large number of calculations.

The example of Gaussian densities is also well suited to illustrate the appropriateness of linear features. Clearly, if the covariance matrices are equal, then (2.25) shows that linear features are optimal. On the other hand, if the expectation vectors are equal and the discriminatory information is in the differences between the covariance matrices, linear feature extraction may still be appropriate. This is shown in the example of Figure 2.10(b) where the covariance matrices are eccentric, differing only in their orientations. However, in the example shown in Figure 2.10(a) (concentric circles) linear features seem to be inappropriate. In practical situations, the covariance matrices will often differ in both shape and orientations. Linear feature extraction is likely to lead to a reduction of the dimension, but this reduction may be less than what is feasible with nonlinear feature extraction.

Linear feature extraction may also improve the ability to generalize. If, in the Gaussian case with unequal covariance matrices, the number of samples in the training set is in the same order as that of the number of parameters, KN^2, overfitting is likely to occur. But linear feature extraction – provided that $D << N$ – helps to improve the generalization ability.

We assume the availability of a training set and a suitable performance measure J(). The design of a feature extraction method boils down to finding the matrix $\mathbf{W}$ that – for the given training set – optimizes the performance measure.

The performance measure of a feature vector $\mathbf{y} = \mathbf{W}\mathbf{z}$ is denoted by $J(\mathbf{y})$ or $J(\mathbf{W}\mathbf{z})$. With this notation, the optimal feature extraction is:

$$\mathbf{W} = \underset{\mathbf{W}}{\operatorname{argmax}}\{J(\mathbf{W}\mathbf{z})\} \tag{6.32}$$

Under the condition that $J(\mathbf{W}\mathbf{z})$ is continuously differentiable in $\mathbf{W}$, the solution of (6.32) must satisfy:

$$\frac{\partial J(\mathbf{W}\mathbf{z})}{\partial \mathbf{W}} = 0 \tag{6.33}$$

Finding a solution of either (6.32) or (6.33) gives us the optimal linear feature extraction. The search can be accomplished numerically using the training set. Alternatively, the search can also be done analytically assuming parameterized conditional densities. Substitution of estimated parameters (using the training set) gives the matrix $\mathbf{W}$.

In the remaining part of this section, the last approach is worked out for two particular cases: feature extraction for two-class problems with Gaussian densities and feature extraction for multi-class problems based on the inter/intra distance measure. The former case will be based on the Bhattacharyya distance.

6.3.1 Feature extraction based on the Bhattacharyya distance with Gaussian distributions

In the two-class case with Gaussian conditional densities a suitable performance measure is the Bhattacharyya distance. In equation (6.19) J_{BHAT} implicitly gives the Bhattacharyya distance as a function of the parameters of the Gaussian densities of the measurement vector $\mathbf{z}$. These parameters are the conditional expectations $\boldsymbol{\mu}_k$ and covariance matrices $\mathbf{C}_k$. Substitution of $\mathbf{y} = \mathbf{W}\mathbf{z}$ gives the expectation vectors and covariance matrices of the feature vector, i.e. $\mathbf{W}\boldsymbol{\mu}_k$ and $\mathbf{W}\mathbf{C}_k\mathbf{W}^T$, respectively. For the sake of brevity, let $\mathbf{m}$ be the difference between expectations of $\mathbf{z}$:

$$\mathbf{m} = \boldsymbol{\mu}_1 - \boldsymbol{\mu}_2 \tag{6.34}$$

Then, substitution of $\mathbf{m}$, $\mathbf{W}\boldsymbol{\mu}_k$ and $\mathbf{W}\mathbf{C}_k\mathbf{W}^T$ in (6.19) gives the Bhattacharyya distance of the feature vector:

$$\begin{aligned} J_{BHAT}(\mathbf{Wz}) = & \frac{1}{4}(\mathbf{Wm})^T\left[\mathbf{W}\mathbf{C}_1\mathbf{W}^T + \mathbf{W}\mathbf{C}_2\mathbf{W}^T\right]^{-1}\mathbf{Wm} \\ & + \frac{1}{2}\ln\left[\frac{|\mathbf{W}\mathbf{C}_1\mathbf{W}^T + \mathbf{W}\mathbf{C}_2\mathbf{W}^T|}{2^D\sqrt{|\mathbf{W}\mathbf{C}_1\mathbf{W}^T||\mathbf{W}\mathbf{C}_2\mathbf{W}^T|}}\right] \end{aligned} \tag{6.35}$$

The first term corresponds to the discriminatory information of the expectation vectors; the second term to the discriminatory information of covariance matrices.

Equation (6.35) is in such a form that an analytic solution of (6.32) is not feasible. However, if one of the two terms in (6.35) is dominant, a solution close to the optimal one is attainable. We consider the two extreme situations first.

Equal covariance matrices

In the case where the conditional covariance matrices are equal, i.e. $\mathbf{C} = \mathbf{C}_1 = \mathbf{C}_2$, we have already seen that classification based on the Mahalanobis distance (see (2.41)) is optimal. In fact, this classification uses a $1 \times N$ dimensional feature extraction matrix given by:

$$\mathbf{W} = \mathbf{m}^T\mathbf{C}^{-1} \tag{6.36}$$

To prove that this equation also maximizes the Bhattacharyya distance is left as an exercise for the reader.

Equal expectation vectors

If the expectation vectors are equal, $\mathbf{m} = 0$, the first term in (6.35) vanishes and the second term simplifies to:

$$J_{BHAT}(\mathbf{Wz}) = \frac{1}{2}\ln\left[\frac{|\mathbf{W}\mathbf{C}_1\mathbf{W}^T + \mathbf{W}\mathbf{C}_2\mathbf{W}^T|}{2^D\sqrt{|\mathbf{W}\mathbf{C}_1\mathbf{W}^T||\mathbf{W}\mathbf{C}_2\mathbf{W}^T|}}\right] \tag{6.37}$$

The $D \times N$ matrix $\mathbf{W}$ that maximizes the Bhattacharyya distance can be derived as follows.

The first step is to apply a whitening operation (Appendix C.3.1) on $\mathbf{z}$ with respect to class ω_1. This is accomplished by the linear operation $\Lambda^{-1/2}\mathbf{V}^T\mathbf{z}$. The matrices $\mathbf{V}$ and Λ follow from factorization of the covariance matrix: $\mathbf{C}_1 = \mathbf{V}\Lambda\mathbf{V}^T$. $\mathbf{V}$ is an orthogonal matrix consisting of the eigenvectors of $\mathbf{C}_1$. Λ is the diagonal matrix containing the corresponding eigenvalues. The process is illustrated in Figure 6.8. The figure shows

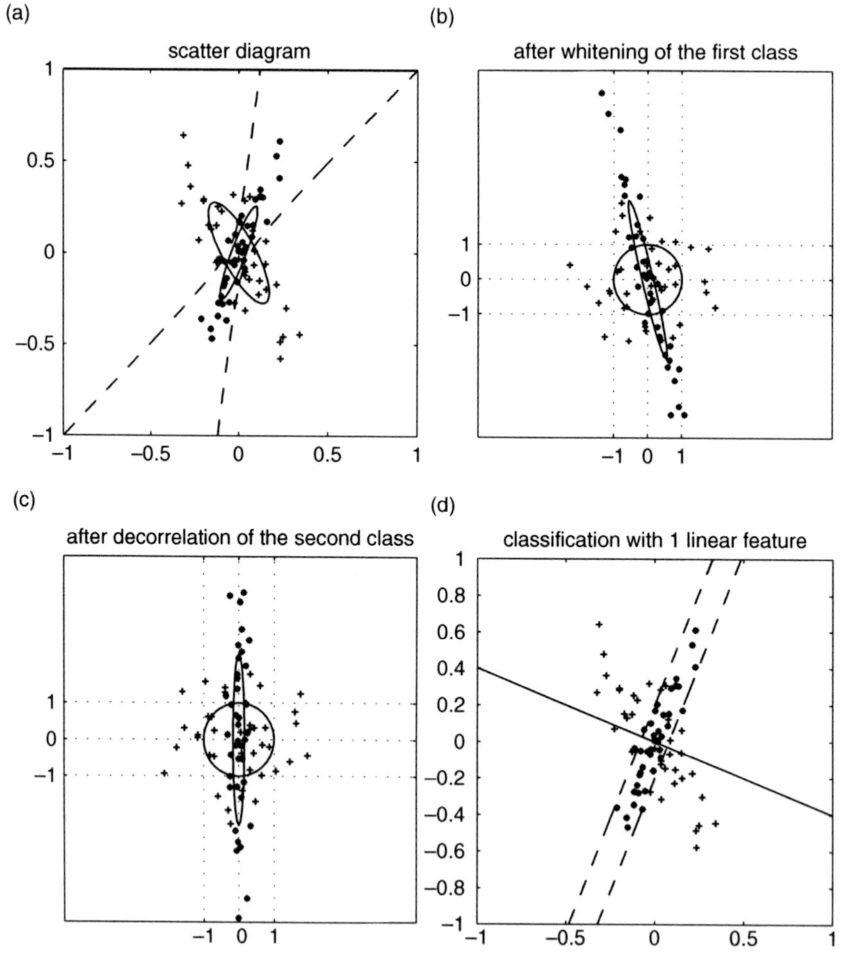

Figure 6.8 Linear feature extraction with equal expectation vectors. (a) Covariance matrices with decision function. (b) Whitening of ω_1 samples. (c) Decorrelation of ω_2 samples. (d) Decision function based on one linear feature

a two-dimensional measurement space with samples from two classes. The covariance matrices of both classes are depicted as ellipses. Figure 6.8(b) shows the result of the operation $\Lambda^{-1/2}\mathbf{V}^T$. The operation $\mathbf{V}^T$ corresponds to a rotation of the coordinate system such that the ellipse of class ω_1 lines up with the axes. The operation $\Lambda^{-1/2}$ corresponds to a scaling of the axes such that the ellipse of ω_1 degenerates into a circle. The figure also shows the resulting covariance matrix belonging to class ω_2.

The result of the operation $\Lambda^{-1/2}\mathbf{V}^T$ on $\mathbf{z}$ is that the covariance matrix associated with ω_1 becomes $\mathbf{I}$ and the covariance matrix associated with ω_2 becomes $\Lambda^{-1/2}\mathbf{V}^T\mathbf{C}_2\mathbf{V}\Lambda^{-1/2}$. The Bhattacharyya distance in the transformed domain is:

$$J_{BHAT}(\Lambda^{-\frac{1}{2}}\mathbf{V}^T\mathbf{z}) = \frac{1}{2}\ln\left[\frac{|\mathbf{I}+\Lambda^{-\frac{1}{2}}\mathbf{V}^T\mathbf{C}_2\mathbf{V}\Lambda^{-\frac{1}{2}}|}{2^N\sqrt{|\Lambda^{-\frac{1}{2}}\mathbf{V}^T\mathbf{C}_2\mathbf{V}\Lambda^{-\frac{1}{2}}|}}\right] \tag{6.38}$$

The second step consists of decorrelation with respect to ω_2. Suppose that $\mathbf{U}$ and Γ are matrices containing the eigenvectors and eigenvalues of the covariance matrix $\Lambda^{-1/2}\mathbf{V}^T\mathbf{C}_2\mathbf{V}\Lambda^{-1/2}$. Then, the operation $\mathbf{U}^T\Lambda^{-1/2}\mathbf{V}^T$ decorrelates the covariance matrix with respect to class ω_2. The covariance matrices belonging to the classes ω_1 and ω_2 transform into $\mathbf{U}^T\mathbf{I}\mathbf{U} = \mathbf{I}$ and Γ, respectively. Figure 6.8(c) illustrates the decorrelation. Note that the covariance matrix of ω_1 (being white) is not affected by the orthonormal operation $\mathbf{U}^T$.

The matrix Γ is a diagonal matrix. The diagonal elements are denoted $\gamma_i = \Gamma_{i,i}$. In the transformed domain $\mathbf{U}^T\Lambda^{-1/2}\mathbf{V}^T\mathbf{z}$, the Bhattacharyya distance is:

$$J_{BHAT}(\mathbf{U}^T\Lambda^{-\frac{1}{2}}\mathbf{V}^T\mathbf{z}) = \frac{1}{2}\ln\left[\frac{|\mathbf{I}+\Gamma|}{2^N\sqrt{|\Gamma|}}\right] = \frac{1}{2}\sum_{i=0}^{N-1}\ln\frac{1}{2}\left(\sqrt{\gamma_i}+\frac{1}{\sqrt{\gamma_i}}\right) \tag{6.39}$$

The expression shows that in the transformed domain the contribution to the Bhattacharyya distance of any element is independent. The contribution of the i-th element is:

$$\frac{1}{2}\ln\frac{1}{2}\left(\sqrt{\gamma_i}+\frac{1}{\sqrt{\gamma_i}}\right) \tag{6.40}$$

Therefore, if in the transformed domain D elements have to be selected, the selection with maximum Bhattacharyya distance is found as the set with the largest contributions. If the elements are sorted according to:

$$\sqrt{\gamma_0} + \frac{1}{\sqrt{\gamma_0}} \geq \sqrt{\gamma_1} + \frac{1}{\sqrt{\gamma_1}} \geq \cdots \geq \sqrt{\gamma_{N-1}} + \frac{1}{\sqrt{\gamma_{N-1}}} \tag{6.41}$$

then the first D elements are the ones with optimal Bhattacharyya distance. Let $\mathbf{U}_D$ be an $N \times D$ submatrix of $\mathbf{U}$ containing the D corresponding eigenvectors of $\Lambda^{-1/2}\mathbf{V}^T\mathbf{C}_2\mathbf{V}\Lambda^{-1/2}$. The optimal linear feature extractor is:

$$\mathbf{W} = \mathbf{U}_D^T \Lambda^{-\frac{1}{2}} \mathbf{V}^T \tag{6.42}$$

and the corresponding Bhattacharyya distance is:

$$J_{BHAT}(\mathbf{Wz}) = \frac{1}{2} \sum_{i=0}^{D-1} \ln \frac{1}{2} \left(\sqrt{\gamma_i} + \frac{1}{\sqrt{\gamma_i}} \right) \tag{6.43}$$

Figure 6.8(d) shows the decision function following from linear feature extraction backprojected in the two-dimensional measurement space. Here, the linear feature extraction reduces the measurement space to a one-dimensional feature space. Application of Bayes classification in this space is equivalent to a decision function in the measurement space defined by two linear, parallel decision boundaries. In fact, the feature extraction is a projection onto a line orthogonal to these decision boundaries.

The general Gaussian case

If both the expectation vectors and the covariance matrices depend on the classes, an analytic solution of the optimal linear extraction problem is not easy. A suboptimal method, i.e. a method that hopefully yields a reasonable solution without the guarantee that the optimal solution will be found, is the one that seeks features in the subspace defined by the differences in covariance matrices. For that, we use the same simultaneous decorrelation technique as in the previous section. The Bhattacharyya distance in the transformed domain is:

$$J_{BHAT}(\mathbf{U}^T \Lambda^{-\frac{1}{2}} \mathbf{V}^T \mathbf{z}) = \frac{1}{4} \sum_{i=0}^{N-1} \frac{d_i^2}{1 + \gamma_i} + \frac{1}{2} \sum_{i=0}^{N-1} \ln \frac{1}{2} \left(\sqrt{\gamma_i} + \frac{1}{\sqrt{\gamma_i}} \right) \tag{6.44}$$

where d_i are the elements of the transformed difference of expectation, i.e. $\mathbf{d} = \mathbf{U}^T\Lambda^{-1/2}\mathbf{V}^T\mathbf{m}$.

Equation (6.44) shows that in the transformed space the optimal features are the ones with the largest contributions, i.e. with the largest

$$\frac{1}{4}\frac{d_i^2}{1+\gamma_i} + \frac{1}{2}\ln\frac{1}{2}\left(\sqrt{\gamma_i} + \frac{1}{\sqrt{\gamma_i}}\right) \tag{6.45}$$

The extraction method is useful especially when most class information is contained in the covariance matrices. If this is not the case, then the results must be considered cautiously. Features that are appropriate for differences in covariance matrices are not necessarily also appropriate for differences in expectation vectors.

Listing 6.2
PRTools code for calculating a Bhattacharryya distance feature extractor.

```
z=gendatl([200 200],0.2);          % Generate a dataset
J=bhatm(z,0);                      % Calculate criterion values
figure; clf; plot(J, 'r.-');       % and plot them
w=bhatm(z,1);                      % Extract one feature
figure; clf; scatterd(z);          % Plot original data
figure; clf; scatterd(z*w);        % Plot mapped data
```

6.3.2 Feature extraction based on inter/intra class distance

The inter/intra class distance, as discussed in Section 6.1.1, is another performance measure that may yield suitable feature extractors. The starting point is the performance measure given in the space defined by $\mathbf{y} = \Lambda^{-1/2}\mathbf{V}^T\mathbf{z}$. Here, Λ is a diagonal matrix containing the eigenvalues of $\mathbf{S}_w$, and $\mathbf{V}$ a unitary matrix containing the corresponding eigenvectors. In the transformed domain the performance measure is expressed as (6.10):

$$J_{INTER/INTRA} = trace(\Lambda^{-\frac{1}{2}}\mathbf{V}^T\mathbf{S}_b\mathbf{V}\Lambda^{-\frac{1}{2}}) \tag{6.46}$$

A further simplification occurs when a second unitary transform is applied. The purpose of this transform is to decorrelate the between-scatter matrix. Suppose that Γ is a diagonal matrix whose diagonal elements $\gamma_i = \Gamma_{i,i}$ are the eigenvalues of the transformed between-scatter matrix $\Lambda^{-1/2}\mathbf{V}^T\mathbf{S}_b\mathbf{V}\Lambda^{-1/2}$. Let $\mathbf{U}$ be a unitary matrix containing the eigenvectors corresponding to Γ. Then, in the transformed domain defined by:

$$\mathbf{y} = \mathbf{U}^T\Lambda^{-\frac{1}{2}}\mathbf{V}^T\mathbf{z} \tag{6.47}$$

the performance measure becomes:

$$J_{INTER/INTRA} = trace(\Gamma) = \sum_{i=0}^{N-1} \gamma_i \tag{6.48}$$

The operation $\mathbf{U}^T$ corresponds to a rotation of the coordinate system such that the between-scatter matrix lines up with the axes. Figure 6.9 illustrates this.

The merit of (6.48) is that the contributions of the elements add up independently. Therefore, in the space defined by $\mathbf{y} = \mathbf{U}^T\Lambda^{-1/2}\mathbf{V}^T\mathbf{z}$ it is easy to select the best combination of D elements. It suffices to determine the D elements from $\mathbf{y}$ whose eigenvalues γ_i are largest. Suppose that the eigenvalues are sorted according to $\gamma_i \geq \gamma_{i+1}$, and that the eigenvectors

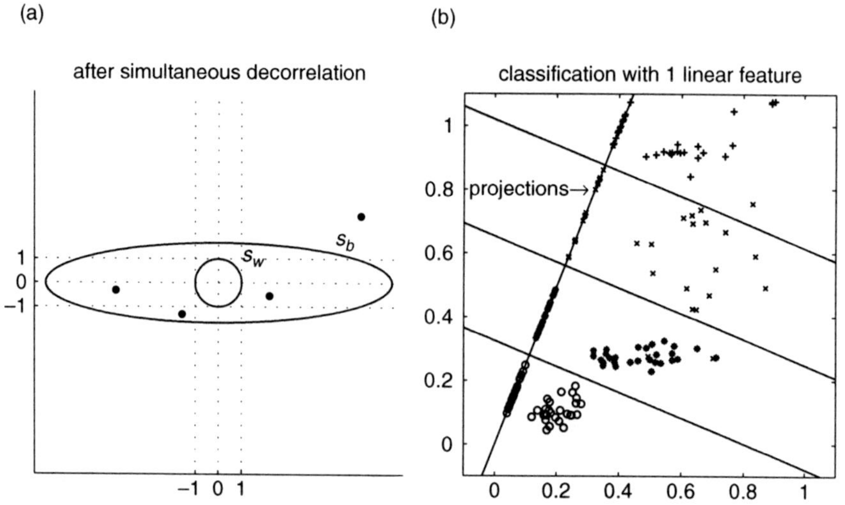

Figure 6.9 Feature extraction based on the interclass/intraclass distance (see Figure 6.2). (a) The within and between scatters after simultaneous decorrelation. (b) Linear feature extraction

corresponding to the D largest eigenvalues are collected in $\mathbf{U}_D$, being an $N \times D$ submatrix of $\mathbf{U}$. Then, the linear feature extraction becomes:

$$\mathbf{W} = \mathbf{U}_D^T \Lambda^{-1/2} \mathbf{V}^T \tag{6.49}$$

The feature space defined by $\mathbf{y} = \mathbf{Wz}$ can be thought of as a linear subspace of the measurement space. This subspace is spanned by the D row vectors in $\mathbf{W}$. The performance measure associated with this feature space is:

$$J_{INTER/INTRA}(\mathbf{Wz}) = \sum_{i=0}^{D-1} \gamma_i \tag{6.50}$$

Example 6.3 Feature extraction based on inter/intra distance
Figure 6.9(a) shows the within-scattering and between-scattering of Example 6.1 after simultaneous decorrelation. The within-scattering has been whitened. After that, the between-scattering is rotated such that its ellipse is aligned with the axes. In this figure, it is easy to see which axis is the most important. The eigenvalues of the between-scatter matrix are $\gamma_0 = 56.3$ and $\gamma_1 = 2.8$, respectively. Hence, omitting the second feature does not deteriorate the performance much.

The feature extraction itself can be regarded as an orthogonal projection of samples on this subspace. Therefore, decision boundaries defined in the feature space correspond to hyperplanes orthogonal to the linear subspace, i.e. planes satisfying equations of the type $\mathbf{Wz} = constant$.

A characteristic of linear feature extraction based on $J_{INTER/INTRA}$ is that the dimension of the feature space found will not exceed $K - 1$, where K is the number of classes. This follows from expression (6.7), which shows that $\mathbf{S}_b$ is the sum of K outer products of vectors (of which one vector linearly depends on the others). Therefore, the rank of $\mathbf{S}_b$ cannot exceed $K - 1$. Consequently, the number of nonzero eigenvalues of $\mathbf{S}_b$ cannot exceed $K - 1$ either. Another way to put this into words is that the K conditional means $\boldsymbol{\mu}_k$ span a $(K - 1)$ dimensional linear subspace in $\mathbb{R}^N$. Since the basic assumption of the inter/intra distance is that within-scattering does not convey any class information, any feature extractor based on that distance can only find class information within that subspace.

Example 6.4 License plate recognition (continued)
In the license plate application, discussed in Example 6.2, the measurement space (consisting of 15×11 bitmaps) is too large with respect to the size of the training set. Linear feature extraction based on maximization of the inter/intra distance reduces this space to at most $D_{max} = K - 1 = 35$ features. Figure 6.10(a) shows how the inter/intra distance depends on D. It can be seen that at about $D = 24$ the distance has almost reached its maximum. Therefore, a reduction to 24 features is possible without losing much information.

Figure 6.10(b) is a graphical representation of the transformation matrix $\mathbf{W}$. The matrix is 24×165. Each row of the matrix serves as a vector on which the measurement vector is projected. Therefore, each row can be depicted as a 15×11 image. The figure is obtained by means of MATLAB code that is similar to Listing 6.3.

Listing 6.3
PRTools code for creating a linear feature extractor based on maximization of the inter/intra distance. The function for calculating the mapping is `fisherm`. The result is an affine mapping, i.e. a mapping of the type $\mathbf{Wz} + \mathbf{b}$. The additive term $\mathbf{b}$ shifts the overall mean of the features to the origin. In this example, the measurement vectors come directly from bitmaps. Therefore, the mapping can be visualized by images. The listing also shows how `fisherm` can be used to get a cumulative plot of $J_{INTER/INTRA}$, as depicted in Figure 6.10(a). The precise call to `fisherm` is discussed in more detail in Exercise 5.

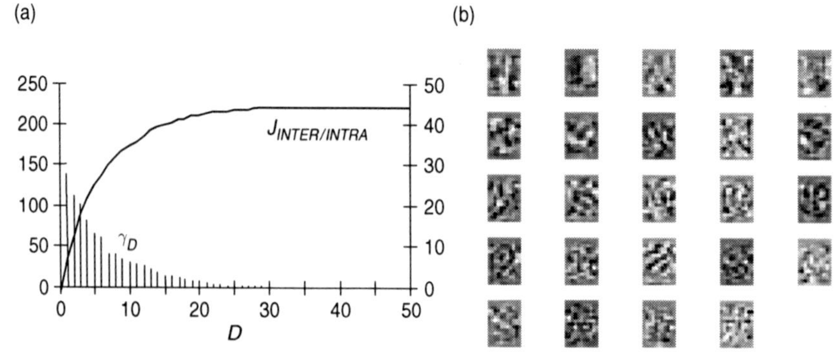

Figure 6.10 Feature extraction in the license plate application. (a) The inter/intra distance as a function of D. (b) First 24 eigenvectors in $\mathbf{W}$ depicted as 15×11 pixel images

```
load license_plates.mat           % Load dataset
figure; clf; show(z);             % Display it
J=fisherm(z,0);                   % Calculate criterion values
figure; clf; plot(J, 'r.-');      % and plot them
w=fisherm(z,24,0.9);              % Calculate the feature extractor
figure; clf; show(w);             % Show the mappings as images
```

The two-class case, Fisher's linear discriminant

$$\begin{aligned} S_b &= \frac{1}{N_S}\left(N_1(\hat{\mu}_1 - \hat{\mu})(\hat{\mu}_1 - \hat{\mu})^T + N_2(\hat{\mu}_2 - \hat{\mu})(\hat{\mu}_2 - \hat{\mu})^T\right) \\ &= \alpha(\hat{\mu}_1 - \hat{\mu}_2)(\hat{\mu}_1 - \hat{\mu}_2)^T \end{aligned} \tag{6.51}$$

where α is a constant that depends on N_1 and N_2. In the transformed space, S_b becomes $\alpha\Lambda^{-1/2}V^T(\hat{\mu}_1 - \hat{\mu}_2)(\hat{\mu}_1 - \hat{\mu}_2)^T V\Lambda^{-1/2}$. This matrix has only one eigenvector with a nonzero eigenvalue $\gamma_0 = \alpha(\hat{\mu}_1 - \hat{\mu}_2)^T S_b^{-1}\ (\hat{\mu}_1 - \hat{\mu}_2)$. The corresponding eigenvector is $\mathbf{u} = \Lambda^{-1/2}V^T(\hat{\mu}_1 - \hat{\mu}_2)$. With that, the feature extractor evolves into (see (6.49)):

$$\begin{aligned} W &= \mathbf{u}^T\Lambda^{-\frac{1}{2}}V^T \\ &= \left(\Lambda^{-\frac{1}{2}}V^T(\hat{\mu}_1 - \hat{\mu}_2)\right)^T \Lambda^{-\frac{1}{2}}V^T \\ &= (\hat{\mu}_1 - \hat{\mu}_2)^T S_w^{-1} \end{aligned} \tag{6.52}$$

This solution – known as *Fisher's linear discriminant* (Fisher, 1936) – is similar to the Bayes classifier for Gaussian random vectors with class-independent covariance matrices; i.e. the Mahalanobis distance classifier. The difference with expression (2.41) is that the true covariance matrix and the expectation vectors have been replaced with estimates from the training set. The multi-class solution (6.49) can be regarded as a generalization of Fisher's linear discriminant.

6.4 REFERENCES

Bhattacharyya, A., On a measure of divergence between two statistical populations defined by their probability distributions. *Bulletin of the Calcutta Mathematical Society*, **35**, 99–110, 1943.

Chernoff, H., A measure of asymptotic efficiency for tests of a hypothesis based on the sum of observations. *Annals of Mathematical Statistics*, **23**, 493–507, 1952.

Devijver, P.A. and Kittler, J., *Pattern Recognition, a Statistical Approach*. Prentice-Hall, London, UK, 1982.

Fisher, R.A., The use of multiple measurements in taxonomic problems. *Annals of Eugenics*, 7, 179–88, 1936.

6.5 EXERCISES

1. Prove equation (6.4). (*)
   Hint: use $\mathbf{z}_{k,n} - \mathbf{z}_{l,m} = (\mathbf{z}_{k,n} - \hat{\boldsymbol{\mu}}_k) + (\hat{\boldsymbol{\mu}}_k - \hat{\boldsymbol{\mu}}) + (\hat{\boldsymbol{\mu}} - \hat{\boldsymbol{\mu}}_l) + (\hat{\boldsymbol{\mu}}_l - \mathbf{z}_{l,m})$.
2. Develop an algorithm that creates a tree structure like in Figure 6.4. Can you adapt that algorithm such that the tree becomes minimal (thus, without the superfluous twigs)? (0)
3. Under what circumstances would it be advisable to use forward selection, or plus-l-takeaway-r selection with $l > r$? And backward selection, or plus-l-takeaway-r selection with $l < r$? (0)
4. Prove that $\mathbf{W} = \mathbf{m}^T\mathbf{C}^{-1}$ is the feature extractor that maximizes the Bhattacharyyaa distance in the two-class Gaussian case with equal covariance matrices. (**)
5. In Listing 6.3, `fisherm` is called with `0.9` as its third argument. Why do you think this is used? Try the same routine, but leave out the third argument (i.e. use `w = fisherm(z, 24)`). Can you explain what you see now? (*)
6. Find an alternative method of preventing the singularities you saw in Exercise 6.5. Will the results be the same as those found using the original Listing 6.3? (**)
7. What is the danger of optimizing the parameters of the feature extraction or selection stage, such as the number of features to retain, on the training set? How could you circumvent this? (0)

7

Unsupervised Learning

In the previous chapter we discussed methods for reducing the dimension of the measurement space in order to decrease the cost of classification and to improve the ability to generalize. In these procedures it was assumed that for all training objects, class labels were available. In many practical applications, however, the training objects are not labelled, or only a small fraction of them are labelled. In these cases it can be worth while to let the data speak for itself. The structure in the data will have to be discovered without the help of additional labels.

An example is colour-based pixel classification. In video-based surveillance and safety applications, for instance, one of the tasks is to track the foreground pixels. Foreground pixels are pixels belonging to the objects of interest, e.g. cars on a parking place. The RGB representation of a pixel can be used to decide whether a pixel belongs to the foreground or not. However, the colours of neither the foreground nor the background are known in advance. Unsupervised training methods can help to decide which pixels of the image belong to the background, and which to the foreground.

Another example is an insurance company which might want to know if typical groups of customers exist, such that it can offer suitable insurance packages to each of these groups. The information provided by an insurance expert may introduce a significant bias. Unsupervised methods can then help to discover additional structures in the data.

In unsupervised methods, we wish to transform and reduce the data such that a specific characteristic in the data is highlighted. In this

Classification, Parameter Estimation and State Estimation: An Engineering Approach using MATLAB
F. van der Heijden, R.P.W. Duin, D. de Ridder and D.M.J. Tax
 ISBN: 0-470-09013-8

chapter we will discuss two main characteristics which can be explored: the subspace structure of data and its clustering characteristics. The first tries to summarize the objects using a smaller number of features than the original number of measurements; the second tries to summarize the data set using a smaller number of objects than the original number. Subspace structure is often interesting for visualization purposes. The human visual system is highly capable of finding and interpreting structure in 2D and 3D graphical representations of data. When higher dimensional data is available, a transformation to 2D or 3D might facilitate its interpretation by humans. Clustering serves a similar purpose, interpretation, but also data reduction. When very large amounts of data are available, it is often more efficient to work with cluster representatives instead of the whole data set. In Section 7.1 we will treat feature reduction, in Section 7.2 we discuss clustering.

7.1 FEATURE REDUCTION

The most popular unsupervised feature reduction method is principal component analysis (Jolliffe, 1986). This will be discussed in Section 7.1.1. One of the drawbacks of this method is that it is a linear method, so nonlinear structures in the data cannot be modelled. In Section 7.1.2 multi-dimensional scaling is introduced, which is a nonlinear feature reduction method.

7.1.1 Principal component analysis

The purpose of *principal component analysis (PCA)* is to transform a high dimensional measurement vector $\mathbf{z}$ to a much lower dimensional feature vector $\mathbf{y}$ by means of an operation:

$$\mathbf{y} = \mathbf{W}_D(\mathbf{z} - \bar{\mathbf{z}}) \tag{7.1}$$

such that $\mathbf{z}$ can be reconstructed accurately from $\mathbf{y}$.

$\bar{\mathbf{z}}$ is the expectation of the random vector $\mathbf{z}$. It is constant for all realizations of $\mathbf{z}$. Without loss of generality, we can assume that $\bar{\mathbf{z}} = \mathbf{0}$ because we can always introduce a new measurement vector, $\tilde{\mathbf{z}} = \mathbf{z} - \bar{\mathbf{z}}$, and apply the analysis to this vector. Hence, under that assumption, we have:

$$\mathbf{y} = \mathbf{W}_D\mathbf{z} \tag{7.2}$$

The $D \times N$ matrix $\mathbf{W}_D$ transforms the N-dimensional measurement space to a D-dimensional feature space. Ideally, the transform is such that $\mathbf{y}$ is a good representation of $\mathbf{z}$ despite of the lower dimension of $\mathbf{y}$. This objective is strived for by selecting $\mathbf{W}_D$ such that an (unbiased) linear MMSE estimate[1] $\hat{\mathbf{z}}_{lMMSE}$ for $\mathbf{z}$ based on $\mathbf{y}$ yields a minimum mean square error (see Section 3.1.5):

$$\mathbf{W}_D = \underset{\mathbf{W}}{\arg\min}\left\{\mathrm{E}\left[\|\hat{\mathbf{z}}_{lMMSE}(\mathbf{y}) - \mathbf{z}\|^2\right]\right\} \quad \text{with} \quad \mathbf{y} = \mathbf{W}\mathbf{z} \tag{7.3}$$

It is easy to see that this objective function does not provide a unique solution for $\mathbf{W}_D$. If a minimum is reached for some $\mathbf{W}_D$, then any matrix $\mathbf{A}\mathbf{W}_D$ is another solution with the same minimum (provided that $\mathbf{A}$ is invertible) as the transformation $\mathbf{A}$ will be inverted by the linear MMSE procedure. For uniqueness, we add two requirements. First, we require that the information carried in the individual elements of $\mathbf{y}$ add up individually. With that we mean that if $\mathbf{y}$ is the optimal D dimensional representation of $\mathbf{z}$, then the optimal $D-1$ dimensional representation is obtained from $\mathbf{y}$, simply by deleting its least informative element. Usually, the elements of $\mathbf{y}$ are sorted in decreasing order of importance, so that the least informative element is always the last element. With this convention, the matrix $\mathbf{W}_{D-1}$ is obtained from $\mathbf{W}_D$ simply by deleting the last row of $\mathbf{W}_D$.

The requirement leads to the conclusion that the elements of $\mathbf{y}$ must be uncorrelated. If not, then the least informative element would still carry predictive information about the other elements of $\mathbf{y}$ which conflicts with our requirement. Hence, the covariance matrix $\mathbf{C}_\mathbf{y}$ of $\mathbf{y}$ must be a diagonal matrix, say Λ. If $\mathbf{C}_\mathbf{z}$ is the covariance matrix of $\mathbf{z}$, then:

$$\mathbf{C}_\mathbf{y} = \mathbf{W}_D\mathbf{C}_\mathbf{z}\mathbf{W}_D^T = \Lambda_D \tag{7.4}$$

For $D = N$ it follows that $\mathbf{C}_\mathbf{z}\mathbf{W}_N^T = \mathbf{W}_N^T\Lambda_N$ because $\mathbf{W}_N$ is an invertible matrix (in fact, an orthogonal matrix) and $\mathbf{W}_N^T\mathbf{W}_N$ must be a diagonal matrix (see Appendix B.5 and C.3.2). As Λ_N is a diagonal matrix, the columns of $\mathbf{W}_N^T$ must be eigenvectors of $\mathbf{C}_\mathbf{z}$. The diagonal elements of Λ_N are the corresponding eigenvalues.

The solution is still not unique because each element of $\mathbf{y}$ can be scaled individually without changing the minimum. Therefore, the second requirement is that each column of $\mathbf{W}_N^T$ has unit length. Since the

[1] Since $\mathbf{z}$ and $\mathbf{y}$ are zero mean, the unbiased linear MMSE estimator coincides with the linear MMSE estimator.

eigenvectors are orthogonal, this requirement is fulfilled by $\mathbf{W}_N\mathbf{W}_N^T = \mathbf{I}$ with $\mathbf{I}$ the $N \times N$ unit matrix. With that, $\mathbf{W}_N$ establishes a rotation on $\mathbf{z}$. The rows of the matrix $\mathbf{W}_N$, i.e. the eigenvectors, must be sorted such that the eigenvalues form a non-ascending sequence. For arbitrary D, the matrix $\mathbf{W}_D$ is constructed from $\mathbf{W}_N$ by deleting the last $N - D$ rows.

The interpretation of this is as follows (see Figure 7.1). The operator $\mathbf{W}_N$ performs a rotation on $\mathbf{z}$ such that its orthonormal basis aligns with the principal axes of the ellipsoid associated with the covariance matrix of $\mathbf{z}$. The coefficients of this new representation of $\mathbf{z}$ are called the *principal components*. The axes of the ellipsoid point in the *principal directions*. The MMSE approximation of $\mathbf{z}$ using only D coefficients is obtained by nullifying the principal components with least variances. Hence, if the principal components are ordered according to their variances, the elements of $\mathbf{y}$ are formed by the first D principal components. The linear MMSE estimate is:

$$\hat{\mathbf{z}}_{lMMSE}(\mathbf{y}) = \mathbf{W}_D^T\mathbf{y} = \mathbf{W}_D^T\mathbf{W}_D\mathbf{z}.$$

PCA can be used as a first step to reduce the dimension of the measurement space. In practice, the covariance matrix is often replaced by the sample covariance estimated from a training set. See Section 5.2.3.

Unfortunately, PCA can be counter-productive for classification and estimation problems. The PCA criterion selects a subspace of the feature space, such that the variance of $\mathbf{z}$ is conserved as much as possible. However, this is done regardless of the classes. A subspace with large variance is not necessarily one in which classes are well separated.

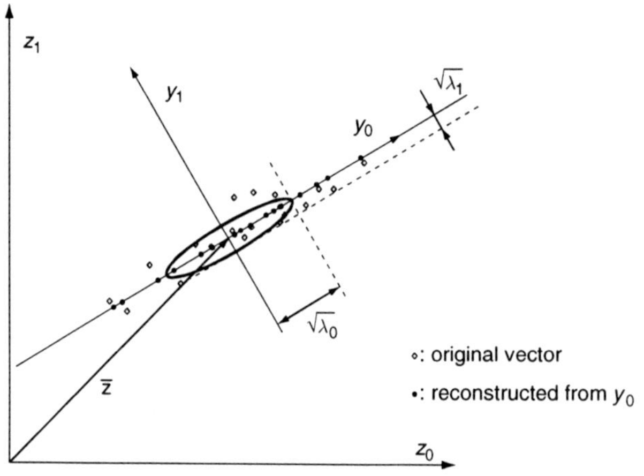

Figure 7.1 Principal component analysis

A second drawback of PCA is that the results are not invariant to the particular choice of the physical units of the measurements. Each element of $\mathbf{z}$ is individually scaled according to its unit in which it is expressed. Changing a unit from, for instance, m (meter) to μm (micrometer) may result in dramatic changes of the principal directions. Usually this phenomenon is circumvented by scaling the vector $\mathbf{z}$ such that the numerical values of the elements all have unit variance.

Example 7.1 Image compression
Since PCA aims at the reduction of a measurement vector in such a way that it can be reconstructed accurately, the technique is suitable for image compression. Figure 7.2 shows the original image. The image plane is partitioned into 32×32 regions each having 8×8 pixels.

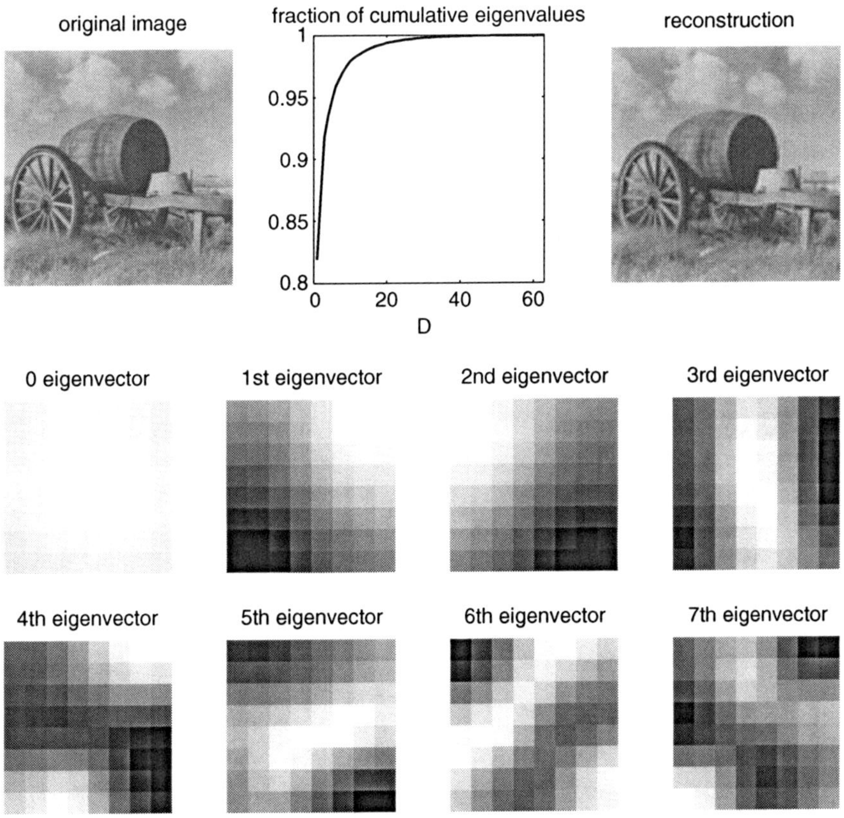

Figure 7.2 Application of PCA to image compression

The grey levels of the pixels are stacked in 64 dimensional vectors $\mathbf{z}_n$ from which the sample mean and the sample covariance matrix have been estimated. The figure also shows the fraction of the cumulative eigenvalues, i.e. $\sum_{i=0}^{D-1} \gamma_i / trace(\Gamma)$ where γ_i is the i-th diagonal element of Γ. The first eight eigenvectors are depicted as 8×8 images. The reconstruction based on these eight eigenvectors is also shown. The compression is about 96%. PRTools code for this PCA compression algorithm is given in Listing 7.1.

Listing 7.1

PRTools code for finding a set of PCA basis vectors for image compression and producing output similar to Figure 7.2.

```
im=double(imread('car.tif'));       % Load image
figure; clf; imshow(im, [0 255]);   % Display image
x=im2col(im,[8 8],'distinct');      % Extract 8x8 windows
z=dataset(x');                      % Create dataset
z.featsize=[8 8];                   % Indicate window size

% Plot fraction of cumulative eigenvalues
v=pca(z,0); figure; clf; plot(v);

% Find 8D PCA mapping and show basis vectors
w=pca(z,8); figure; clf; show(w)

% Reconstruct image and display it
z_hat=z*w*w';
im_hat=col2im (+z_hat',[8 8], size(im),'distinct');
figure; clf; imshow (im_hat, [0 255]);
```

The original image blocks are converted into vectors, a data set is created and a PCA base is found. This base is then shown and used to reconstruct the image. Note that this listing also uses the MATLAB Image Processing toolbox, specifically the functions `im2col` and `col2im`. The function `pca` is used for the actual analysis.

7.1.2 Multi-dimensional scaling

Principal component analysis is limited to finding linear combinations of features to map the original data to. This is sufficient for many applications, such as discarding low-variance directions in which only noise is suspected to be present. However, if the goal of a mapping is to inspect data in a two or three dimensional projection, PCA might discard too much information. For example, it can project two distinct groups in the data on top of each other; or it might not show whether data is distributed nonlinearly in the original high dimensional measurement space.

To retain such information, a nonlinear mapping method is needed. As there are many possible criteria to fit a mapping, there are many different methods available. Here we will discuss just one: *multidimensional scaling* (*MDS*) (Kruskal and Wish, 1977) or *Sammon mapping* (Sammon Jr, 1969). The self-organizing map, discussed in Section 7.2.5, can be considered as another nonlinear projection method.

MDS is based on the idea that the projection should preserve the distances between objects as well as possible. Given a data set containing N dimensional measurement vectors (or feature vectors) $\mathbf{z}_i$ $i = 1, \ldots, N_S$, we try to find new D dimensional vectors $\mathbf{y}_i$ $i = 1, \ldots N_S$ according to this criterion. Usually, $D \ll N$; if the goal is to visualize the data, we choose $D = 2$ or $D = 3$. If δ_{ij} denotes the known distance between objects $\mathbf{z}_i$ and $\mathbf{z}_j$, and d_{ij} denotes the distance between projected objects $\mathbf{y}_i$ and $\mathbf{y}_j$, then distances can be preserved as well as possible by placing the $\mathbf{y}_i$ such that the *stress* measure

$$E_S = \frac{1}{\sum_{i=1}^{N_S} \sum_{j=i+1}^{N_S} \delta_{ij}^2} \sum_{i=1}^{N_S} \sum_{j=i+1}^{N_S} \left(\delta_{ij} - d_{ij}\right)^2 \tag{7.5}$$

is minimized. To do so, we take the derivative of (7.5) with respect to the objects $\mathbf{y}$. It is not hard to derive that, when Euclidean distances are used, then

$$\frac{\partial E_S}{\partial \mathbf{y}_i} = -\frac{2}{\sum_{j=1}^{N_S} \sum_{k=j+1}^{N_S} \delta_{jk}^2} \sum_{\substack{j=1 \\ j \neq i}}^{N_S} \frac{\delta_{ij} - d_{ij}}{d_{ij}} (\mathbf{y}_i - \mathbf{y}_j) \tag{7.6}$$

There is no closed-form solution setting this derivative to zero, but a gradient descent algorithm can be applied to minimize (7.5). The MDS algorithm then becomes:

Algorithm 7.1: Multi-dimensional scaling

1. *Initialization*: Randomly choose an initial configuration of the projected objects $\mathbf{y}^{(0)}$. Alternatively, initialize $\mathbf{y}^{(0)}$ by projecting the original data set on its first D principal components. Set $t = 0$.

2. *Gradient descent*

- For each object $\mathbf{y}_i^{(t)}$, calculate the gradient according to (7.6).
- Update: $\mathbf{y}_i^{(t+1)} = \mathbf{y}_i^{(t)} - \alpha \partial E_S / \partial \mathbf{y}_i^{(t)}$, where α is a learning rate.
- As long as E_S significantly decreases, set $t = t + 1$ and go to step 2.

Figures 7.3(a) to (c) show examples of two-dimensional MDS mapping. The data set, given in Table 7.1, consists of the geodesic distances of 13 world cities. These distances can only be fully brought in accordance with the true three-dimensional geographical positions of the cities if the spherical surface of the earth is accounted for. Nevertheless, MDS has found two-dimensional mappings that resemble the usual Mercator projection of the earth surface on the tangent plane at the North Pole.

Since distances are invariant to translation, rotation and mirroring, MDS can result in arbitrarily shifted, rotated and mirrored mappings.

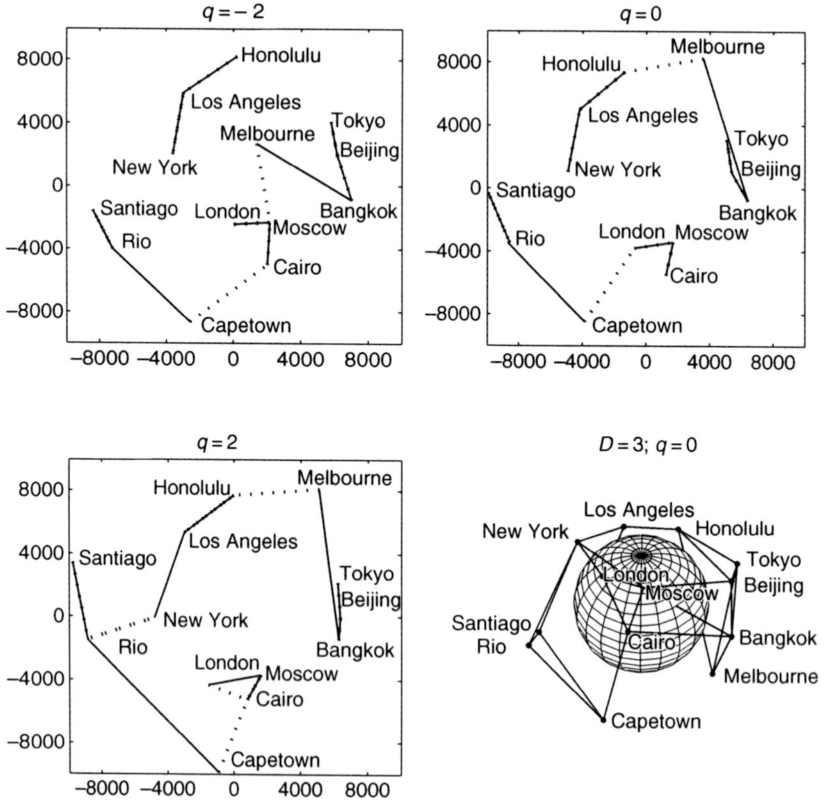

Figure 7.3 MDS applied to a matrix of geodesic distances of world cities

Table 7.1 Distance matrix of 13 cities in (km)

0	3290	7280	10100	10600	9540	13300	7350	7060	13900	16100	17700	4640	Bangkok
	0	7460	12900	8150	8090	10100	9190	5790	11000	17300	19000	2130	Beijing
		0	7390	14000	3380	12100	14000	2810	8960	9950	12800	9500	Cairo
			0	18500	9670	16100	10300	10100	12600	6080	7970	14800	Capetown
				0	11600	4120	8880	11200	8260	13300	11000	6220	Honolulu
					0	8760	16900	2450	5550	9290	11700	9560	London
						0	12700	9740	3930	10100	9010	8830	Los Angeles
							0	14400	16600	13200	11200	8200	Melbourne
								0	7510	11500	14100	7470	Moscow
									0	7770	8260	10800	New York
										0	2730	18600	Rio
											0	17200	Santiago
												0	Tokyo

The mapped objects in Figures 7.3 have been rotated such that New York and Beijing lie on a horizontal line with New York on the left. Furthermore, the vertical axis is mirrored such that London is situated below the line New York–Beijing.

The reason for the preferred projection on the tangent plane near the North Pole is that most cities in the list are situated in the Northern hemisphere. The exceptions are Santiago, Rio, Capetown and Melbourne. The distances for just these cities are least preserved. For instance, the true geodesic distance between Capetown and Melbourne is about 10 000 (km), but they are mapped opposite to each other with a distance of about 18 000 (km).

For specific applications, it might be fruitful to focus more on local structure or on global structure. A way of doing so is by using the more general stress measure, where an additional free parameter q is introduced:

$$E_S^q = \frac{1}{\sum_{i=1}^{N_S} \sum_{j=i+1}^{N_S} \delta_{ij}^{(q+2)}} \sum_{i=1}^{N_S} \sum_{j=i+1}^{N_S} \delta_{ij}^q (\delta_{ij} - d_{ij})^2 \tag{7.7}$$

For $q = 0$, (7.7) is equal to (7.5). However, as q is decreased, the $(\delta_{ij} - d_{ij})^2$ term remains constant but the δ_{ij}^q term will weigh small distances heavier than large ones. In other words, local distance preservation is emphasized, whereas global distance preservation is less important. This is demonstrated in Figure 7.3(a) to (c), where $q = -2$, $q = 0$ and $q = +2$ have been used, respectively. Conversely, if q is increased, the resulting stress measure will emphasize preserving global distances over local ones. When the stress measure is used with $q = 1$, the resulting mapping is also called a *Sammon mapping*.

The effects of using different values of q are demonstrated in Table 7.2. Here, for each city in the list, the nearest and the furthest cities are given. For each pair, the true geodesic distance from Table 7.1 is shown together with the corresponding distances in the MDS mappings. Clearly, for short distances, $q = -2$ is more accurate. For long distances, $q = +2$ is favourable. In Figure 7.3(a) to (c) the nearest cities (according to the geodesic distances) are indicated by solid thin lines. In addition, the nearest cities according to the MDS maps are indicated by dotted thick lines, Clearly, if $q = -2$, the nearest neighbours are best preserved. This is expected because the relations between nearest neighbours are best preserved if the local structure is best preserved.

Figure 7.3 also shows a three-dimensional MDS mapping of the world cities. For clarity, a sphere is also included. MDS is able to find a 2D to 3D mapping such that the resulting map resembles the true geographic position on the earth surface. The positions are not exactly on the surface of a sphere because the input of the mapping is a set of geodesic distances, while the output map measures distances according to a three-dimensional Euclidean metric.

The MDS algorithm has the same problems any gradient-based method has: it is slow and it cannot be guaranteed to converge to a global optimum. Another problem the MDS algorithm in particular suffers from is the fact that the number of distances in a data set grows quadratically with N_S. If the data sets are very large, then we need to select a subset of the points to map in order to make the application tractable. Finally, a major problem specific to MDS is that there is no clear way of finding the projection $\mathbf{y}_{new}$ of a new, previously unseen point $\mathbf{z}_{new}$. The only way of making sure that the new configuration $\{\mathbf{y}_i, i = 1, \ldots N_S; \mathbf{y}_{new}\}$ minimizes (7.5) is to recalculate the entire MDS mapping. In practice, this is infeasible. A crude approximation is to use triangulation: to map down to D dimensions, search for the $D + 1$ nearest neighbours of $\mathbf{z}_{new}$ among the training points $\mathbf{z}_i$. A mapping $\mathbf{y}_{new}$ for $\mathbf{z}_{new}$ can then be found by preserving the distances to the projections of these neighbours exactly. This method is fast, but inaccurate: because it uses nearest neighbours, it will not give a smooth interpolation of originally mapped points. A different, but more complicated solution is to use an MDS mapping to train a neural network to predict the $\mathbf{y}_i$ for given $\mathbf{z}_i$. This will give smoother mappings, but brings with it all the problems inherent in neural network training. An example of how to use MDS in PRTools is given in Listing 7.2.

Table 7.2 Results of MDS mapping

	Nearest cities					Furthest cities				
	City	Geodesic distance	Differences between geodesic and MDS distance			City	Geodesic distance	Differences between geodesic and MDS distance		
			$q=-2$	$q=0$	$q=+2$			$q=-2$	$q=0$	$q=+2$
Bangkok	Beijing	3290	213	1156	1928	Santiago	17700	−146	−52	−18
Beijing	Tokyo	2130	−8	37	−168	Santiago	19000	1797	2307	1468
Cairo	Moscow	2810	104	626	983	Melbourne	14000	6268	−19	−237
Capetown	Rio	6080	−1066	−1164	−5750	Honolulu	18500	1310	2387	864
Honolulu	Los Angeles	4120	−225	183	−62	Capetown	18500	1310	2387	864
London	Moscow	2450	−39	−221	−1277	Melbourne	16900	11528	3988	2315
Los Angeles	New York	3930	−49	−124	−1924	Capetown	16100	1509	2605	670
Melbourne	Bangkok	7350	−75	−2196	−2327	London	16900	11528	3988	2315
Moscow	London	2450	−39	−221	−1277	Melbourne	14400	9320	2563	1952
New York	Los Angeles	3930	−49	−124	−1924	Melbourne	16600	10794	4878	2559
Rio	Santiago	2730	−72	−701	−1658	Tokyo	18600	1518	2296	1643
Santiago	Rio	2730	−72	−701	−1658	Beijing	19000	1797	2307	1468
Tokyo	Beijing	2130	−8	37	−168	Rio	18600	1518	2296	1643

Listing 7.2
PRTools code for performing an MDS mapping.

```
load worldcities;                      % Load dataset D
options.q=2;
w=mds(D,2,options);                    % Map to 2D with q=2
figure; clf; scatterd(D*w,'both');     % Plot projections
```

7.2 CLUSTERING

Instead of reducing the number of features, we now focus on reducing the number of objects in the data set. The aim is to detect 'natural' clusters in the data, i.e. clusters which agree with our human interpretation of the data. Unfortunately, it is very hard to define what a natural cluster is. In most cases, a cluster is defined as a subset of objects for which the resemblance between the objects within the subset is larger than the resemblance with other objects in other subsets (clusters).

This immediately introduces the next problem: how is the resemblance between objects defined? The most important cue for the resemblance of two objects is the distance between the objects, i.e. their dissimilarity. In most cases the Euclidean distance between objects is used as a dissimilarity measure, but there are many other possibilities. The L_p norm is well known (see Appendix A.1.1 and A.2):

$$d_p(\mathbf{z}_i, \mathbf{z}_j) = \left(\sum_{n=1}^{N} (z_{i,n} - z_{j,n})^p \right)^{\frac{1}{p}} \tag{7.8}$$

The *cosine distance* uses the angle between two vectors as a dissimilarity measure. It is often used in the automatic clustering of text documents:

$$d(\mathbf{z}_i, \mathbf{z}_j) = 1 - \frac{\mathbf{z}_i^T \mathbf{z}_j}{||\mathbf{z}_i||_2 ||\mathbf{z}_j||_2} \tag{7.9}$$

In PRTools the basic distance computation is implemented in the function `proxm`. Several methods for computing distances (and similarities) are defined. Next to the two basic distances mentioned above, also some similarity measures are defined, like the inner product between vectors, and the Gaussian kernel. The function is implemented as a mapping. When a mapping **w** is trained on some data **z**, the application y∗w then computes the distance between any vector in **z** and any vector in **y**. For the squared Euclidean distances a short-cut function `distm` is defined.

Listing 7.3
PRTools code for defining and applying a proximity mapping.

```
z=gendatb(3);              % Create some train data
y=gendats(5);              % and some test data
w=proxm(z,'d',2);          % Squared Euclidean distance to z
D=y*w;                     % 5 x 3 distance matrix
D=distm(y,z);              % The same 5 x 3 distance matrix
w=proxm(z,'o');            % Cosine distance to z
D=y*w;                     % New 5 x 3 distance matrix
```

The distance between objects should reflect the important structures in the data set. It is assumed in all clustering algorithms that distances between objects are informative. This means that when objects are close in the feature space, they should also resemble each other in the real world. When the distances are not defined sensibly, and remote objects in the feature space correspond to similar real-world objects, no clustering algorithm will be able to give acceptable results without extra information from the user. In these sections it will therefore be assumed that the features are scaled such that the distances between objects are informative. Note that a cluster does not necessarily correspond directly to a class. A class can consist of multiple clusters, or multiple classes may form a single cluster (and will therefore probably be hard to discriminate between).

By the fact that clustering is unsupervised, it is very hard to evaluate a clustering result. Different clustering methods will yield a different set of clusters, and the user has to decide which clustering is to be preferred. A quantitative measure of the quality of the clustering is the average distance of the objects to their respective cluster centre. Assume that the objects $\mathbf{z}_i$ ($i = 1, \ldots, N_S$) are clustered in K clusters, C_k ($k = 1, \ldots, K$) with cluster centre $\boldsymbol{\mu}_k$ and to each of the clusters N_k objects are assigned:

$$\mathrm{J} = \frac{1}{N_S} \sum_{k=1}^{K} \frac{1}{N_k} \sum_{\mathbf{z} \in C_k} (\mathbf{z} - \boldsymbol{\mu}_k)^2 \tag{7.10}$$

Other criteria, such as the ones defined in Chapter 5 for supervised learning, can also be applied.

With these error criteria, different clustering results can be compared provided that K is kept constant. Clustering results found for varying K cannot be compared. The pitfall of using criteria like (7.10) is that the optimal number of clusters is $K = N_S$, i.e. the case in which each object belongs to its own cluster. In the average-distance criterion it will result in the trivial solution: $J = 0$.

The choice of the number of clusters K is a very fundamental problem. In some applications, an expected number of clusters is known beforehand. Using this number for the clustering does not always yield optimal results for all types of clustering methods: it might happen that, due to noise, a local minimum is reached. Using a slightly larger number of clusters is therefore sometimes preferred. In most cases the number of clusters to look for is one of the research questions in the investigation of a data set. Often, the data is repeatedly clustered using a range of values of K, and the clustering criterion values are compared. When a significant decrease in a criterion value appears, a 'natural' clustering is probably found. Unfortunately, in practice, it is very hard to objectively say when a significant drop occurs. True automatic optimization of the number of clusters is possible in just a very few situations, when the cluster shapes are given.

In the coming sections we will discuss four methods for performing a clustering: hierarchical clustering, K-means clustering, mixtures of Gaussians and finally self-organizing maps.

7.2.1 Hierarchical clustering

The basic idea of hierarchical clustering (Johnson, 1967) is to collect objects into clusters by combining the closest objects and clusters to larger clusters until all objects are in one cluster. An important advantage is that the objects are not just placed in K distinct groups, but are placed in a hierarchy of clusters. This gives more information about the structure in the data set, and shows which clusters are similar or dissimilar. This makes it possible to detect subclusters in large clusters, or to detect outlier clusters.

Figure 7.4 provides an example. The distance matrix of the 13 world cities given in Table 7.1 are used to find a hierarchical cluster structure. The figure shows that at the highest level the cities in the southern hemisphere are separated from the ones in the northern part. At a distance level of about 5000 km we see the following clusters: 'East Asiatic cities', 'European/Arabic cities', 'North American cities', 'African city', 'South American cities', 'Australian city'.

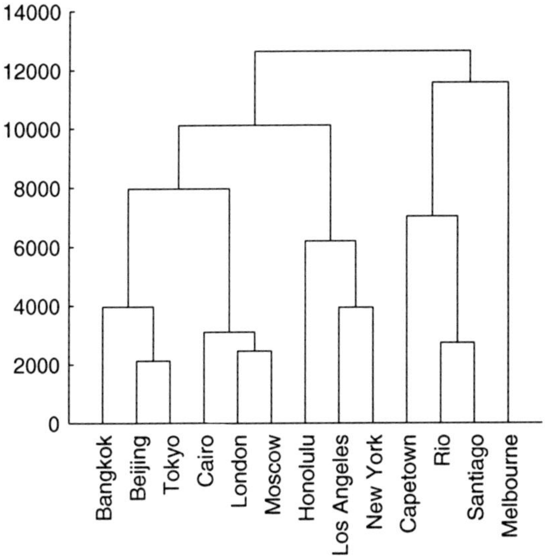

Figure 7.4 Hierarchical clustering of the data in Table 7.1

Given a set of N_S objects $\mathbf{z}_i$ to be clustered, and an $N_S \times N_S$ distance matrix between these objects, the hierarchical clustering involves the following steps:

Algorithm 7.2: Hierarchical clustering

1. Assign each object to its own cluster, resulting in N_S clusters, each containing just one object. The initial distances between all clusters are therefore just the distances between all objects.
2. Find the closest pair of clusters and merge them into a single cluster, so that the number of clusters reduces by one.
3. Compute the distance between the new cluster and each of the old clusters, where the distance between two clusters can be defined in a number of ways (see below).
4. Repeat steps 2 and 3 until all items are clustered into a single cluster of size N_S, or until a predefined number of clusters K is achieved.

In step 3 it is possible to compute the distances between clusters in several different ways. We can distinguish single-link clustering, average-link clustering and complete-link clustering. In single-link clustering, the distance between two clusters C_i and C_j is defined as the shortest distance from any object in one cluster to any object in the other cluster:

$$d_{sl}(C_i, C_j) = \min_{\mathbf{x} \in C_i, \mathbf{y} \in C_j} \|\mathbf{x} - \mathbf{y}\|^2 \tag{7.11}$$

For average-link clustering, the minimum operator is replaced by the average distance, and for the complete-link clustering it is replaced by the maximum operator.

In Figure 7.5 the difference between single link and complete link is shown for a very small toy data set ($N_S = 6$). At the start of the clustering, both single-link (left) and complete-link clustering (right) combine the same objects to clusters. When larger clusters appear, in the lower row, different objects are combined. The different definitions for the inter-cluster distances result in different characteristic cluster shapes. For single-link clustering, the clusters tend to become long and spidery, while for complete-link clustering the clusters become very compact.

The user now has to decide on what the most suitable number of clusters is. This can be based on a *dendrogram*. The dendrogram shows at which distances the objects or clusters are grouped together. Examples

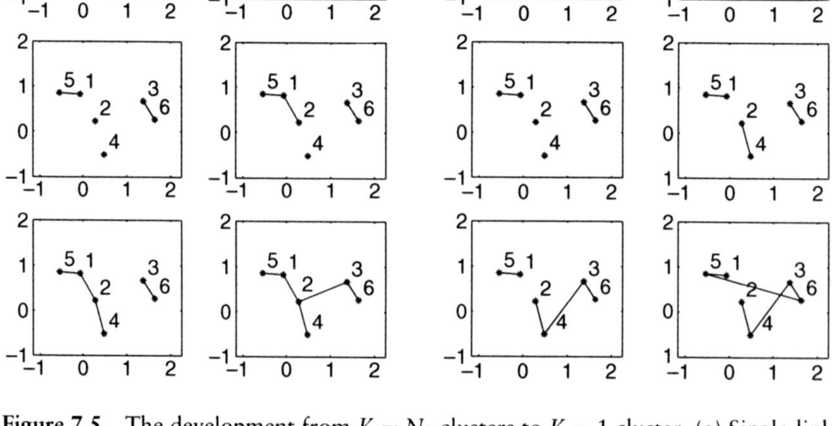

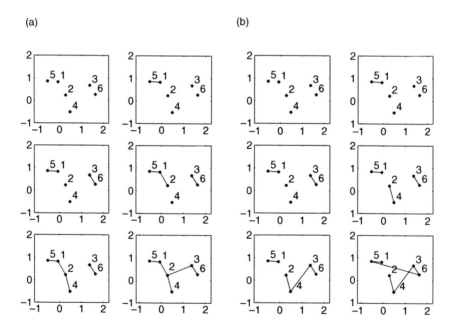

Figure 7.5 The development from $K = N_S$ clusters to $K = 1$ cluster. (a) Single-link clustering. (b) Complete-link clustering

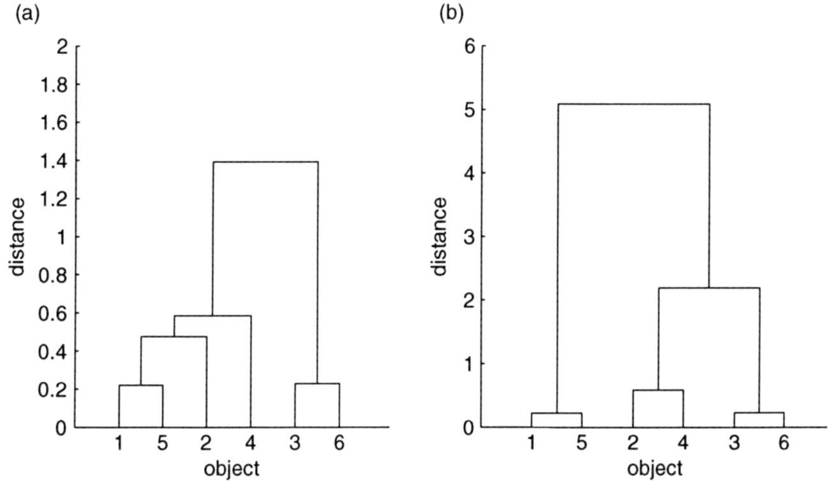

Figure 7.6 Hierarchical clustering with two different clustering types. (a) Single-link clustering. (b) Complete-link clustering

for single- and complete-link clustering are shown in Figure 7.6. At smaller distances, pairs of single objects are combined, at higher distances complete clusters. When there is a large gap in the distances, as can be seen in the single-link dendrogram, it is an indication that the two clusters are far apart. Cutting the dendrogram at height 1.0 will then result in a 'natural' clustering, consisting of two clusters. In many practical cases, the cut is not obvious to define, and the user has to guess an appropriate number of clusters.

Note that this clustering is obtained using a fixed data set. When new objects become available, there is no straightforward way to include it in an existing clustering. In these cases, the clustering will have to be constructed from the beginning using the complete data set.

In PRTools, it is simple to construct a hierarchical clustering; Listing 7.4 shows an example. Note that the clustering operates on a distance matrix rather than the data set. A distance matrix can be obtained with the function `distm`.

Listing 7.4
PRTools code for obtaining a hierarchical clustering.

```
z=gendats(5);                    % Generate some data
figure; clf; scatterd(z);        % and plot it
dendr=hclust(distm(z),'s');      % Single link clustering
figure; clf; plotdg(dendr);      % Plot the dendrogram
```

7.2.2 K-means clustering

K-means clustering (Bishop, 1995) differs in two important aspects from hierarchical clustering. First, K-means clustering requires the number of clusters K beforehand. Second, it is not hierarchical, instead it partitions the data set into K disjoint subsets. Again the clustering is basically determined by the distances between objects. The K-means algorithm has the following structure:

Algorithm 7.3: K-means clustering

1. Assign each object randomly to one of the clusters $k = 1, \ldots K$.
2. Compute the means of each of the clusters:

$$\boldsymbol{\mu}_k = \frac{1}{N_k} \sum_{\mathbf{z}_i \in C_k} \mathbf{z}_i \tag{7.12}$$

3. Reassign each object $\mathbf{z}_i$ to the cluster with the closest mean $\boldsymbol{\mu}_k$.
4. Return to step 2 until the means of the clusters do not change anymore.

The initialization step can be adapted to speed up the convergence. Instead of randomly labelling the data, K randomly chosen objects are taken as cluster means. Then the procedure enters the loop in step 3. Note again that the procedure depends on distances, in this case between the objects $\mathbf{z}_i$ and the means $\boldsymbol{\mu}_k$. Scaling the feature space will here also change the final clustering result. An advantage of K-means clustering is that it is very easy to implement. On the other hand, it is unstable: running the procedure several times will give several different results. Depending on the random initialization, the algorithm will converge to different (local) minima. In particular when a high number of clusters is requested, it often happens that some clusters do not gain sufficient support and are ignored. The effective number of clusters then becomes much less than K.

In Figure 7.7 the result of a K-means clustering is shown for a simple 2D data set. The means are indicated by the circles. At the start of the optimization, i.e. at the start of the trajectory, each mean coincides with a data object. After 10 iteration steps, the solution converged. The result of the last iteration is indicated by '×'. In this case, the number of clusters in the data and the predefined $K = 3$ match. A fairly stable solution is found.

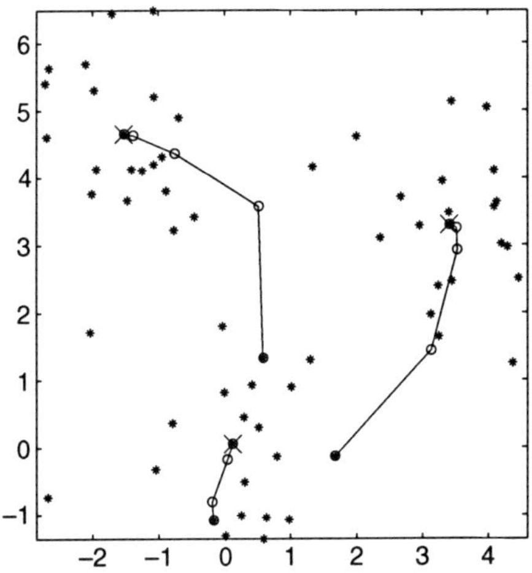

Figure 7.7 The development of the cluster means during 10 update steps of the K-means algorithm

Example 7.2 Classification of mechanical parts, K-means clustering
Two results of the K-means algorithm applied to the unlabelled data set of Figure 5.1(b) are shown in Figure 7.8. The algorithm is called with $K = 4$. The differences between the two results are solely caused by the different realizations of the random initialization of the algorithm. The first result, Figure 7.8(a), is more or less correct (compare with the correct labelling as given in Figure 5.1(a). Unfortunately, the result in Figure 7.8(b) indicates that this success is not reproducible.

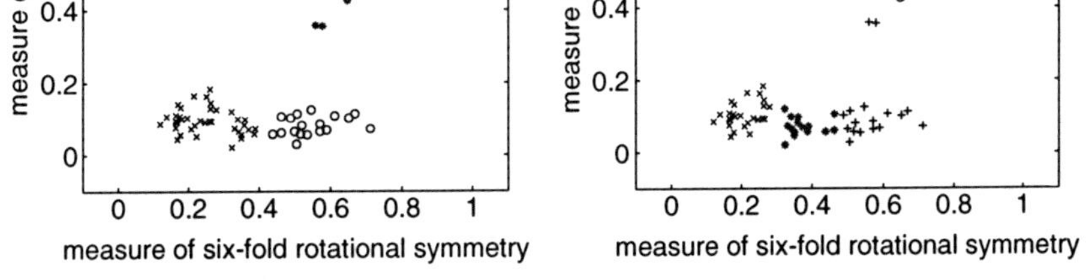

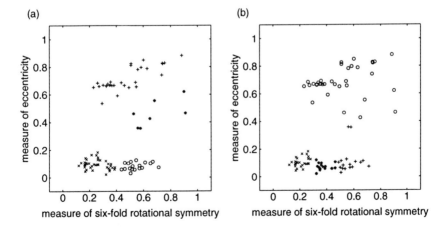

Figure 7.8 Two results of K-means clustering applied to the 'mechanical parts' data set

The algorithm is implemented in PRTools with the function `kmeans`. See Listing 7.5.

Listing 7.5

PRTools code for fitting and plotting a K-means clustering, with $K = 4$.

```
load nutsbolts_unlabeled;  % Load the data set z
lab=kmeans(z,4);           % Perform k-means clustering
y=dataset(z,lab);          % Label by cluster assignment
figure; clf; scatterd(y);  % and plot it
```

7.2.3 Mixture of Gaussians

In the K-means clustering algorithm, spherical clusters were assumed by the fact that the Euclidean distance to the cluster centres $\boldsymbol{\mu}_k$ is computed. All objects on a circle or hypersphere around the cluster centre will have the same resemblance to that cluster. However, in many cases clusters have more structure than that. In the *mixture of Gaussians* model (Dempster *et al.*, 1977; Bishop, 1995), it is assumed that the objects in each of the K clusters are distributed according to a Gaussian distribution. That means that each cluster is not only characterized by a mean $\boldsymbol{\mu}_k$ but also by a covariance matrix $\mathbf{C}_k$. In effect, a complete density estimate of the data is performed, where the density is modelled by:

$$p(\mathbf{z}) = \sum_{k=1}^{K} \pi_k \mathrm{N}(\mathbf{z}|\boldsymbol{\mu}_k, \mathbf{C}_k) \tag{7.13}$$

$\mathrm{N}(\mathbf{z}|\boldsymbol{\mu}_k,\mathbf{C}_k)$ stands for the multivariate Gaussian distribution. π_k are the mixing parameters (for which $\sum_{k=1}^{K} \pi_k = 1$ and $\pi_k \geq 0$). The mixing parameter π_k can be regarded as the probability that $\mathbf{z}$ is produced by a random number generator with probability density $\mathrm{N}(\mathbf{z}|\boldsymbol{\mu}_k, \mathbf{C}_k)$.

The parameters to be estimated are: the number K of mixing components, the mixing parameters $\pi_1, \ldots, \pi_K$, the mean vectors $\boldsymbol{\mu}_k$ and the covariance matrices $\mathbf{C}_k$. We will denote this set of free parameters by $\Psi = \{\pi_k, \boldsymbol{\mu}_k, \mathbf{C}_k | k = 1, \ldots, K\}$. This increase in the number of free parameters, compared to the K-means algorithm, leads to the need for a higher number of training objects in order to reliably estimate these cluster parameters. On the other hand, because more flexible cluster shapes are applied, fewer clusters might have to be used to approximate the structure in the data.

One method for fitting the parameters of (7.13) to the data is to use maximum likelihood estimation (see Section 3.1.4), i.e. to optimize the log-likelihood:

$$L(Z|\Psi) \stackrel{def}{=} \ln p(\mathbf{z}_1, \ldots, \mathbf{z}_{N_S}|\Psi) = \ln \prod_{n=1}^{N_S} p(\mathbf{z}_n|\Psi) \tag{7.14}$$

with $Z = \{\mathbf{z}_1, \ldots, \mathbf{z}_{N_S}\}$. The optimization of $L(Z|\Psi)$ w.r.t. Ψ is complicated and cannot be done directly with this equation. Fortunately, there exists a standard 'hill-climbing' algorithm to optimize the parameters in this equation. The *expectation-maximization* (EM) algorithm is a general procedure which assumes the following problem definition. We have two data spaces: the observed data Z and the so-called *missing data* (*hidden data*) $X = \{\mathbf{x}_1, \ldots, \mathbf{x}_{N_S}\}$. Let Y be the complete data in which each vector $\mathbf{y}_n$ consists of a known part $\mathbf{z}_n$ and a missing part $\mathbf{x}_n$. Thus, the vectors $\mathbf{y}_n$ are defined by $\mathbf{y}_n^T = [\,\mathbf{z}_n^T \quad \mathbf{x}_n^T\,]$. Only $\mathbf{z}_n$ is available in the training set, $\mathbf{x}_n$ is missing, and as such unknown. The EM algorithm uses this model of 'incomplete data' to iteratively estimate the parameters Ψ of the distribution of $\mathbf{z}_n$. It tries to maximize the log-likelihood $L(Z|\Psi)$. In fact, if the result of the i-th iteration is denoted by $\Psi^{(i)}$, then the EM algorithm assures that $L(Z|\Psi^{(i+1)}) \geq L(Z|\Psi^{(i)})$.

The mathematics behind the EM algorithm is rather technical and will be omitted. However, the intuitive idea behind the EM algorithm is as follows. The basic assumption is that with all data available, that is if we had observed the complete data Y, the maximization of $L(Y|\Psi)$ would be simple, and the problem could be solved easily. Unfortunately, only Z is available and X is missing. Therefore, we maximize the expectation $\mathrm{E}[L(Y|\Psi)|Z]$ instead of $L(Y|\Psi)$. The expectation is taken over the complete data Y, but under the condition of the observed data Z. The estimate from the i-th iteration follows from:

$$\begin{aligned} \mathrm{E}[L(Y|\Psi)|Z] &= \int (\ln p(Y|\Psi)) p(Y|Z, \Psi^{(i)}) dY \\ \Psi^{(i+1)} &= \arg\max_{\Psi} \{\mathrm{E}[L(Y|\Psi)|Z]\} \end{aligned} \tag{7.15}$$

The integral extends over the entire Y space. The first step is the E step; the last one is the M step.

In the application of EM optimization to a mixture of Gaussians (with predefined K; see also the discussion on page 228) the missing part $\mathbf{x}_n$, associated with $\mathbf{z}_n$, is a K dimensional vector. $\mathbf{x}_n$ indicates which one of the K Gaussians generated the corresponding object $\mathbf{z}_n$. The $\mathbf{x}_n$ are called *indicator variables*. They use position coding to indicate the Gaussian associated with $\mathbf{z}_n$. In other words, if $\mathbf{z}_n$ is generated by the k-th Gaussian, then $x_{n,k} = 1$ and all other elements in $\mathbf{x}_n$ are zero. With that, the prior probability that $x_{n,k} = 1$ is π_k.

In this case, the complete log-likelihood of Ψ can be written as:

$$\begin{aligned} L(Y|\Psi) &= \ln \prod_{n=1}^{N_S} p(\mathbf{z}_n, \mathbf{x}_n|\Psi) = \ln \prod_{n=1}^{N_S} p(\mathbf{z}_n|\mathbf{x}_n, \Psi) p(\mathbf{x}_n|\Psi) \\ &= \ln \prod_{n=1}^{N_S} \prod_{k=1}^{K} [\mathrm{N}(\mathbf{z}_n|\boldsymbol{\mu}_k, \mathbf{C}_k)\pi_k]^{x_{n,k}} \\ &= \sum_{k=1}^{K} \sum_{n=1}^{N_S} x_{n,k} \ln \mathrm{N}(\mathbf{z}_n|\boldsymbol{\mu}_k, \mathbf{C}_k) + x_{n,k} \ln \pi_k \end{aligned} \tag{7.16}$$

Under the condition of a given Z, the probability density $p(Y|Z,\Psi) = p(X,Z|Z,\Psi)$ can be replaced with the marginal probability $P(X|Z,\Psi)$. Therefore in (7.15), we have:

$$\begin{aligned} \mathrm{E}[L(Y|\Psi)|Z] &= \int (\ln p(Y|\Psi)) p(Y|Z, \Psi^{(i)}) dY \\ &= \sum_{X} L(Y|\Psi) P(X|Z, \Psi^{(i)}) \end{aligned} \tag{7.17}$$

But since in (7.16) $L(Y|\Psi)$ is linear in X, we conclude that

$$\mathrm{E}[L(Y|\Psi)|Z] = L(\overline{X}, Z|\Psi) \tag{7.18}$$

where $\overline{X}$ is the expectation of the missing data under the condition of Z and $\Psi^{(i)}$:

$$\begin{aligned} \bar{x}_{n,k} &= \mathrm{E}[x_{n,k}|\mathbf{z}_n, \Psi^{(i)}] = P(x_{n,k}|\mathbf{z}_n, \Psi^{(i)}) \\ &= \frac{p(\mathbf{z}_n|x_{n,k}, \Psi^{(i)}) P(x_{n,k}|\Psi^{(i)})}{p(\mathbf{z}_n|\Psi^{(i)})} = \frac{\mathrm{N}\left(\mathbf{z}_n|\boldsymbol{\mu}_k^{(i)}, \mathbf{C}_k^{(i)}\right)\pi_k^{(i)}}{\sum_{j=1}^{K} \mathrm{N}\left(\mathbf{z}_n|\boldsymbol{\mu}_j^{(i)}, \mathbf{C}_j^{(i)}\right)\pi_j^{(i)}} \end{aligned} \tag{7.19}$$

The variable $\bar{x}_{n,k}$ is called the *ownership* because it indicates to what degree sample $\mathbf{z}_n$ is attributed to the k-th component.

For the M step, we have to optimize the expectation of the log-likelihood, that is $E[L(Y|\Psi)|Z] = L(\overline{X}, Z|\Psi)$. We do this by substituting $\bar{x}_{n,k}$ for $x_{n,k}$ into equation (7.16), taking the derivative with respect to Ψ and setting the result to zero. Solving the equations will yield expressions for the parameters $\Psi = \{\pi_k, \boldsymbol{\mu}_k, C_k\}$ in terms of the data $\mathbf{z}_n$ and $\bar{x}_{n,k}$.

Taking the derivative of $L(\overline{X}, Z|\Psi)$ with respect to $\boldsymbol{\mu}_k$ gives:

$$\sum_{n=1}^{N_S} \bar{x}_{n,k}(\mathbf{z}_n - \boldsymbol{\mu}_k)^T C_k^{-1} = 0 \tag{7.20}$$

Rewriting this, gives the update rule for $\boldsymbol{\mu}_k$:

$$\hat{\boldsymbol{\mu}}_k = \frac{\sum_{n=1}^{N_S} \bar{x}_{n,k}\mathbf{z}_n}{\sum_{n=1}^{N_S} \bar{x}_{n,k}} \tag{7.21}$$

The estimation of C_k is somewhat more complicated. With the help of (b.39), we can derive:

$$\frac{\partial}{\partial C_k^{-1}} \ln N(\mathbf{z}_n|\hat{\boldsymbol{\mu}}_k, C_k) = \frac{1}{2}C_k - \frac{1}{2}(\mathbf{z}_n - \hat{\boldsymbol{\mu}}_k)(\mathbf{z}_n - \hat{\boldsymbol{\mu}}_k)^T \tag{7.22}$$

This results in the following update rule for C_k:

$$\hat{C}_k = \frac{\sum_{n=1}^{N_S} \bar{x}_{n,k}(\mathbf{z}_n - \hat{\boldsymbol{\mu}}_k)(\mathbf{z}_n - \hat{\boldsymbol{\mu}}_k)^T}{\sum_{n=1}^{N_S} \bar{x}_{n,k}} \tag{7.23}$$

Finally, the parameters π_k cannot be optimized directly because of the extra constraint, namely that $\sum_{k=1}^{K} \pi_k = 1$. This constraint can be enforced by introducing a Lagrange multiplier λ and extending the log-likelihood (7.16) by:

$$L'(Y|\Psi) = \sum_{n=1}^{N_S} \sum_{k=1}^{K} x_{n,k} \ln N(\mathbf{z}_n|\boldsymbol{\mu}_k, C_k) + x_{n,k} \ln \pi_k - \lambda\left(\sum_{k=1}^{K} \pi_k - 1\right) \tag{7.24}$$

Substituting $\bar{x}_{n,k}$ for $x_{n,k}$ and setting the derivative with respect to π_k to zero, yields:

$$\sum_{n=1}^{N_S} \bar{x}_{n,k} = \lambda \pi_k \tag{7.25}$$

Summing equation (7.25) over all clusters, we get:

$$\sum_{k=1}^{K} \sum_{n=1}^{N_S} \bar{x}_{n,k} = \lambda \sum_{k=1}^{K} \pi_k = \lambda \tag{7.26}$$

Further note, that summing $\sum_{k=1}^{K} \sum_{n=1}^{N_S} \bar{x}_{n,k}$ over all clusters gives the total number of objects N_S, thus it follows that $\lambda = N_S$. By substituting this result back into (7.25), the update rule for π_k becomes:

$$\hat{\pi}_k = \frac{1}{N_S} \sum_{n=1}^{N_S} \bar{x}_{n,k} \tag{7.27}$$

Note that with (7.27), the determination of $\boldsymbol{\mu}_k$ and $\mathbf{C}_k$ in (7.21) and (7.23) can be simplified to:

$$\begin{aligned}
\hat{\boldsymbol{\mu}}_k &= \frac{1}{N_S \hat{\pi}_k} \sum_{n=1}^{N_S} \bar{x}_{n,k} \mathbf{z}_n \\
\hat{\mathbf{C}}_k &= \frac{1}{N_S \hat{\pi}_k} \sum_{n=1}^{N_S} \bar{x}_{n,k} (\mathbf{z}_n - \hat{\boldsymbol{\mu}}_k)(\mathbf{z}_n - \hat{\boldsymbol{\mu}}_k)^T
\end{aligned} \tag{7.28}$$

The complete EM algorithm for fitting a mixture of Gaussian model to a data set is as follows.

Algorithm 7.4: EM algorithm for estimating a mixture of Gaussians
Input: The number K of mixing components and the data $\mathbf{z}_n$.

1. *Initialization*: Select randomly (or heuristically) $\Psi^{(0)}$. Set $i = 0$.
2. *Expectation step* (E step): Using the observed data set $\mathbf{z}_n$ and the estimated parameters $\Psi^{(i)}$, calculate the expectations $\bar{\mathbf{x}}_n$ of the missing values $\mathbf{x}_n$ using (7.19).

3. *Maximization step* (M step): Optimize the model parameters $\Psi^{(i+1)} = \{\hat{\pi}_k, \hat{\boldsymbol{\mu}}_k, \hat{\mathbf{C}}_k\}$ by maximum likelihood estimation using the expectation $\bar{\mathbf{x}}_n$ calculated in the previous step. See (7.27) and (7.28).
4. Stop if the likelihood did not change significantly in step 3. Else, increment i and go to step 2.

Clustering by EM results in more flexible cluster shapes than K-means clustering. It is also *guaranteed* to converge to (at least) a local optimum. However, it still has some of the same problems of K-means: the choice of the appropriate number of clusters, the dependence on initial conditions and the danger of convergence to local optima.

Example 7.3 Classification of mechanical parts, EM algorithm for mixture of Gaussians
Two results of the EM algorithm applied to the unlabelled data set of Figure 5.1(b) are shown in Figure 7.9. The algorithm is called with $K = 4$, which is the correct number of classes. Figure 7.9(a) is a correct result. The position and size of each component is appropriate. With random initializations of the algorithm, this result is reproduced with a probability of about 30%. Unfortunately, if the initialization is unlucky, the algorithm can also get stuck in a local minimum, as shown in Figure 7.9(b). Here, the components are at the wrong position.

The EM algorithm is implemented in PRTools by the function `emclust`. Listing 7.6 illustrates its use.

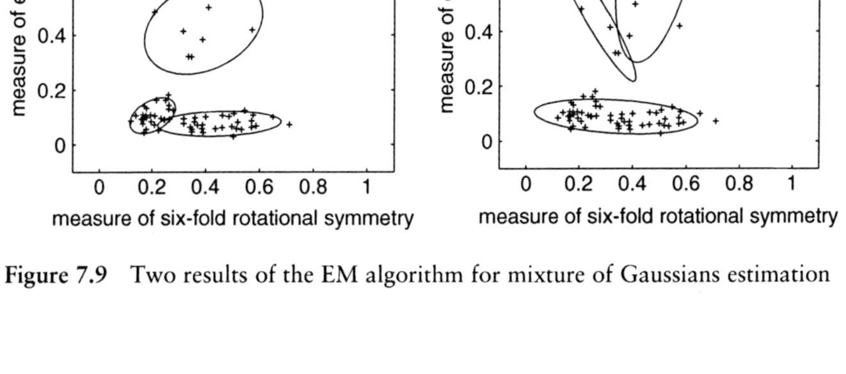

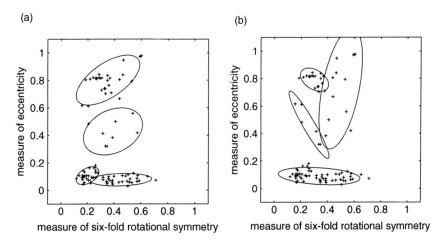

Figure 7.9 Two results of the EM algorithm for mixture of Gaussians estimation

Listing 7.6
MATLAB code for calling the EM algorithm for mixture of Gaussians estimation.

```
load nutsbolts_unlabeled;        % Load the data set z
z = setlabtype(z,'soft');        % Set probabilistic labels
[lab,w] = emclust(z,qdc,4);      % Cluster using EM
figure; clf; scatterd(z);
plotm(w,[],0.2:0.2:1);           % Plot results
```

7.2.4 Mixture of probabilistic PCA

An interesting variant of the mixture of Gaussians is the *mixture of probabilistic principal component analyzers* (Tipping and Bishop, 1999). Each single model is still a Gaussian like in (7.13), but its covariance matrix is constrained:

$$\mathbf{C}'_k = \mathbf{W}_k^T \mathbf{W}_k + \sigma_k^2 \mathbf{I} \tag{7.29}$$

where the $D \times N$ matrix $\mathbf{W}_k$ has the D eigenvectors corresponding to the largest eigenvalues of $\mathbf{C}_k$ as its columns, and the noise level outside the subspace spanned by $\mathbf{W}_k$ is estimated using the remaining eigenvalues:

$$\sigma_k^2 = \frac{1}{N-D} \sum_{m=D+1}^{N} \lambda_m \tag{7.30}$$

The EM algorithm to fit a mixture of probabilistic principal component analyzers proceeds just as for a mixture of Gaussians, using $\mathbf{C}'_k$ instead of $\mathbf{C}_k$. At the end of the M step, the parameters $\mathbf{W}_k$ and σ_k^2 are re-estimated for each cluster k by applying normal PCA to $\mathbf{C}_k$ and (7.30), respectively.

The mixture of probabilistic principal component analyzers introduces a new parameter, the subspace dimension D. Nevertheless, it uses far fewer parameters (when $D \ll N$) than the standard mixture of Gaussians. Still it is possible to model nonlinear data, which cannot be done using normal PCA. Finally, an advantage over normal PCA is that it is a full probabilistic model, i.e. it can be used directly as a density estimate.

In PRTools, probabilistic PCA is implemented in `qdc`: an additional parameter specifies the number of subspace dimensions to use. To train a mixture, one can use `emclust`, as is illustrated in Listing 7.7.

Listing 7.7
MATLAB code for calling the EM algorithm for mixture of probabilistic principal component analyzers estimation.

```
load nutsbolts_unlabeled;                   % Load the data set z
z = setlabtype(z, 'soft');                  % Set probabilistic labels
[lab,w] = emclust(z,qdc([],[],[],1),4);     % Cluster 1D PCAs using EM
figure; clf; scatterd(z);
plotm(w, [],0.2:0.2:1);                     % Plot results
```

7.2.5 Self-organizing maps

The *self-organizing map* (SOM; also known as *self-organizing feature map, Kohonen map*) is an unsupervised clustering and feature extraction method in which the cluster centres are constrained in their placing (Kohonen, 1995). The construction of the SOM is such that all objects in the input space retain as much as possible their distance and neighbourhood relations in the mapped space. In other words, the *topology* is preserved in the mapped space. The method is therefore strongly related to multi-dimensional scaling.

The mapping is performed by a specific type of neural network, equipped with a special learning rule. Assume that we want to map an N-dimensional measurement space to a D-dimensional feature space, where $D \ll N$. In fact, often $D = 1$ or $D = 2$. In the feature space, we define a finite orthogonal grid with $M_1 \times M_2 \times \cdots \times M_D$ grid points. At each grid point we place a neuron. Each neuron stores an N-dimensional vector that serves as a cluster centre. By defining a grid for the neurons, each neuron does not only have a neighbouring neuron in the measurement space, it also has a neighbouring neuron in the grid. During the learning phase, neighbouring neurons in the grid are enforced to also be neighbours in the measurement space. By doing so, the local topology will be preserved.

The SOM is updated using an iterative update rule, which honours the topological constraints on the neurons. The iteration runs over all training objects $\mathbf{z}_n, n = 1, \ldots, N_S$ and at each iteration, the following steps are taken:

Algorithm 7.5: Training a self-organizing map:

1. *Initialization*: Choose the grid size, $M_1, \ldots, M_D$, and initialize the weight vectors $\mathbf{w}_j^{(0)}$ (where $j = 1, \ldots, M_1 \times \cdots M_D$) of each neuron, for instance, by assignment of the values of $M_1 \times \cdots M_D$ different samples from the data set and that are randomly selected. Set $i = 0$.

2. *Iterate*:

   2.1 Find, for each object $\mathbf{z}_n$ in the training set, the most similar neuron $\mathbf{w}_k^{(i)}$.

$$k(\mathbf{z}_n) = \arg\min_j \|\mathbf{z}_n - \mathbf{w}_j^{(i)}\| \tag{7.31}$$

   This is called the best-matching or winning neuron for this input vector.

   2.2 Update the winning neuron and its neighbours using the update rule:

$$\mathbf{w}_j^{(i+1)} = \mathbf{w}_j^{(i)} + \eta^{(i)} h^{(i)}(|k(\mathbf{z}_n) - j|)(\mathbf{z}_n - \mathbf{w}_j^{(i)}) \tag{7.32}$$

   2.3 Repeat 2.1 and 2.2 for all samples $\mathbf{z}_n$ in the data set.

   2.4 If the weights in the previous steps did not change significantly, then stop. Else, increment i and go to step 2.1.

Here $\eta^{(i)}$ is the learning rate and $h^{(i)}(|k(\mathbf{z}_n) - j|)$ is a weighting function. Both can depend on the iteration number i. This weighting function weighs how much a particular neuron in the grid is updated. The term $|k(\mathbf{z}_n) - j|$ indicates the distance between the winning neuron $k(\mathbf{z}_n)$ and neuron j, *measured over the grid*. The winning neuron (for which $j = k(\mathbf{z}_n)$) will get the maximal weight, because $h^{(i)}()$ is chosen such that:

$$h^{(i)}() \leq 1 \text{ and } h^{(i)}(0) = 1 \tag{7.33}$$

Thus, the winning neuron will get the largest update. This update moves the neuron in the direction of $\mathbf{z}_n$ by the term $(\mathbf{z}_n - \mathbf{w}_j)$.

The other neurons in the grid will receive smaller updates. Since we want to preserve the neighbourhood relations only locally, the further the neuron is from the winning neuron, the smaller the update. A commonly used weighting function which satisfies these requirements is the Gaussian function:

$$h^{(i)}(x) = \exp\left(-\frac{x^2}{\sigma_{(i)}^2}\right) \tag{7.34}$$

For this function a suitable scale $\sigma_{(i)}$ over the map should be defined. This weighting function can be used for a grid of any dimension (not just one-dimensional), when we realize that $|k(\mathbf{z}_n) - j|$ means in general the distance between the winning neuron $k(\mathbf{z}_n)$ and neuron j over the grid.

Let us clarify this with an example. Assume we start with data $\mathbf{z}_n$ uniformly distributed in a square in a two-dimensional measurement space, and we want to map this data into a one-dimensional space. Therefore, $K = 15$ neurons are defined. These neurons are ordered, such that neuron $j - 1$ is the left neighbour of neuron j and neuron $j + 1$ is the right neighbour. In the weighting function $\sigma = 1$ is used, in the update rule $\eta = 0.01$; these are not changed during the iterations. The neurons have to be placed as objects in the feature space such that they represent the data as best as possible. Listing 7.8 shows an implementation for training a one-dimensional map in PRTools.

Listing 7.8
PRTools code for training and plotting a self-organizing map.

```
z = rand(100,2);                  % Generate the data set z
w = som(z,15);                    % Train a 1D SOM and show it
figure; clf; scatterd(z); plotsom(w);
```

In Figure 7.10 four scatter plots of this data set with the SOM ($K = 15$) are shown. In the left subplot, the SOM is randomly initialized by picking K objects from the data set. The lines between the neurons indicate the neighbouring relationships between the neurons. Clearly, neighbouring neurons in feature space are not neighbouring in the grid. In the fourth subplot it is visible that after 100 iterations over the data set, the one-dimensional grid has organized itself over the square. This solution does not change in the next 500 iterations.

With one exception, the neighbouring neurons in the measurement space are also neighbouring neurons in the grid. Only where the one-dimensional string crosses, neurons far apart in the grid become close neighbours in feature space. This local optimum, where the map did not unfold completely in the measurement space, is often encountered in SOMs. It is very hard to get out of this local optimum. The solution would be to restart the training with another random initialization.

Many of the unfolding problems, and the speed of convergence, can be solved by adjusting the learning parameter $\eta^{(i)}$ and the characteristic width in the weighting function $h^{(i)}(|k(\mathbf{z}_n) - j|)$ during the iterations. Often, the following functional forms are used:

$$\begin{aligned} \eta^{(i)} &= \eta^{(0)} \exp(-i/\tau_1) \\ \sigma_{(i)} &= \sigma_{(0)} \exp(-i/\tau_2) \end{aligned} \tag{7.35}$$

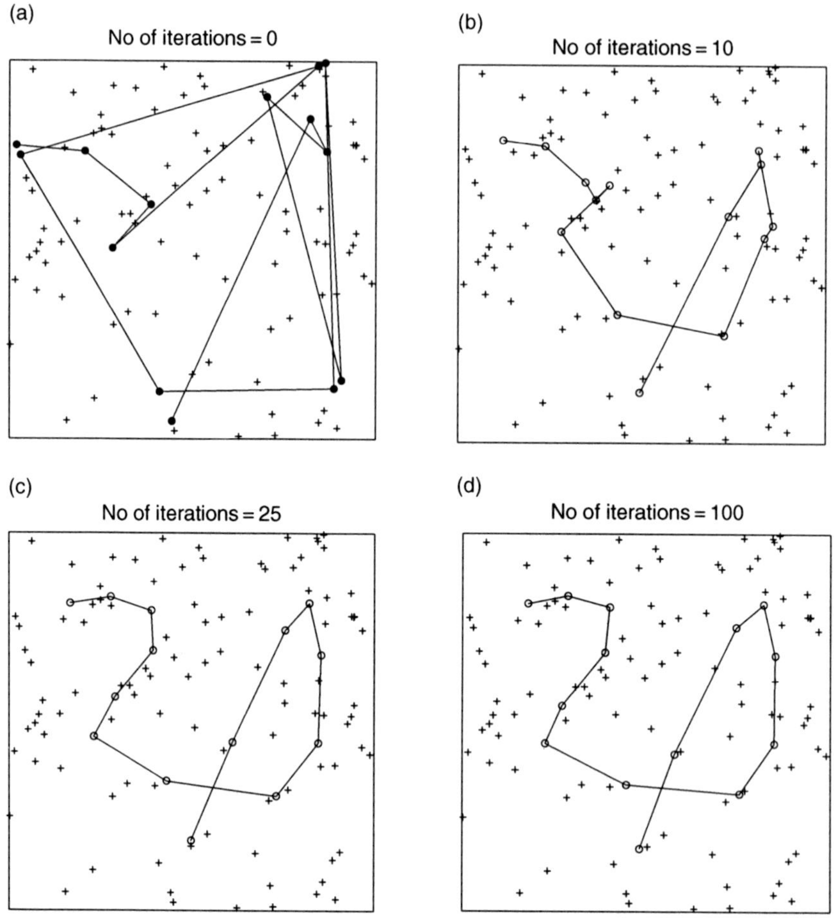

Figure 7.10 The development of a one-dimensional self-organizing map, trained on a two-dimensional uniform distribution: (a) initialization; (b)–(d) after 10, 25 and 100 iterations, respectively

This introduces two additional scale parameters which have to be set by the user.

Although the SOM offers a very flexible and powerful tool for mapping a data set to one or two dimensions, the user is required to make many important parameter choices: the dimension of the grid, the number of neurons in the grid, the shape and initial width of the neighbourhood function, the initial learning rate and the iteration dependencies of the neighbourhood function and learning rate. In many cases a two-dimensional grid is chosen for visualization purposes, but it might not fit the data well. The retrieved data reduction will therefore not reflect the true structure in the data and visual inspection only reveals an artificially induced structure. Training several SOMs with different settings might provide some stable solution.

Example 7.4 SOM of the RGB samples of an illuminated surface
Figure 7.11(a) shows the RGB components of a colour image. The imaged object has a spherical surface that is illuminated by a spot illuminator. The light–material interaction involves two distinct

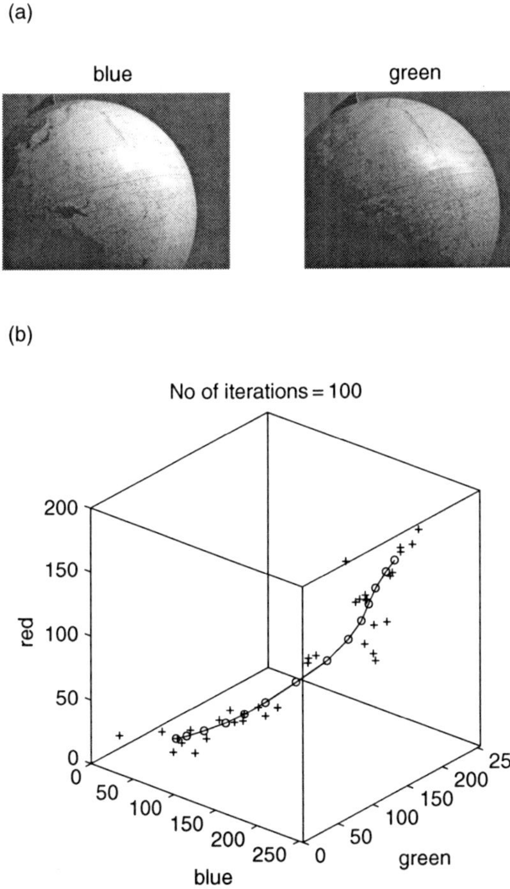

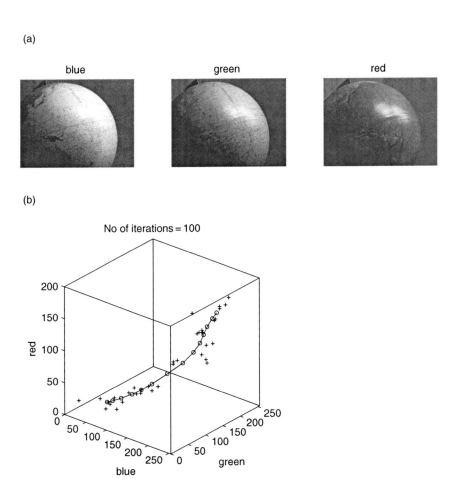

Figure 7.11 A SOM that visualizes the effects of a highlight. (a) RGB image of an illuminated surface with a highlight (=glossy spot). (b) Scatter diagram of RGB samples together with a one-dimensional SOM

physical mechanisms: diffuse reflection and specular (mirror-like) reflection. The RGB values that result from diffuse reflection are invariant to the geometry. Specular reflection only occurs at specific surface orientations determined by the position of the illuminator and the camera. Therefore, specular reflection is seen in the image as a glossy spot, a so-called highlight. The colour of the surface is determined by its spectral properties of the diffuse reflection component. Usually, specular reflection does not depend on the wavelength of the light so that the colour of the highlight is solely determined by the illuminator (usually white light).

Since light is additive, a RGB value $\mathbf{z}$ is observed as a linear combination of the diffuse component and the off-specular component: $\mathbf{z} = \alpha \mathbf{z}_{diff} + \beta \mathbf{z}_{spec}$. The variables α and β depend on the geometry. The estimation of $\mathbf{z}_{diff}$ and $\mathbf{z}_{spec}$ from a set of samples of $\mathbf{z}$ is an interesting problem. Knowledge of $\mathbf{z}_{diff}$ and $\mathbf{z}_{spec}$ would, for instance, open the door to 'highlight removal'.

Here, a one-dimensional SOM of the data is helpful to visualize the data. Figure 7.11(b) shows such a map. The figure suggests that the data forms a one-dimensional manifold, i.e. a curve in the three-dimensional RGB space. The manifold has the shape of an elbow. In fact, the orientation of the lower part of the elbow corresponds to data from an area with only diffuse reflection, i.e. to $\alpha \mathbf{z}_{diff}$. The upper part corresponds to data from an area with both diffuse and specular reflection, i.e. to $\alpha \mathbf{z}_{diff} + \beta \mathbf{z}_{spec}$.

7.2.6 Generative topographic mapping

We conclude this chapter with a probabilistic version of the self-organizing map, the *generative topographic mapping* (or GTM) (Bishop *et al.*, 1998). The goal of the GTM is the same as that of the SOM: to model the data by clusters with the constraint that neighbouring clusters in the original space are also neighbouring clusters in the mapped space. Contrary to the SOM, a GTM is fully probabilistic and it can be trained using the EM algorithm.

The GTM starts from the idea that the density $p(\mathbf{z})$ can be represented in terms of a D-dimensional *latent variable* $\mathbf{q}$, where in general $D \ll N$. For this, a function $\mathbf{z} = \boldsymbol{\phi}(\mathbf{q}; \mathbf{W})$ with weights $\mathbf{W}$ has to be found. The functions maps a point $\mathbf{q}$ into a corresponding object $\mathbf{z} = \boldsymbol{\phi}(\mathbf{q}; \mathbf{W})$. Because in reality the data will not be mapped perfectly, we assume some Gaussian noise with variance σ^2 in all directions. The full model for the probability of observing a vector $\mathbf{z}$ is:

$$p(\mathbf{z}|\mathbf{q},\mathbf{W},\sigma^2) = \frac{1}{2\pi^{\frac{D}{2}}\sigma^D}\exp\left(-\frac{\|\boldsymbol{\phi}(\mathbf{q};\mathbf{W})-\mathbf{z}\|^2}{\sigma^2}\right) \tag{7.36}$$

In general, the distribution of $\mathbf{z}$ in the high dimensional space can be found by integration over the latent variable $\mathbf{q}$:

$$p(\mathbf{z}|\mathbf{W},\sigma^2) = \int p(\mathbf{z}|\mathbf{q},\mathbf{W},\sigma^2)p(\mathbf{q})d\mathbf{q} \tag{7.37}$$

In order to allow an analytical solution of this integral, a simple grid-like probability model is chosen for $p(\mathbf{q})$, just like in the SOM:

$$p(\mathbf{q}) = \frac{1}{K}\sum_{k=1}^{K}\delta(\mathbf{q}-\mathbf{q}_k) \tag{7.38}$$

i.e. a set of Dirac functions centred on grid nodes $\mathbf{q}_k$. The log-likelihood of the complete model can then be written as:

$$\ln L(\mathbf{W},\sigma^2) = \sum_{n=1}^{N_S}\ln\left(\frac{1}{K}\sum_{k=1}^{K}p(\mathbf{z}_n|\mathbf{q}_k,\mathbf{W},\sigma^2)\right) \tag{7.39}$$

Still, the functional form for the mapping function $\boldsymbol{\phi}(\mathbf{q};\,\mathbf{W})$ has to be defined. This function maps the low dimensional grid to a manifold in the high dimensional space. Therefore, its form controls how nonlinear the manifold can become. In the GTM, a regression on a set of fixed basis functions is used:

$$\boldsymbol{\phi}(\mathbf{q};\mathbf{W}) = \mathbf{W}\boldsymbol{\gamma}(\mathbf{q}) \tag{7.40}$$

$\boldsymbol{\gamma}(\mathbf{q})$ is a vector containing the output of M basis functions, which are usually chosen to be Gaussian with means on the grid points and a fixed width σ_φ. $\mathbf{W}$ is a $N \times M$ weight matrix.

Given settings for K, M and σ_φ, the EM algorithm can be used to estimate $\mathbf{W}$ and σ^2. Let the complete data be $\mathbf{y}_n^T = [\mathbf{z}_n^T \quad \mathbf{x}_n^T]$, with $\mathbf{x}_n$ the hidden variables. $\mathbf{x}_n$ is a K-dimensional vector. The element $x_{n,k}$ codes the membership of $\mathbf{z}_n$ with respect to the k-th cluster. After proper initialization, the EM algorithm proceeds as follows:

Algorithm 7.6: EM algorithm for training a GTM:

1. *E step*: estimate the missing data:

$$\bar{x}_{n,k} = p(\mathbf{q}_k|\mathbf{z}_n, \hat{\mathbf{W}}, \hat{\sigma}^2) = \frac{p(\mathbf{z}_n|\mathbf{q}_k, \hat{\mathbf{W}}, \hat{\sigma}^2)}{\sum_{k=1}^{K} p(\mathbf{z}_n|\mathbf{q}_k, \hat{\mathbf{W}}, \hat{\sigma}^2)} \tag{7.41}$$

using Bayes' theorem.

2. *M step*: re-estimate the parameters:

$$\hat{\mathbf{W}}^T = (\Gamma^T \mathbf{G} \Gamma + \lambda \mathbf{I})^{-1} \Gamma^T \overline{\mathbf{X}}^T \mathbf{Z} \tag{7.42}$$

$$\hat{\sigma}^2 = \frac{1}{N_S N} \sum_{n=1}^{N_S} \sum_{k=1}^{K} \bar{x}_{n,k} \|\hat{\mathbf{W}} \boldsymbol{\gamma}(\mathbf{q}_k) - \mathbf{z}_n\|^2 \tag{7.43}$$

3. Repeat 1 and 2 until no significant changes occur.

G is a $K \times K$ diagonal matrix with $G_{kk} = \sum_{n=1}^{N_S} z_{n,k}$ as elements. Γ is a $K \times M$ matrix containing the basis functions: $\Gamma_{k,m} = \gamma_m(\mathbf{q}_k)$. $\overline{\mathbf{X}}$ is a $N_S \times K$ matrix with the $\bar{x}_{n,k}$ as elements. $\mathbf{Z}$ is a $N_S \times N$ matrix. It contains the data vectors $\mathbf{z}_n^T$ as its rows. Finally, λ is a regularization parameter, which is needed in cases where $\Gamma^T \mathbf{G} \Gamma$ becomes singular and the inverse cannot be computed.

The GTM can be initialized by performing PCA on the data set, projecting the data to D dimensions and finding $\mathbf{W}$ such that the resulting latent variables $\mathbf{q}_n$ approximate as well as possible the projected data. The noise variance σ^2 can be found as the average of the $N - D$ smallest eigenvalues, as in probabilistic PCA.

Training the GTM is usually quick and converges well. However, it suffers from the standard problem of EM algorithms in that it may converge to a local optimum. Furthermore, its success depends highly on the choices for K (the number of grid points), M (the number of basis functions for the mapping), σ_φ (the width of these basis functions) and

λ (the regularization term). For inadequate choices, the GTM may reflect the structure in the data very poorly. The SOM has the same problem, but needs to estimate somewhat less parameters.

An advantage of the GTM over the SOM is that the parameters that need to be set by the user have a clear interpretation, unlike in the SOM where unintuitive parameters as learning parameters and time parameters have to be defined. Furthermore, the end result is a probabilistic model, which can easily be compared to other models (e.g. in terms of likelihood) or combined with them.

Figure 7.12 shows some examples of GTMs ($D = 1$) trained on uniformly distributed data. Figure 7.12(a) clearly shows how the GTM can be overtrained if too many basis functions – here, 10 – are used. In Figure 7.12(b), less basis functions are used and the manifold found is much smoother. Another option is to use regularization, which also gives a more smooth result as shown in Figure 7.12(c) but cannot completely prevent extreme nonlinearities.

Listing 7.9 shows how a GTM can be trained and displayed in PRTools.

Listing 7.9

PRTools code for training and plotting generative topographic mapping.

```
z=rand(100,2);      % Generate the data set z
w=gtm(z,15);        % Train a 1D GTM and show it
figure; clf; scatterd(z); plotgtm(w);
```

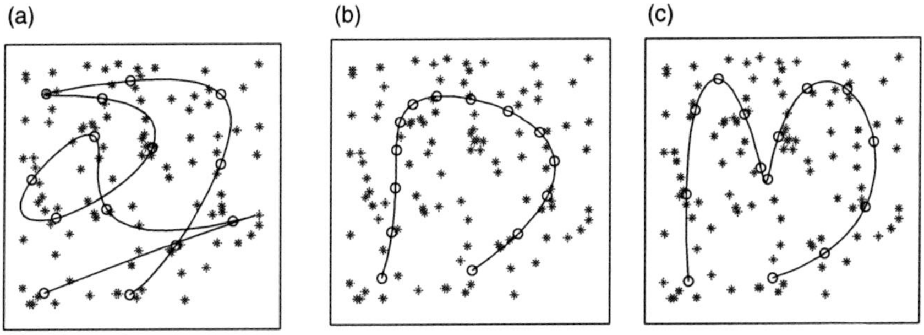

Figure 7.12 Trained generative topographic mappings. (a) $K = 14, M = 10, \sigma_\varphi = 0.2$ and $\lambda = 0$. (b) $K = 14, M = 5, \sigma_\varphi = 0.2$ and $\lambda = 0$. (c) $K = 14, M = 10, \sigma_\varphi = 0.2$ and $\lambda = 0.01$

7.3 REFERENCES

Bishop, C.M., *Neural Networks for Pattern Recognition*, Oxford University Press, Oxford, UK, 1995.

Bishop, C.M., Svensén, M. and Williams, C.K.I., GTM: the generative topographic mapping. *Neural Computation*, **10**(1), 215–34, 1998.

Dempster, N.M., Laird, A.P. and Rubin, D.B., Maximum likelihood from incomplete data via the EM algorithm. *Journal of the Royal Statististical Society B*, **39**, 185–197, 1977.

Johnson, S.C., Hierarchical clustering schemes. *Psychometrika*, **2**, 241–54, 1967.

Jolliffe, I.T., *Principal Component Analysis*, Springer-Verlag, New York, 1986.

Kohonen, T., *Self-organizing Maps*, Springer-Verlag, Heidelberg, Germany, 1995.

Kruskal, J.B. and Wish, M., *Multidimensional scaling*, Sage Publications, Beverly Hills, CA, 1977.

Sammon, Jr, J.W., A nonlinear mapping for data structure analysis. *IEEE Transactions on Computers*, **C-18**, 401–9, 1969.

Tipping, M.E. and Bishop, C.M., Mixtures of probabilistic principal component analyzers. *Neural Computation*, **11**(2), 443–82, 1999.

7.4 EXERCISES

1. Generate a two-dimensional data set **z** uniformly distributed between 0 and 1. Create a second data set **y** uniformly distributed between −1 and 2. Compute the (Euclidean) distances between **z** and **y**, and find the objects in **y** which have distance smaller than 1 to an object in **z**. Make a scatter plot of these objects. Using a large number of objects in **z**, what should be the shape of the area of objects with distance smaller than 1? What would happen if you change the distance definition to the city-block distance (Minkowski metric with $q = 1$)? And what would happen if the cosine distance is used? (0)
2. Create a data set `z = gendatd(50,50,4,2);` . Make a scatter plot of the data. Is the data separable? Predict what would happen if the data is mapped to one dimension. Check your prediction by mapping the data using `pca(z,1)`, and training a simple classifier on the mapped data (such as `ldc`). (0)
3. Load the `worldcities` data set and experiment with using different values for q in the MDS criterion function. What is the effect? Can you think of another way of treating close sample pairs different from far-away sample pairs? (0)
4. Derive equations (7.21), (7.23) and (7.27). (∗)
5. Discuss in which data sets it can be expected that the data is distributed in a subspace, or in clusters. In which cases will it not be useful to apply clustering or subspace methods? (∗)
6. What is a desirable property of a clustering when the same algorithm is run multiple times on the same data set? Develop an algorithm that uses this notion to estimate the number of clusters present in the data. (∗∗)
7. In terms of scatter matrices (see the previous chapter), what does the K-means algorithm minimize? (0)

8. Under which set of assumptions does the EM algorithm degenerate to the K-means algorithm? (**)
9. What would be a simple way to lower the risk of ending up in a poor local minimum with the K-means and EM algorithms? (0)
10. Which parameter(s) control(s) the generalization ability of the self-organizing map (for example, its ability to predict the locations of previously unseen samples)? And that of the generative topographic mapping? (*)

8

State Estimation in Practice

Chapter 4 discussed the theory needed for the design of a state estimator. The current chapter addresses the practical issues related to the design. Usually, the engineer cycles through a number of design stages of which some are depicted in Figure 8.1.

One of the first steps in the design process is *system identification*. The purpose is to formulate a mathematical model of the system of interest. As stated in Chapter 4, the model is composed of two parts: the state space model of the physical process and the measurement model of the sensory system. Using these models, the theory from Chapter 4 provides us with the mathematical expressions of the optimal estimator.

The next questions in the design process are the issues of *observability* (can all states of the process be estimated from the given set of measurements?) and *stability*. If the system is not observable or not stable, either the model must be revised or the sensory system must be redesigned.

If the design passes the observability and the stability tests, the attention is focussed at the computational issues. Due to finite arithmetic precision, there might be some pitfalls. Since in state estimation the measurements are processed sequentially, the effects of round-off errors may accumulate and may cause inaccurate results. The estimator may even completely fail to work due to numerical instabilities. Although the optimal solution of an estimation problem is often unique, there are a number of different implementations which are

Classification, Parameter Estimation and State Estimation: An Engineering Approach using MATLAB
F. van der Heijden, R.P.W. Duin, D. de Ridder and D.M.J. Tax
 ISBN: 0-470-09013-8

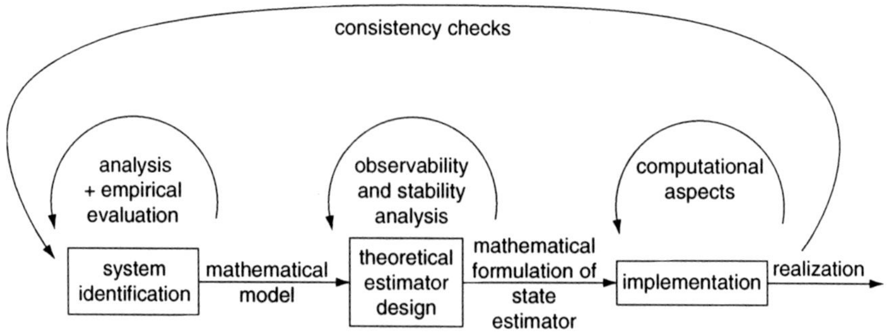

Figure 8.1 Design stages for state estimators

all mathematically equivalent, and thus all representing the same solution, but with different sensitivities to round-off errors. Thus, in this stage of the design process the appropriate implementation must be selected.

As soon as the estimator has been realized, *consistency checks* must be performed to see whether the estimator behaves in accordance with the expectations. If these checks fail, it is necessary to return to an earlier stage, i.e. refinements of the models, selection of another implementation, etc.

Section 8.1 presents a short introduction to system identification. The topic is a discipline on its own and will certainly not be covered here in its full length. For a full treatment we refer to the pertinent literature (Box and Jenkins, 1976; Eykhoff, 1974, Ljung and Glad, 1994; Ljung, 1999; Söderström and Stoica, 1989). Section 8.2 discusses the observability and the dynamic stability of an estimator. Section 8.3 deals with the computational issues. Here, several implementations are given each with its own sensitivities to numerical instabilities. Section 8.4 shows how consistency checks can be accomplished. Finally, Section 8.5 deals with extensions of the discrete Kalman filter. These extensions make the estimator applicable to a wider class of problems, i.e. non-white/cross-correlated noise sequences and offline estimation.

Some aspects of state estimator design are not discussed in this book; for instance *sensitivity analysis* and *error budgets* (Gelb *et al.*, 1974). These techniques are systemic methods for the identification of the most vulnerable parts of the design.

Most topics in this chapter concern Kalman filtering as introduced in Section 4.2.1, though some are also of relevance for extended Kalman filtering. For the sake of convenience, the equations are repeated here. The point of departure in Kalman filtering is a linear-Gaussian model of the physical process:

$$
\begin{aligned}
\mathbf{x}(i+1) &= \mathbf{F}(i)\mathbf{x}(i) + \mathbf{L}(i)\mathbf{u}(i) + \mathbf{w}(i) \quad i = 0, 1, \ldots && \text{(state equation)} \\
\mathbf{z}(i) &= \mathbf{H}(i)\mathbf{x}(i) + \mathbf{v}(i) && \text{(measurement model)}
\end{aligned}
\tag{8.1}
$$

$\mathbf{x}(i)$ is the state vector with dimension M. $\mathbf{z}(i)$ is the measurement vector with dimension N. The process noise $\mathbf{w}(i)$ and measurement noise $\mathbf{v}(i)$ are white Gaussian noise sequences, zero mean, and with covariance matrix $\mathbf{C_w}(i)$ and $\mathbf{C_v}(i)$, respectively. Process noise and measurement noise are uncorrelated: $\mathbf{C_{wv}}(i) = \mathbf{0}$. The prior knowledge is that $\mathbf{x}(0)$ has a Gaussian distribution with expectation $\mathrm{E}[\mathbf{x}(0)]$ and covariance matrix $\mathbf{C_x}(0)$.

The MMSE solution to the online estimation problem is developed in Section 4.2.1, and is known as the discrete Kalman filter. The solution is an iterative scheme. Each iteration cycles through (4.27) and (4.28), which are repeated here for convenience:

$$
\begin{aligned}
&\text{update:} \\
&\quad \hat{\mathbf{z}}(i) = \mathbf{H}(i)\bar{\mathbf{x}}(i|i-1) && \text{(predicted measurement)} \\
&\quad \mathbf{S}(i) = \mathbf{H}(i)\mathbf{C}(i|i-1)\mathbf{H}^T(i) + \mathbf{C_v}(i) && \text{(innovation matrix)} \\
&\quad \mathbf{K}(i) = \mathbf{C}(i|i-1)\mathbf{H}^T(i)\mathbf{S}^{-1}(i) && \text{(Kalman gain matrix)} \\
&\quad \bar{\mathbf{x}}(i|i) = \bar{\mathbf{x}}(i|i-1) + \mathbf{K}(i)(\mathbf{z}(i) - \hat{\mathbf{z}}(i)) && \text{(updated estimate)} \\
&\quad \mathbf{C}(i|i) = \mathbf{C}(i|i-1) - \mathbf{K}(i)\mathbf{S}(i)\mathbf{K}^T(i) && \text{(error covariance matrix)} \\
&\text{prediction:} \\
&\quad \bar{\mathbf{x}}(i+1|i) = \mathbf{F}(i)\bar{\mathbf{x}}(i|i) + \mathbf{L}(i)\mathbf{u}(i) && \text{(prediction)} \\
&\quad \mathbf{C}(i+1|i) = \mathbf{F}(i)\mathbf{C}(i|i)\mathbf{F}^T(i) + \mathbf{C_w}(i) && \text{(predicted state covariance)}
\end{aligned}
\tag{8.2}
$$

The iterative procedure is initiated with the prediction for $i = 0$ set equal to the prior:

$$
\bar{\mathbf{x}}(0|-1) \stackrel{def}{=} \mathrm{E}[\mathbf{x}(0)] \text{ and } \mathbf{C}(0|-1) \stackrel{def}{=} \mathbf{C_x}(0).
$$

8.1 SYSTEM IDENTIFICATION

System identification is the act of formulating a mathematical model of a given dynamic system based on input and output measurements of that system, and on general knowledge of the physical process at hand. The discipline of system identification not only finds application in estimator design, but also monitoring, fault detection and diagnosis (e.g. for maintenance of machines) and design of control systems.

A dichotomy of models exists between parametric models and nonparametric models. The nonparametric models describe the system by means of tabulated data of, for instance, the Fourier transfer function(s) or the edge response(s). Various types of parametric models exist, e.g. state space models, poles-zeros models and so on. In our case, state space models are the most useful, but other parametric models can also be used since most of these models can be converted to a state space.

The identification process can roughly be broken down into four parts: structuring, experiment design, estimation, evaluation and selection.

8.1.1 Structuring

The first activity is structuring. The structure of the model is settled by addressing the following questions. What is considered part of the system and what is environment? What are the (controllable) input variables? What are the possible disturbances (process noise)? What are the state variables? What are the output variables? What are the physical laws that relate the physical variables? Which parameters of these laws are known, and which are unknown? Which of these parameters can be measured directly?

Usually, there is not a unique answer to all these questions. In fact, the result of structuring is a set of candidate models.

Example 8.1 Candidate models describing a simple hydraulic system
The hydraulic system depicted in Figure 8.2 consists of two identical tanks connected by a pipeline with flow $q_1(t)$. The input flow $q_0(t)$ is acting on the first tank. $q_2(t)$ is the output flow from the second tank.

The relation between the level and the net input flow of a tank is $C\partial h = q\partial t$. C is the capacity of the tank. If the horizontal cross-sections of the tanks are constant, the capacity does not depend on h. In the present example, the capacity of both tanks is $C = 420$ (cm^2). The order of the system is at least two; the two states being the levels $h_1(t)$ and $h_2(t)$.

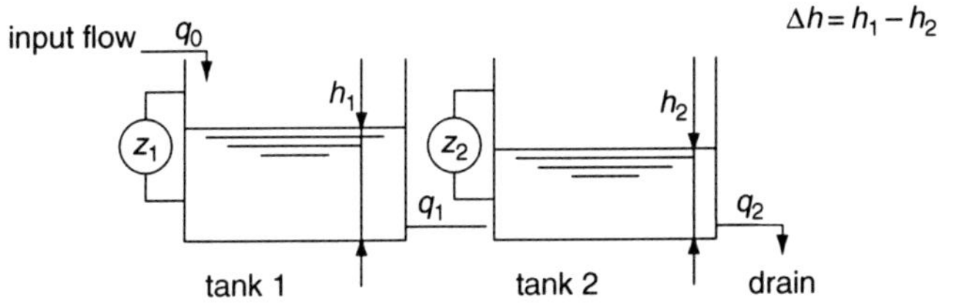

Figure 8.2 A simple hydraulic system consisting of two connected tanks

For the model of the flow through the pipelines we consider three possibilities, each leading to a different structure of the model.

Candidate model I: Frictionless liquids; Torricelli's law
For a frictionless liquid, Torricelli's law states that when a tank leaks, the sum of potential and kinetic energy is constant: $q^2 = 2A^2gh$. A is the area of the hole. Application of this law gives rise to the following second order, nonlinear model (g is the gravitational constant):

$$\begin{aligned} \dot{h}_1 &= -\frac{1}{C}\sqrt{2A_1^2 g(h_1 - h_2)} + \frac{1}{C}q_0 \\ \dot{h}_2 &= +\frac{1}{C}\sqrt{2A_1^2 g(h_1 - h_2)} - \frac{1}{C}\sqrt{2A_2^2 g h_2} \end{aligned} \tag{8.3}$$

Candidate model II: Linear friction
Here, we assume that the difference of pressure on both sides of a pipeline holds a linear relation with the flow: $\Delta p = Rq$. The parameter R is the resistance. Since $\Delta p = \rho g \Delta h$ (ρ is the mass density), the assumption brings the following linear, second order model:

$$\begin{aligned} \dot{h}_1 &= \frac{\rho g}{R_1 C}(h_2 - h_1) + \frac{1}{C}q_0 \\ \dot{h}_2 &= \frac{\rho g}{R_1 C}(h_1 - h_2) + \frac{-\rho g}{R_2 C}h_2 \end{aligned} \tag{8.4}$$

Candidate model III: Linear friction and hydraulic inertness
A liquid within a pipeline with a length ℓ and cross-section A experiences a force $\ell\rho\dot{q}$ (second law of Newton: $F = ma$). This force is induced by the difference of pressure $F = A\Delta p = A\rho g \Delta h$. Thus,

$\ell\dot{q} = Ag\Delta h$. With friction, the equation is $ARq + \ell\rho\dot{q} = A\rho g\Delta h$. With that, we arrive at the following linear, fourth order model:

$$\begin{bmatrix} \ddot{h}_1 \\ \ddot{h}_2 \end{bmatrix} = \frac{1}{C}\begin{bmatrix} -m_1 & m_1 \\ m_1 & -m_1 - m_2 \end{bmatrix}\begin{bmatrix} h_1 \\ h_2 \end{bmatrix} + \begin{bmatrix} -r_1 & 0 \\ r_1 - r_2 & -r_2 \end{bmatrix}\begin{bmatrix} \dot{h}_1 \\ \dot{h}_2 \end{bmatrix}$$
$$+\frac{1}{C}\begin{bmatrix} r_1 & 1 \\ r_2 - r_1 & 0 \end{bmatrix}\begin{bmatrix} q_0 \\ \dot{q}_0 \end{bmatrix} \qquad (8.5)$$

where
$m_1 = gA_1/\ell_1$, $m_2 = gA_2/\ell_2$, $r_1 = R_1A_1/(\ell_1\rho)$ and $r_2 = R_2A_2/(\ell_2\rho)$.

8.1.2 Experiment design

The purpose of the experimentation is to enable the determination of the unknown parameters and the evaluation and selection of the candidate models. The experiment comes down to the enforcement of some input test signals to the system at hand and the acquisition of measured data. The design aspects of the experiments are the choice of input signals, the choice of the sensors and the sensor locations and the preprocessing of the data.

The input signals should be such that the relevant aspects of the system at hand are sufficiently covered. The bandwidth of the input signal must match the bandwidth of interest. The signal magnitude must be in a range as required by the application. Furthermore, the signal energy should be sufficiently large so as to achieve suitable signal-to-noise ratios at the output of the sensors. Here, a trade-off exists between the bandwidth, the registration interval and the signal magnitude.

For instance, a very short input pulse covers a large bandwidth and permits a low registration interval, but requires a tremendous (perhaps excessive) magnitude in order to have sufficient signal-to-noise ratios. At the other side of the extremes, a single sinusoidal burst with a long duration covers only a very narrow bandwidth, and the signal magnitude can be kept low yet offering enough signal energy. A signal, often used, is the pseudorandom binary signal.

The choice of sensors is another issue. First, the set of identifiable, unknown parameters of the model must be determined. For instance, if in equation (8.4) ρ and R_1 are unknown, then these parameters are not separately identifiable from the level measurements because these parameters always occur in the combination ρ/R_1. Thus, in this case we can treat ρ/R_1 as one identifiable parameter. Sometimes, it might be necessary to temporarily use additional (often expensive) sensors so as to enable the estimation of all relevant parameters. As soon as the system identification is satisfactorily accomplished, these additional sensors can be removed from the system.

Often, the acquired data need preprocessing before the parameter estimation and evaluation take place. Reasons for doing so are, for instance:

- If the bandwidth of the noise is larger than the bandwidth of interest, filtering can be applied to suppress the noise in the unimportant frequency ranges.
- If a linearized model is strived for, the unimportant offsets should be removed (offset correction, baseline removal). Often, this is done by subtraction of the average from the signal.
- Sudden peaks (spikes) in the data are probably caused by disturbances such as mechanical shocks and electrical inferences due to insufficient shielding. These peaks should be removed.

In order to prevent overfitting, it might be useful to split the data according to two time intervals. The data in the first interval is used for parameter estimation. The second interval is used for model evaluation. Cross-evaluation might also be useful.

Example 8.2 Experimental data from the hydraulic system
Figure 8.3 shows data obtained from the hydraulic system depicted in Figure 8.2. The data is obtained using two level sensors that measure the levels h_1 and h_2. The sample period is $\Delta = 5$ (s). The standard deviation of the sensor noise is about $\sigma_v = 0.04$ (cm).

The measured levels in Figure 8.3 correspond to the free response of the system obtained with zero input and with an initial condition in which both tanks are completely filled, i.e. $h_1(0) = h_2(0) = 25$ (cm). Such an experiment is useful if it is envisaged that in the application this kind of level swings can occur.

8.1.3 Parameter estimation

Suppose that all unknown parameters are gathered in one parameter vector $\boldsymbol{\alpha}$. The discrete system equation is denoted by $\mathbf{f}(\mathbf{x}, \mathbf{w}, \boldsymbol{\alpha})$. The

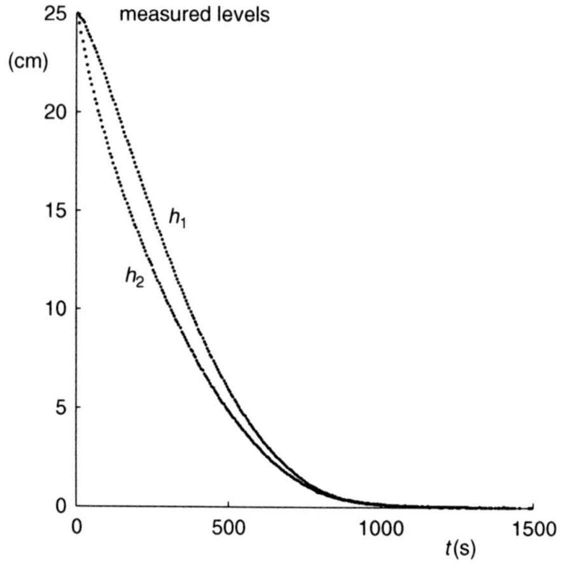

Figure 8.3 Experimental data obtained from the hydraulic system

measurement system is modelled, as before, by $\mathbf{z} = \mathbf{h}(\mathbf{x},\mathbf{v})$. We then have the sequence of measurements according to the following recursions:

$$\left.\begin{aligned} \mathbf{z}(i) &= \mathbf{h}(\mathbf{x}(i), \mathbf{v}(i)) \\ \mathbf{x}(i+1) &= \mathbf{f}(\mathbf{x}(i), \mathbf{w}(i), \boldsymbol{\alpha}) \end{aligned}\right\} \quad \text{for} \quad i = 0, 1, \ldots, I-1 \tag{8.6}$$

I is the length of the sequence. $\mathbf{x}(0)$ is the initial condition (which may be known or unknown). $\mathbf{v}(i)$ and $\mathbf{w}(i)$ are the measurement noise and the process noise, respectively.

One possibility for estimating $\boldsymbol{\alpha}$ is to process the sequence $\mathbf{z}(i)$ in batches. For that purpose, we stack all measurement vectors to one $I \times N$ dimensional vector, say $\mathbf{Z}$. Equation (8.6) defines the conditional probability density $p(\mathbf{Z}|\boldsymbol{\alpha})$. The stochastic nature of $\mathbf{Z}$ is due to the randomness of $\mathbf{w}(i)$, $\mathbf{v}(i)$ and possibly $\mathbf{x}(0)$. Equation (8.6) shows how this randomness propagates to $\mathbf{Z}$. Once the conditional density $p(\mathbf{Z}|\boldsymbol{\alpha})$ has been settled, the complete estimation machinery from Chapter 3 applies, thus providing the optimal solution of $\boldsymbol{\alpha}$. Especially, maximum likelihood estimation is popular since (8.6) can be used to calculate the (log-)likelihood of $\boldsymbol{\alpha}$. A numerical optimization procedure must provide the solution.

Working in batches soon becomes complicated due to the (often) nonlinear nature of the relations involved in (8.6). Many alternative techniques have been developed. For linear systems, processing in the frequency domain may be advantageous. Another possibility is to process the measurements sequentially. The trick is to regard the parameters as state vectors $\boldsymbol{\alpha}(i)$. Static parameters do not change in time. So, the corresponding state equation is $\boldsymbol{\alpha}(i+1) = \boldsymbol{\alpha}(i)$. Sometimes, it is useful to allow slow variations in $\boldsymbol{\alpha}(i)$. This is helpful in order to model drift phenomena, but also to improve the convergence properties of the procedure. A simple model would be a process that is similar to random walk (Section 4.2.1): $\boldsymbol{\alpha}(i+1) = \boldsymbol{\alpha}(i) + \boldsymbol{\omega}(i)$. The white noise sequence $\boldsymbol{\omega}(i)$ is the driving force for the changes. Its covariance matrix $\mathbf{C}_{\omega}$ should be small in order to prevent a too wild behaviour of $\boldsymbol{\alpha}(i)$.

Using this model, equation (8.6) transforms into:

$$\begin{bmatrix} \mathbf{x}(i+1) \\ \boldsymbol{\alpha}(i+1) \end{bmatrix} = \begin{bmatrix} \mathbf{f}(\mathbf{x}(i), \mathbf{w}(i), \boldsymbol{\alpha}(i)) \\ \boldsymbol{\alpha}(i) + \boldsymbol{\omega}(i) \end{bmatrix} \quad \text{(state equation)}$$
$$\mathbf{z}(i) = \mathbf{h}(\mathbf{x}(i), \mathbf{v}(i)) \quad \text{(measurement equation)} \tag{8.7}$$

The original state vector $\mathbf{x}(i)$ has been augmented with $\boldsymbol{\alpha}(i)$. The new state equation can be written as $\boldsymbol{\xi}(i+1) = \boldsymbol{\phi}(\boldsymbol{\xi}(i), \mathbf{w}(i), \boldsymbol{\omega}(i))$ with $\boldsymbol{\xi}(i) \stackrel{def}{=} [\, \mathbf{x}^T(i) \; \boldsymbol{\alpha}^T(i) \,]^T$. This opens the door to simultaneous online estimation of both $\mathbf{x}(i)$ and $\boldsymbol{\alpha}(i)$ using the techniques discussed in Chapter 4. However, the new state function $\boldsymbol{\phi}(\cdot)$ is nonlinear and for online estimation we must resort to estimators that can handle these nonlinearities, e.g. extended Kalman filtering (Section 4.2.2), or particle filtering (Section 4.4).

Note that if $\boldsymbol{\omega}(i) \equiv 0$, then $\boldsymbol{\alpha}(i)$ is a random constant and (hopefully) its estimate $\hat{\boldsymbol{\alpha}}(i)$ converges to a constant. If we allow $\boldsymbol{\omega}(i)$ to deviate from zero by setting $\mathbf{C}_{\omega}$ to some (small) nonzero diagonal matrix, the estimator becomes *adaptive*. It has the potential to keep track of parameters that drift.

Example 8.3 Parameter estimation for the hydraulic system

In order to estimate the parameters of the three models of the hydraulic system using the data from the previous example the following procedure was applied. First, a particle filter was executed in order to get a rough indication of the magnitudes of the parameters. Figure 8.4(a) shows the results of the filter applied to the Torricelli model. In this model, there are two parameters A_1 and A_2 which were both initiated with a uniform distribution between 0 and 2(cm^2). The parameters were modelled with $A(i+1) = A(i) + \omega(i)$ where $\omega(i)$ is white noise with a standard deviation of 0.004(cm^2).

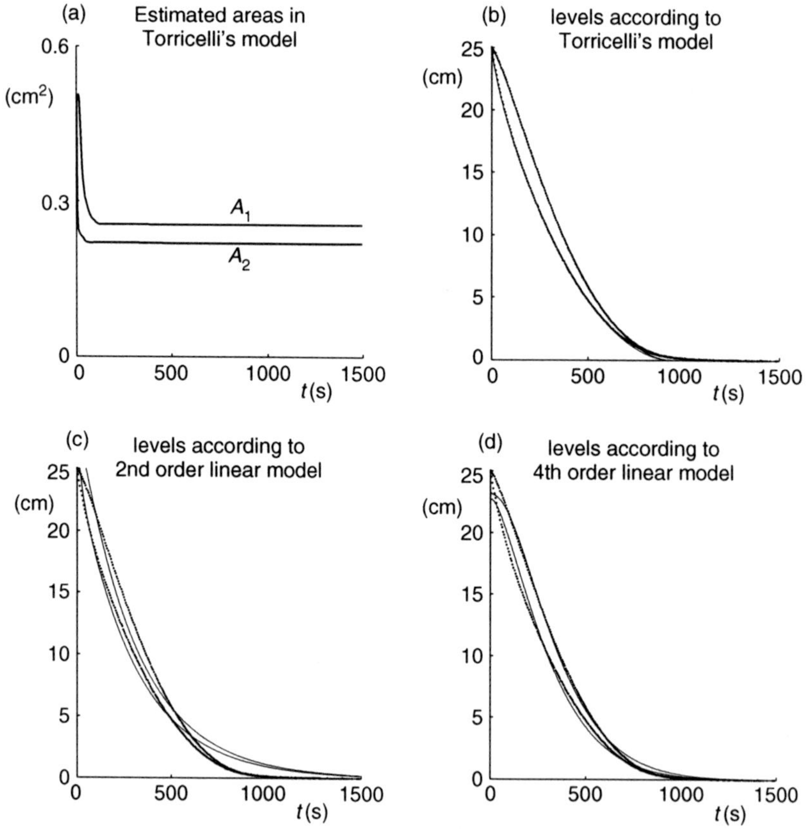

Figure 8.4 Results of the estimation of the parameters of the hydraulic models. The dotted lines are the measurements. The solid lines are results from the model

The second stage of the estimation procedure is a refinement of the parameters based on maximum likelihood estimation. Equation (8.6) was used to numerically evaluate the log-likelihood as a function of the parameters. A normal distribution of $\mathbf{v}(i)$ was assumed. Therefore, instead of the log-likelihood we can equivalently well calculate the sum of squared Mahalanobis distances:

$$J(\hat{\mathbf{x}}(0), \boldsymbol{\alpha}) = \sum_{i=0}^{I} (\mathbf{z}(i) - \hat{\mathbf{x}}(i))^T \mathbf{C}_\mathbf{v}^{-1} (\mathbf{z}(i) - \hat{\mathbf{x}}(i))$$
$$\text{with: } \hat{\mathbf{x}}(i+1) = \mathbf{f}(\hat{\mathbf{x}}(i), \boldsymbol{\alpha}) \text{ for } i = 0, 1, \ldots, I-1 \tag{8.8}$$

The minimalization of $J(\hat{\mathbf{x}}(0), \alpha)$ using the MATLAB function `fminsearch` from the Optimization Toolbox gives the final result. The numerical optimization is initiated with the parameters obtained from particle filtering.

Figures 8.4(b), (c) and (d) show the estimated levels $\hat{\mathbf{x}}(i)$ obtained with minimal J() for the three considered models.

8.1.4 Evaluation and model selection

The last step is the evaluation of the candidate models and the final selection. For that purpose, we select a quality measure and evaluate and compare the various models. Popular quality measures are the log-likelihood and the sum of squares of the residuals (the difference between measurements and model-predicted measurements).

Models with a low order are preferable because the risk of overfitting is minimized and the estimators are less sensitive to estimation errors of the parameters. Therefore, it might be beneficial to consider not only the model with the best quality measure, but also the models which score slightly less, but with a lower order. In order to evaluate these models, other tests are useful. Ideally, the cross correlation between the residuals and the input signal is zero. If not, there is still part of the behaviour of the system that is not explained by the model.

Example 8.4 Model selection for the hydraulic system
Qualitative evaluation of the three models using the log-likelihood as a quality measure (or equivalently, the sum of squared Mahalanobis distances) yields the following results:

Torricelli's model:	$J = 4875$
Second order linear model:	$J = 281140$
Fourth order linear model:	$J = 62595$

Here, there is wide gap between the nonlinear Torricelli's model and the two linear models. Yet, the quality measure $J = 4875$ is still too large to ascribe it fully to the measurement noise. Without the modelling errors, the mean value of the sum of squared Mahalanobis distances $(I = 300, N = 2)$ is 600. Inspection of Figure 8.4(b) reveals that the feet of the curves cause the discrepancy. Perhaps the Torricelli model should be extended with a small linear friction term.

8.1.5 Identification of linear systems with a random input

There is a rich literature devoted to the problem of explaining a random sequence $x(i)$ by means of a linear system driven by white noise (Box, 1976). An example of such a model is the autoregressive model introduced in Section 4.2.1. The Mth order AR model is:

$$x(i) = \sum_{n=1}^{M} \alpha_n x(i-n) + w(i) \tag{8.9}$$

This type of model is easily cast into a state space model. As such it can be used to describe non-white process noise. More general schemes are the autoregressive moving average (ARMA) models and the autoregressive integrating moving average (ARIMA) models. The discussion here is only introductory and is restricted to AR models. For a full treatment we refer to the pertinent literature.

The identification of an AR model from an observed sequence $x(i)$ boils down to the determination of the order M, and the estimation of the parameters α_n and σ_w^2. Assuming that the system is in the steady state, the estimation can be done by solving the *Yule–Walker* equations. These equations arise if we multiply (8.9) on both sides by $x(i-1), \ldots, x(i-M)$, and take expectations:

$$\begin{aligned} \mathrm{E}[x(i)x(i-k)] &= \sum_{n=1}^{M} \alpha_n \mathrm{E}[x(i-n)x(i-k)] \\ &\quad + \mathrm{E}[w(i)x(i-k)] \quad \text{for} \quad k = 1, \ldots, M \end{aligned} \tag{8.10}$$

Since $\mathrm{E}[x(i-k)w(i)] = 0$ and $r_k \stackrel{def}{=} \mathrm{E}[x(i)(i-k)]/\sigma_x^2$, equation (8.10) defines the following systems of linear relations (see also equation (4.21)):

$$\begin{bmatrix} r_1 \\ r_2 \\ r_3 \\ \vdots \\ r_M \end{bmatrix} = \begin{bmatrix} 1 & r_1 & r_2 & \cdots & \cdots & r_{M-1} \\ r_1 & 1 & r_1 & r_2 & \cdots & r_{M-2} \\ r_2 & r_1 & 1 & r_1 & \cdots & r_{M-3} \\ \vdots & & & \ddots & \ddots & \vdots \\ r_{M-1} & r_{M-2} & \cdots & & \cdots & 1 \end{bmatrix} \begin{bmatrix} \alpha_1 \\ \alpha_2 \\ \alpha_3 \\ \\ \alpha_M \end{bmatrix} \tag{8.11}$$

The parameters α_n are found by estimating the correlation coefficients r_k and solving (8.11).

The parameter σ_w^2 is obtained by multiplying (8.9) by $x(i)$ and taking expectations:

$$\sigma_x^2 = \sum_{n=1}^{M} a_n \sigma_x^2 r_k + \sigma_w^2 \tag{8.12}$$

Estimation of σ_x^2 and solving (8.12) gives us the estimate of σ_w^2.

The order of the system can be retrieved by a concept called the *partial autocorrelation function*. Suppose that an AR sequence $x(i)$ has been observed with unknown order M. The procedure for the identification of this sequence is to first estimate the correlation coefficients r_k yielding estimates $\hat{r}_k$. Then, for a number of hypothesized orders $\hat{M} = 1, 2, 3, \ldots$ we estimate the AR coefficients $\hat{\alpha}_{k,\hat{M}}$ for $k = 1, \ldots, \hat{M}$ (the subscript $\hat{M}$ has been added to discriminate between coefficients of different orders). From these coefficients, the last one of each sequence, i.e. $\hat{\alpha}_{\hat{M},\hat{M}}$, is called the partial autocorrelation function. It can be proven that:

$$\alpha_{\hat{M},\hat{M}} = 0 \quad \text{for} \quad \hat{M} > M \tag{8.13}$$

Thus, the order M is determined by checking where $\hat{\alpha}_{\hat{M},\hat{M}}$ drops down to near zero.

Example 8.5 AR model of a pseudorandom binary sequence
Figure 8.5(a) shows a realization of a zero mean, pseudorandom binary sequence. A discrete Markov model, given in terms of transition probabilities (see Section 4.3.1), would be an appropriate model for this type of signal. However, sometimes it is useful to describe the sequence with a linear, AR model. This occurs, for instance, when the sequence is an observation of process noise in an (otherwise) linear plant. The application of a Kalman filter requires the availability of a linear model, and thus the process noise must be modelled as an AR process.

Figure 8.5(b) shows the partial autocorrelation function obtained from a registration of $x(i)$ consisting of 4000 samples (Figure 8.5(a) only shows the first 500 samples). The plot has been made using MATLAB's function `aryule` from the Signal Processing Toolbox. Clearly, the partial autocorrelation function drops down at $\hat{M} = 2$,

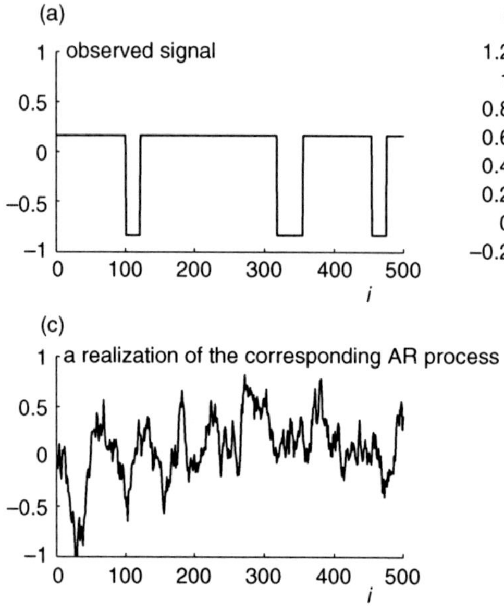

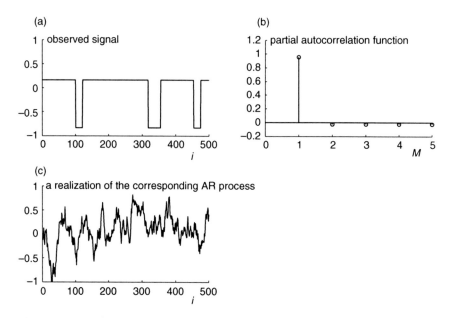

Figure 8.5 Modelling a pseudorandom binary signal by an AR process

so the estimated order is $\hat{M} = 1$, i.e. the best AR model is of first order. Figure 8.5(c) shows a realization of such a process.

8.2 OBSERVABILITY, CONTROLLABILITY AND STABILITY

8.2.1 Observability

We consider a deterministic linear system:

$$\begin{aligned} \mathbf{x}(i+1) &= \mathbf{F}(i)\mathbf{x}(i) + \mathbf{L}(i)\mathbf{u}(i) \\ \mathbf{z}(i) &= \mathbf{H}(i)\mathbf{x}(i) \end{aligned} \tag{8.14}$$

The system is called *observable* if with known $\mathbf{F}(i)$, $\mathbf{L}(i)\mathbf{u}(i)$ and $\mathbf{H}(i)$ the state $\mathbf{x}(i)$ (with fixed i) can be solved from a sequence $\mathbf{z}(i), \mathbf{z}(i+1), \ldots$ of measurements. The system is called *completely observable* if it is observable for any i. In the following, we assume $\mathbf{L}(i)\mathbf{u}(i) = \mathbf{0}$. This is without any loss of generality since the influence to $\mathbf{z}(i), \mathbf{z}(i+1), \ldots$ of a $\mathbf{L}(i)\mathbf{u}(i)$ not being zero can be neutralized easily. Hence, the observability of a system solely depends on $\mathbf{F}(i)$ and $\mathbf{H}(i)$.

An approach to find out whether the system is observable is to construct the *observability Gramian* (Bar-Shalom and Li, 1993). From (8.14):

$$\begin{bmatrix} \mathbf{z}(i) \\ \mathbf{z}(i+1) \\ \mathbf{z}(i+2) \\ \vdots \end{bmatrix} = \begin{bmatrix} \mathbf{H}(i)\mathbf{x}(i) \\ \mathbf{H}(i+1)\mathbf{x}(i+1) \\ \mathbf{H}(i+2)\mathbf{x}(i+2) \\ \vdots \end{bmatrix} = \begin{bmatrix} \mathbf{H}(i) \\ \mathbf{H}(i+1)\mathbf{F}(i) \\ \mathbf{H}(i+2)\mathbf{F}(i)\mathbf{F}(i+1) \\ \vdots \end{bmatrix} \mathbf{x}(i) \tag{8.15}$$

Equation (8.15) is of the type $z = H\mathbf{x}$. The least squares estimate is $\hat{\mathbf{x}} = (H^T H)^{-1} H^T z$. See Section 3.3.1. The solution exists if and only if the inverse of $H^T H$ exists, or in other words, if the rank of $H^T H$ is equal to the dimension M of the state vector. Equivalent conditions are that $H^T H$ is positive definite (i.e. $\mathbf{y}^T H^T H \mathbf{y} > 0$ for every $\mathbf{y} \neq 0$), or that the eigenvalues of $H^T H$ are all positive. See Appendix B.5.

Translated to the present case, the requirement is that for at least one $n \geq 0$ the observability Gramian $\mathfrak{G}$, defined by:

$$\mathfrak{G} = \mathbf{H}^T(i)\mathbf{H}(i) + \sum_{j=1}^{n} \left(\mathbf{H}(i+j) \prod_{k=0}^{j-1} \mathbf{F}(i+k) \right)^T \left(\mathbf{H}(i+j) \prod_{k=0}^{j-1} \mathbf{F}(i+k) \right) \tag{8.16}$$

has rank equal to M. Equivalently we check whether the Gramian is positive definite. For time-invariant systems, F and H do not depend on time, and the Gramian simplifies to:

$$\mathfrak{G} = \sum_{j=0}^{n} (\mathbf{H}\mathbf{F}^j)^T (\mathbf{H}\mathbf{F}^j) \tag{8.17}$$

If the system F is stable (the magnitude of all eigenvalues of F are less than one), we can set $n \to \infty$ to check whether the system is observable.

A second approach to determine the observability of a time-invariant, deterministic system is to construct the *observability matrix*:

$$\mathbf{M} = \begin{bmatrix} \mathbf{H} \\ \mathbf{H}\mathbf{F} \\ \mathbf{H}\mathbf{F}^2 \\ \vdots \\ \mathbf{H}\mathbf{F}^{M-1} \end{bmatrix} \tag{8.18}$$

According to (8.15), $\mathbf{x}(i)$ can be retrieved from a sequence $\mathbf{z}(i), \ldots, \mathbf{z}(i+M-1)$ if $\mathbf{M}$ is invertible; that is, if the rank of $\mathbf{M}$ equals M.

The advantage of using the observability Gramian instead of the observability matrix is that the former is more stable. Modelling errors and round-off errors in the coefficients in both $\mathbf{F}$ and $\mathbf{H}$ could make the difference between an invertible $\mathfrak{G}$ or $\mathbf{M}$ and a noninvertible one. However, $\mathfrak{G}$ is less prone to small errors than $\mathbf{M}$ is.

A more quantitative measure of the observability is obtained by using the eigenvalues of the Gramian $\mathfrak{G}$. A suitable measure is the ratio between the smallest eigenvalue and the largest eigenvalue. The system is less observable as this ratio tends to zero. A likewise result can be obtained by using the singular values of the matrix $\mathbf{M}$ (see singular value decomposition in Appendix B.6).

Example 8.6 Observability of a second order system
Consider the system $(\mathbf{F}, \mathbf{H})$ given by:

$$\mathbf{F} = \begin{bmatrix} 0.66 & 0.12 \\ 0.32 & 0.74 \end{bmatrix} \quad \mathbf{H} = \begin{bmatrix} \frac{1}{3} & \frac{1}{4} \end{bmatrix}$$

The rank of both the Gramian $\mathfrak{G}$ and the observability matrix $\mathbf{M}$ appear to be one, indicating that the system is not observable. However, if the coefficients of $\mathbf{F}$ and $\mathbf{H}$ are represented in single precision IEEE floating point format, the relative round-off error is in the order of 10^{-8}. These round-off errors cause the ratio of eigenvalues of $\mathfrak{G}$ to be in the order of 10^{-16} instead of zero. The comparable ratio of singular values of $\mathbf{M}$ is in the order of 10^{-9}. Clearly, both ratios indicate that the observability is poor. However, the one of $\mathbf{M}$ is overoptimistic. In fact, if MATLAB's function `rank()` is applied to $\mathfrak{G}$ and $\mathbf{M}$, the former returns one, while the latter (erroneously) yields two. The corresponding MATLAB code, given in Listing 8.1, uses functions from the Control System Toolbox. Especially, the function `ss()` is of interest. It creates a state space model, i.e. a special structure array containing all the matrices of a linear time-invariant system.

Listing 8.1
Two methods of obtaining an observability measure of a linear time-invariant system.

```
F=[0.66 0.12; 0.32 0.74];   % Define the system
H=[1/3 1/4];
Fs=double(single(F));       % Round-off to 32 bits
Hs=double(single(H));
B=[1; 0]; D=0;
sys=ss(Fs,B,Hs,D,-1);       % Create state-space model
M=obsv(Fs,Hs);              % Get observability matrix
G=gram(sys,'o');            % and Gramian
eigG=eig(G);                % Calculate eigenvalues
svdM=svd(M);                % and singular values
disp('ratio of eigenvalues of Gramian:');
min(eigG)/max(eigG)
disp('ratio of singular values of observability matrix:');
min(svdM)/max(svdM)
```

The concept of observability can also be extended such that the influence of measurement noise is incorporated. The *stochastic observability* has a strong connection with a particular implementation of the Kalman filter, known as the *information filter*. The details of this extension will follow in Section 8.3.3.

8.2.2 Controllability

In control theory, the concept of controllability usually refers to the ability that for any state $\mathbf{x}(i)$ at a given time i a finite input sequence $\mathbf{u}(i), \mathbf{u}(i+1), \ldots, \mathbf{u}(i+n-1)$ exists that can drive the system to an arbitrary final state $\mathbf{x}(i+n)$. If this is possible for any time, the system is called *completely controllable*.[1] As with observability, the controllability of a system can be revealed by checking the rank of a Gramian. For time-invariant systems, the controllability can also be analysed by means of the *controllability matrix*. This matrix arises from the following equation:

[1] Some authors use the word 'reachability' instead, and reserve the word 'controllability' for a system that can be driven to zero (but not necessarily to an arbitrary state). See Åmström and Wittenmark (1990).

$$\begin{aligned}
\mathbf{x}(i+1) &= \mathbf{F}\mathbf{x}(i) + \mathbf{L}\mathbf{u}(i) \\
\mathbf{x}(i+2) &= \mathbf{F}^2\mathbf{x}(i) + \mathbf{F}\mathbf{L}\mathbf{u}(i) + \mathbf{L}\mathbf{u}(i+1) \\
\mathbf{x}(i+3) &= \mathbf{F}^3\mathbf{x}(i) + \mathbf{F}^2\mathbf{L}\mathbf{u}(i) + \mathbf{F}\mathbf{L}\mathbf{u}(i+1) + \mathbf{L}\mathbf{u}(i+2) \\
&\vdots \\
\mathbf{x}(i+n) &= \mathbf{F}^n\mathbf{x}(i) + \sum_{j=0}^{n-1} \mathbf{F}^j\mathbf{L}\mathbf{u}(i+j)
\end{aligned} \tag{8.19}$$

or:

$$\begin{bmatrix} \mathbf{L} & \mathbf{F}\mathbf{L} & \ldots & \mathbf{F}^{n-1}\mathbf{L} \end{bmatrix} \begin{bmatrix} \mathbf{u}(i) \\ \mathbf{u}(i+1) \\ \vdots \\ \mathbf{u}(i+n-1) \end{bmatrix} = \mathbf{x}(i+n) - \mathbf{F}^n\mathbf{x}(i) \tag{8.20}$$

The minimum number of steps, n, is at most equal to M, the dimension of the state vector. Therefore, in order to test the controllability of the system (**F**, **L**) it suffices to check whether the controllability matrix $[\mathbf{L} \quad \mathbf{F}\mathbf{L} \quad \ldots \quad \mathbf{F}^{M-1}\mathbf{L}]$ has rank M.

The MATLAB functions for creating the controllability matrix and Gramian are `ctrb()` and `gram()`, respectively.

8.2.3 Dynamic stability and steady state solutions

The term *stability* refers to the ability of a system to resist to and recover from disturbances acting on this system. A state estimator has to face three different causes of instabilities: sensor instability, numerical instability and dynamic instability.

Apart from the usual sensor noise and sensor linearity errors, a sensor may produce unusual glitches and other errors caused by hard to predict phenomena, such as radio interference, magnetic interference, thermal drift, mechanical shocks and so on. This kind of behaviour is sometimes denoted by *sensor instability*. Its early detection can be done using consistency checks (to be discussed in Section 8.4).

Numerical instabilities originate from round-off errors. Particularly, the inversion of a near singular matrix may cause large errors due to its sensitivity to round-off errors. A careful implementation of the design must prevent these errors. See Section 8.3.

The third cause for instability lies in the dynamics of the state estimator itself. In order to study the dynamic stability of the state estimator it is necessary to consider the estimator as a dynamic system with as inputs the measurements $\mathbf{z}(i)$ and the control vectors $\mathbf{u}(i)$. See Appendix D. The output consists of the estimates $\bar{\mathbf{x}}(i|i)$. In linear systems, the stability does not depend on the input sequences. For the stability analysis it suffices to assume zero $\mathbf{z}(i)$ and $\mathbf{u}(i)$. The equations of interest are derived from (8.2):

$$\bar{\mathbf{x}}(i+1|i+1) = (\mathbf{I} - \mathbf{K}(i)\mathbf{H}(i))\mathbf{F}(i-1)\bar{\mathbf{x}}(i|i) \tag{8.21}$$

with:

$$\mathbf{K}(i) = \mathbf{P}(i)\mathbf{H}^T(i)\left(\mathbf{H}(i)\mathbf{P}(i)\mathbf{H}^T(i) + \mathbf{C}_\mathbf{v}(i)\right)^{-1}$$

$\mathbf{P}(i)$ is the covariance matrix $\mathbf{C}(i+1|i)$ of the predicted state $\bar{\mathbf{x}}(i+1|i)$. $\mathbf{P}(i)$ is recursively defined by the *discrete Ricatti equation*:

$$\begin{aligned}\mathbf{P}(i+1) = {} & \mathbf{F}(i)\mathbf{P}(i)\mathbf{F}^T(i) + \mathbf{C}_\mathbf{w}(i) \\ & - \mathbf{F}(i)\mathbf{P}(i)\mathbf{H}^T(i)\left(\mathbf{H}(i)\mathbf{P}(i)\mathbf{H}^T(i) + \mathbf{C}_\mathbf{v}(i)\right)^{-1}\mathbf{H}(i)\mathbf{P}^T(i)\mathbf{F}^T(i)\end{aligned} \tag{8.22}$$

The recursion starts with $\mathbf{P}(0) \stackrel{def}{=} \mathbf{C}_\mathbf{x}(0)$. The first term in (8.22) represents the absorption of uncertainty due to the dynamics of the system during each time step (provided that $\mathbf{F}$ is stable; otherwise it represents the growth of uncertainty). The second term represents the additional uncertainty at each time step due to the process noise. The last term represents the reduction of uncertainty thanks to the measurements.

For the stability analysis of a Kalman filter it is of interest to know whether a sequence of process noise, $\mathbf{w}(i)$, can influence each element of the state vector independently. The answer to this question is found by writing the covariance matrix $\mathbf{C}_\mathbf{w}(i)$ of the process noise as $\mathbf{C}_\mathbf{w}(i) = \mathbf{G}(i)\mathbf{G}^T(i)$. Here, $\mathbf{G}(i)$ can be obtained by an eigenvalue/eigenvector diagonalization of $\mathbf{C}_\mathbf{w}(i)$. That is $\mathbf{C}_\mathbf{w}(i) = \mathbf{V}_\mathbf{w}(i)\boldsymbol{\Lambda}_\mathbf{w}(i)\mathbf{V}_\mathbf{w}^T(i)$ and $\mathbf{G}(i) = \mathbf{V}_\mathbf{w}(i)\boldsymbol{\Lambda}_\mathbf{w}^{1/2}(i)$. See Appendix B.5 and C.3. $\boldsymbol{\Lambda}_\mathbf{w}(i)$ is a $K \times K$ diagonal matrix where K is the number of nonzero eigenvalues of

$\mathbf{C}_\mathbf{w}(i)$. The matrices $\mathbf{G}(i)$ and $\mathbf{V}_\mathbf{w}(i)$ are $M \times K$ matrices. The introduction of $\mathbf{G}(i)$ allows us to write the system equation as:

$$\mathbf{x}(i+1) = \mathbf{F}(i)\mathbf{x}(i) + \mathbf{L}(i)\mathbf{u}(i) + \mathbf{G}(i)\tilde{\mathbf{w}}(i) \tag{8.23}$$

where $\tilde{\mathbf{w}}(i)$ is a K dimensional Gaussian white noise vector with covariance matrix $\mathbf{I}$. The noise sequence $\mathbf{w}(i)$ influences each element of $\mathbf{x}(i)$ independently if the system $(\mathbf{F}(i), \mathbf{G}(i))$ is controllable by the sequence $\tilde{\mathbf{w}}(i)$. We then have the following theorem (Bar-Shalom, 1993), which provides a sufficient (but not a necessary) condition for stability:

> If a time invariant system $(\mathbf{F}, \mathbf{H})$ is completely observable and the system $(\mathbf{F}, \mathbf{G})$ is completely controllable, then for any initial condition $\mathbf{P}(0) = \mathbf{C}_\mathbf{x}(0)$ the solution of the Ricatti equation converges to a unique, finite, invertible steady state covariance matrix $\mathbf{P}(\infty)$.

The observability of the system assures that the sequence of measurements contains information about the complete state vector. Therefore, the observability guarantees that the prediction covariance matrix $\mathbf{P}(\infty)$ is bounded from above. Consequently, the steady state Kalman filter is BIBO stable. See Appendix D.3.2. If one or more state elements are not observable, the Kalman gains for these elements will be zero, and the estimation of these elements is purely prediction driven (model based without using measurements). If the system $\mathbf{F}$ is unstable for these elements, then so is the Kalman filter.

The controllability with respect $\tilde{\mathbf{w}}(i)$ assures that $\mathbf{P}(\infty)$ is unique (does not depend on $\mathbf{C}_\mathbf{x}(0)$), and that its inverse exists. If the system is not controllable, then $\mathbf{P}(\infty)$ might depend on the specific choice of $\mathbf{C}_\mathbf{x}(0)$. If for the non-controllable state elements the system $\mathbf{F}$ is stable, then the corresponding eigenvalues of $\mathbf{P}(\infty)$ are zero. Consequently, the Kalman gains for these elements are zero, and the estimation of these elements is again purely prediction driven.

The discrete algebraic Ricatti equation, $\mathbf{P}(i+1) = \mathbf{P}(i)$, mentioned in Chapter 4, provides the steady state solution for the Ricatti equation:

$$\mathbf{P} = \mathbf{F}\mathbf{P}\mathbf{F}^T + \mathbf{C}_\mathbf{w} - \mathbf{F}\mathbf{P}\mathbf{H}^T\left(\mathbf{H}\mathbf{P}\mathbf{H}^T + \mathbf{C}_\mathbf{v}\right)^{-1}\mathbf{H}\mathbf{P}^T\mathbf{F}^T \tag{8.24}$$

The steady state is reached due to the balance between the growth of uncertainty (the second term and possibly the first term) and the reduction of uncertainty (the third term and possibly the first term).

A steady state solution of the Ricatti equation gives rise to a constant Kalman gain. The corresponding Kalman filter, derived from (8.2) and (8.21):

$$\bar{\mathbf{x}}(i+1|i+1) = (\mathbf{I} - \mathbf{KH})\mathbf{F}\,\bar{\mathbf{x}}(i|i) + (\mathbf{I} - \mathbf{KH})\mathbf{Lu}(i) + \mathbf{Kz}(i+1) \quad (8.25)$$

with:

$$\mathbf{K}(i) = \mathbf{PH}^T\left(\mathbf{HPH}^T + \mathbf{C_v}\right)^{-1}$$

becomes time invariant.

Example 8.7 Stability of a system that is not observable
Consider the system $(\mathbf{F}, \mathbf{H}, \mathbf{C_w}, \mathbf{C_v})$ given by:

$$\mathbf{F} = \begin{bmatrix} 0.66 & 0.32 \\ 0.12 & 0.74 \end{bmatrix} \quad \mathbf{H} = \begin{bmatrix} 1.2 & 2.4 \end{bmatrix} \quad \mathbf{C_w} = \begin{bmatrix} 1 & 0 \\ 0 & 1 \end{bmatrix} \qquad \mathbf{C_v} = [1]$$

This system is controllable, but not observable. Yet, the Ricatti equation is asymptotically stable with steady state solution:

$$\mathbf{P} = \begin{bmatrix} 1.3110 & -0.0822 \\ -0.0822 & 1.1293 \end{bmatrix} \text{and eigenvalues 1.0976 and 1.3427}$$

The eigenvalues of the corresponding steady state Kalman filter, i.e. the eigenvalues of $(\mathbf{I} - \mathbf{KH})\mathbf{F}$, are 0.5 and 0.101. Thus the Kalman filter is stable.

Listing 8.2 provides the MATLAB code for this example. The function `dlqe()` returns the Kalman gain matrix together with the prediction covariance, the error covariance and the eigenvalues of the Kalman filter. Alternatively, we use the function `kalman()` that creates the estimator as a state space model. Internally, it uses the function `dare()` which solves the discrete algebraic Ricatti equation. The functions are from the Control System Toolbox.

The explanation for the stability in the last example is as follows. Suppose that the measurements are switched off, that is, $\mathbf{K} = 0$. In that case, the estimates just follow the dynamics of the system: $\bar{\mathbf{x}}(i+1|i+1) = \mathbf{F}\,\bar{\mathbf{x}}(i|i) + \mathbf{Lu}(i)$. Since $\mathbf{F}$ is stable, the initial uncertainty $\mathbf{C_x}(0)$ will be absorbed. In the steady state, the final uncertainty is given by the balance:

$$\mathbf{C_x}(\infty) = \mathbf{FC_x}(\infty)\mathbf{F}^T + \mathbf{C_w} \quad (8.26)$$

(the *discrete Lyapunov equation*). Since the Kalman filter is optimal, the steady state solution of the Ricatti equation is bounded from above by $\mathbf{C_x}(\infty)$, and thus asymptotical stable. The MATLAB function `dlyap()` returns the solution of the discrete Lyapunov equation.

Example 8.8 Stability of a system that is not observable (continued)
A further inspection of the situation confirms the statement made above. The solution of the discrete Lyapunov equation is:

$$\mathbf{C_x}(\infty) = \begin{bmatrix} 4.0893 & 2.3094 \\ 2.3094 & 3.2472 \end{bmatrix} \text{with eigenvalues 1.32 and 6.02}$$

Indeed, $\mathbf{P} \leq \mathbf{C_x}(\infty)$ (meaning that the difference $\mathbf{C_x}(\infty) - \mathbf{P}$ is positive semidefinite; i.e. possesses only non-negative eigenvalues). The eigenvalues of $\mathbf{F}$ are 0.5 and 0.9. The eigenvalue of 0.5 corresponds to the state in the diagonalized system (Appendix D.3.1) that is not observed by the measurements. However, this state is stable. The Kalman gain for this state is zero, and thus the steady state Kalman filter copies this eigenvalue.

If the second eigenvalue of $\mathbf{F}$ is increased from 0.9 to, say, 1.5, the system is not stable anymore, nevertheless the steady state solution of the Ricatti equation exists. The corresponding Kalman filter is stable. However, if the first eigenvalue of $\mathbf{F}$ is increased from 0.5 to 1.5, the system is again not stable. But this time, the corresponding Kalman filter isn't stable either. The Ricatti equation is not stable anymore.

Example 8.9 Stability of a system that is not controllable
Consider the system $(\mathbf{F}, \mathbf{H}, \mathbf{C_w}, \mathbf{C_v})$ given by:

$$\mathbf{F} = \begin{bmatrix} 0.66 & 0.32 \\ 0.12 & 0.74 \end{bmatrix} \quad \mathbf{C_w} = \begin{bmatrix} 0.1111 & 0.0833 \\ 0.0833 & 0.0625 \end{bmatrix}$$

$$\mathbf{H} = [1 \quad 1] \qquad \mathbf{C_v} = [1]$$

This system is observable. The covariance matrix of the process noise can be written as: $\mathbf{C_w} = \mathbf{G}\mathbf{G}^T$ with $\mathbf{G}^T = [\,0.333 \quad 0.25\,]$. The system $(\mathbf{F}, \mathbf{G})$ is not controllable as a simple test can show. The steady state solution of the Ricatti equation is:

$$\mathbf{P} = \begin{bmatrix} 0.2166 & 0.1624 \\ 0.1624 & 0.1218 \end{bmatrix} \text{with eigenvalues 0 and 0.338}$$

P is not invertible. The eigenvalues of **F** are 0.9 and 0.5. The Kalman filter, with eigenvalues 0.5411 and 0.5, is stable.

The explanation of this behaviour is as follows. The diagonalized system has one controllable state (corresponding to an eigenvalue of 0.9). For this state, the Kalman filter behaves regularly. The second state (with eigenvalue 0.5) is not controllable. This state is not affected by the process noise. It is a stable state, and thus, the initial uncertainty fades out. The zero variance of this state causes a zero eigenvalue in $\mathbf{C_x}(\infty)$. With that, the Kalman gain for that state also becomes zero because without uncertainty there is no need for measurements. Consequently, the eigenvalue of the system repeats itself in the Kalman filter. The zero eigenvalue of $\mathbf{C_x}(\infty)$ causes a corresponding zero eigenvalue in **P**. Thus, this matrix is not invertible.

If the second eigenvalue of **F** is increased from 0.5 to 1.5, the initial condition $\mathbf{C_x}(0)$ influences the long term behaviour of $\mathbf{C_x}(i)$. If $\mathbf{C_x}(0) = \mathbf{0}$, then $\mathbf{C_x}(i)$ converges to a constant. But this solution is not stable. A small perturbation of $\mathbf{C_x}(0)$ causes $\mathbf{C_x}(i)$ to diverge to infinity. Small perturbations of $\mathbf{C_x}(0)$ trigger $\mathbf{P}(i)$ to follow quite different trajectories, but they finally converge to a nonzero steady state for which the Kalman filter is stable.

The last example shows that if a system $(\mathbf{F}, \mathbf{G})$ is not controllable, some eigenvalues of the prediction covariance matrix may become zero. The matrix **P** is positive semidefinite. Such a situation does not contribute to the numerical stability.

Listing 8.2

Steady state solution of a system that is not observable.

```
lambda=diag([0.9 0.5]);        % Define a system with
V=[1/3 -1; 1/4 1/2];           % eigenvalues 0.9 and 0.5
F=V*lambda*inv(V);
H=inv(V); H(2,:)=[];           % Define a measurement matrix
                               % that only observes one state
Cv=eye(1); Cw=eye(2);          % Define covariance matrices
% Discrete steady state Kalman filter:
[M,P,Z,E]=dlqe(F,eye(2),H,Cw,Cv);
Cx_inf=dlyap(F,Cw)             % Solution of discrete Lyapunov equation
```

```
disp('Kalman gain matrix');                          disp(M);
disp('Eigenval. of Kalman filter');                  disp(E);
disp('Error covariance');                            disp(Z);
disp('Prediction covariance');                       disp(P);
disp('Eigenval. of prediction covariance');          disp(eig(P));
disp('Solution of discrete Lyapunov equation');      disp(Cx_inf);
disp('Eigenval. of sol.
  of discrete Lyapunov eq.');                        disp(eig
                                                       (Cx_inf));
```

8.3 COMPUTATIONAL ISSUES

A straightforward implementation of the time-variant Kalman filter may result in too large estimation errors. The magnitudes of these errors are not compatible with the error covariance matrices. The filter may even completely diverge even though theoretically the filter should be stable. This anomalous behaviour is due to a number representation with limited precisions. In order to find where round-off errors have the largest impact it is instructive to rephrase the Kalman equations in (8.2) as follows:

Ricatti loop:

$$\begin{aligned} \mathbf{C}(i|i) &= \mathbf{C}(i|i-1) - \mathbf{C}(i|i-1)\mathbf{H}^T(\mathbf{H}\mathbf{C}(i|i-1)\mathbf{H}^T + \mathbf{C}_\mathbf{v})^{-1}\,\mathbf{H}\mathbf{C}(i|i-1) \\ \mathbf{C}(i+1|i) &= \mathbf{F}\mathbf{C}(i|i)\mathbf{F}^T + \mathbf{C}_\mathbf{w} \end{aligned}$$

$$\begin{aligned} &\downarrow \\ \mathbf{K}(i) &= \mathbf{C}(i|i-1)\mathbf{H}^T\mathbf{S}^{-1}(i) \\ \mathbf{S}(i) &= \mathbf{H}\mathbf{C}(i|i-1)\mathbf{H}^T + \mathbf{C}_\mathbf{v} \\ &\downarrow \end{aligned} \tag{8.27}$$

estimation loop:

$$\begin{aligned} \bar{\mathbf{x}}(i|i) &= \bar{\mathbf{x}}(i|i-1) + \mathbf{K}(i)(\mathbf{z}(i) - \mathbf{H}\bar{\mathbf{x}}(i|i-1)) \\ \bar{\mathbf{x}}(i+1|i) &= \mathbf{F}\,\bar{\mathbf{x}}(i|i) + \mathbf{L}\mathbf{u}(i) \end{aligned}$$

For simplicity, the system (**F**, **L**, **H**) is written without the time index.

As can be seen, the filter consists of two loops. If the Kalman filter is stable, the second loop (the estimation loop) is usually not that sensitive to round-off errors. Possible induced errors are filtered in much the same way as the measurement noise. However, the loop depends on the Kalman gains $\mathbf{K}(i)$. Large errors in $\mathbf{K}(i)$ may cause the filter to become unstable. These gains come from the first loop.

In the first loop (the Ricatti loop) the prediction covariance matrix $\mathbf{P}(i) = \mathbf{C}(i+1|i)$ is recursively calculated. As can be seen, the recursion involves nonlinear matrix operations including a matrix inversion. Especially, the representation of these matrices (and through this the Kalman gain) may be sensitive to the effects of round-off errors.

The sensitivity to round-off errors becomes apparent if an eigenvalue–eigenvector decomposition (Appendix B.5 and C.3.2) is applied to the covariance matrix $\mathbf{P}$:

$$\mathbf{P} = \sum_{m=1}^{M} \lambda_m \mathbf{v}_m \mathbf{v}_m^T \tag{8.28}$$

λ_m are the eigenvalues of $\mathbf{P}$ and $\mathbf{v}_m$ the corresponding eigenvectors. The eigenvalues of a properly behaving covariance matrix are all positive (the matrix is positive definite and non-singular). However, the range of the eigenvalues may be very large. This finds expression in the condition number $\lambda_{max}/\lambda_{min}$ of the matrix. Here, $\lambda_{max} = \max(\lambda_m)$ and $\lambda_{min} = \min(\lambda_m)$. A large condition number indicates that if the matrix is inverted, the propagation of round-off errors will be large:

$$\mathbf{P}^{-1} = \sum_{m=1}^{M} \frac{1}{\lambda_m} \mathbf{v}_m \mathbf{v}_m^T \tag{8.29}$$

In a floating point representation of $\mathbf{P}$, the exponents are largely determined by λ_{max}. Therefore, the round-off error in λ_{min} is proportional to λ_{max}, and may be severe. It will result in large errors in $1/\lambda_{min}$.

Another operation with a large sensitivity to round-off errors is the subtraction of two similar matrices.

These errors can result in a loss of symmetry in the covariance matrices and a loss of positive definiteness. In some cases, the eigenvalues of the covariance matrices can even become negative. If this occurs, the errors may accumulate during each recursion, and the process may diverge.

As an example, consider the following Ricatti loop:

$$\begin{aligned} \mathbf{S}(i) &= \mathbf{HC}(i|i-1)\mathbf{H}^T + \mathbf{C_v} \\ \mathbf{K}(i) &= (\mathbf{HC}(i|i-1))^T\mathbf{S}^{-1}(i) \\ \mathbf{C}(i|i) &= \mathbf{C}(i|i-1) - \mathbf{K}(i)\mathbf{HC}(i|i-1) \\ \mathbf{C}(i+1|i) &= \mathbf{FC}(i|i)\mathbf{F}^T + \mathbf{C_w} \end{aligned} \tag{8.30}$$

This loop is mathematically equivalent to (8.2), but is computationally less expensive because the factor $\mathbf{HC}(i|i-1)$ appears at several places and can be reused. However, the form is prone to round-off errors:

- The part $\mathbf{K}(i)\mathbf{H}$ can easily introduce asymmetries in $\mathbf{C}(i|i)$.
- The subtraction $\mathbf{I} - \mathbf{K}(i)\mathbf{H}$ can introduce negative eigenvalues.

The implementation should be used cautiously. The preferred implementation is (8.2):

$$\mathbf{C}(i|i) = \mathbf{C}(i|i-1) - \mathbf{K}(i)\mathbf{S}(i)\mathbf{K}^T(i) \tag{8.31}$$

It requires more operations than expression (8.30), but it is more balanced. Therefore, the risk of introducing asymmetries is much lower. Still, this implementation does not guarantee non-negative eigenvalues.

Listing 8.3 implements the Kalman filter using (8.31). The functions `acquire_measurement_vector()` and `get_control_vector()` are placeholders for the interfacing to the measurement system and a possible control system. These functions should also take care of the timing of the estimation process. The required number of operations (using MATLAB's standard matrix arithmetic) of the update step is about $2M^2N + 3MN^2 + N^3$. The number of the operations of the prediction step is about $2M^3$.

Listing 8.3
Conventional time variant Kalman filter for a time invariant system.

```
load linsys;          % Load a system: F,H,Cw,Cv,Cx0,x0,L
C=Cx0;                % Initialize prior uncertainty
xprediction=x0;       % and prior mean
while (1)             % Endless loop
  % Update:
  S=H*C*H'+Cv;        % Innovation matrix
  K=C*H'*S^-1;        % Kalman gain matrix
```

```
  z=acquire_measurement_vector();
  innovation=z - H*xprediction;
  xestimation=xprediction+K*innovation;
  C=C - K*S*K';
  % Prediction:
  u=get_control_vector();
  xprediction=F*xestimation+L*u;
  C=F*C*F'+Cw;
end
```

Example 8.10 Numerical instability of the conventional Kalman filter
Consider the system $(\mathbf{F}, \mathbf{H}, \mathbf{C}_\mathbf{w}, \mathbf{C}_\mathbf{v}, \mathbf{C}_\mathbf{x}(0))$ given by:

$$\mathbf{F} = \begin{bmatrix} 0.9698 & 0.1434 & -0.1101 \\ -0.1245 & 0.9547 & 0.2557 \\ 0.1312 & -0.2455 & 0.9557 \end{bmatrix} \qquad \mathbf{C}_\mathbf{x}(0) = \begin{bmatrix} 100 & 0 & 0 \\ 0 & 100 & 0 \\ 0 & 0 & 100 \end{bmatrix}$$

$$\mathbf{C}_\mathbf{w} = \mathbf{A}\mathbf{A}^T \;\; \text{with} \;\; \mathbf{A} = \begin{bmatrix} 8.1174 \\ 3.9092 \\ 4.3388 \end{bmatrix}$$

$$\mathbf{H} = \begin{bmatrix} 0.0812 & 0.0391 & 0.0434 \\ -0.4339 & 0.9010 & 0 \\ -0.3909 & -0.1883 & 0.9010 \end{bmatrix} \qquad \mathbf{C}_\mathbf{v} = \begin{bmatrix} 10^{-6} & 0 & 0 \\ 0 & 10^{-6} & 0 \\ 0 & 0 & 10^{-6} \end{bmatrix}$$

The eigenvalues of $\mathbf{F}$ are $0.999 \exp(-0.1j\pi), 0.999 \exp(0.1j\pi)$ and 0.98 where $j = \sqrt{-1}$. The magnitude of the first two eigenvalues are close to unit, and therefore, the system is just stable. A diagonalization of the system would reveal that the process noise only affects the state that corresponds with the eigenvalue 0.98. Hence, the system is not controllable by the process noise. The measurement matrix is such that it measures the diagonalized states.

An implementation of (8.30) for this system yields results as shown in Figure 8.6. The results are obtained by means of a simulation of the system and a 32-bit IEEE floating point implementation of the Kalman filter (including the Ricatti loop). The relative round-off errors are of the order 10^{-8}. The Kalman filter appears to be unstable due to the negative eigenvalues of the covariance matrix. The same implementation using 64-bit double precision (relative error is around 10^{-16}) yields stable results. Application of the code in Listing 8.3, but performed with 32 bit precision gives the results shown in Figure 8.7. Although the filter is stable now, it still doesn't work properly.

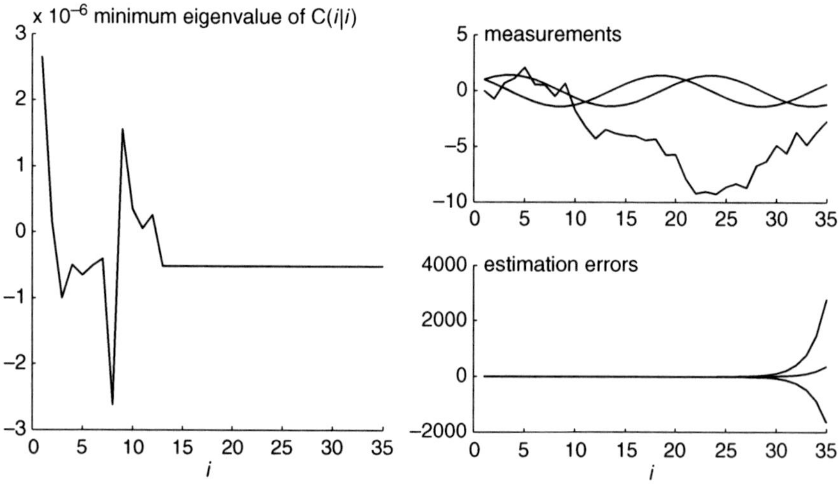

Figure 8.6 Results of a computationally efficient implementation of the conventional Kalman filter. The filter is unstable due to an eigenvalue of **P** that remains negative

The symmetry of a matrix **P** can be enforced by the assignment: $\mathbf{P} := (\mathbf{P} + \mathbf{P}^T)/2$. The addition of this type of statement at some critical locations in the code helps a little, but often not enough. The true remedy is to use a proper implementation combined with sufficient number precision. The remaining part of this section discusses a number of different implementations.

8.3.1 The linear-Gaussian MMSE form

In Section 3.1.3 we derived the MMSE estimator for static variables in the linear-Gaussian case. In section 3.1.5, the unbiased linear MMSE estimator was derived. Since the MMSE solution, expressed in equation (3.20), appeared to be linear and unbiased, the conclusion was drawn that this solution is identical to the unbiased linear MMSE solution (given in (3.33) and (3.45)). However, the two solutions have different forms. We denote the first solution by the (linear-Gaussian) MMSE form. The second solution is the Kalman form. The equivalence between the two solutions can be shown by using the matrix inversion lemma, (b.10).

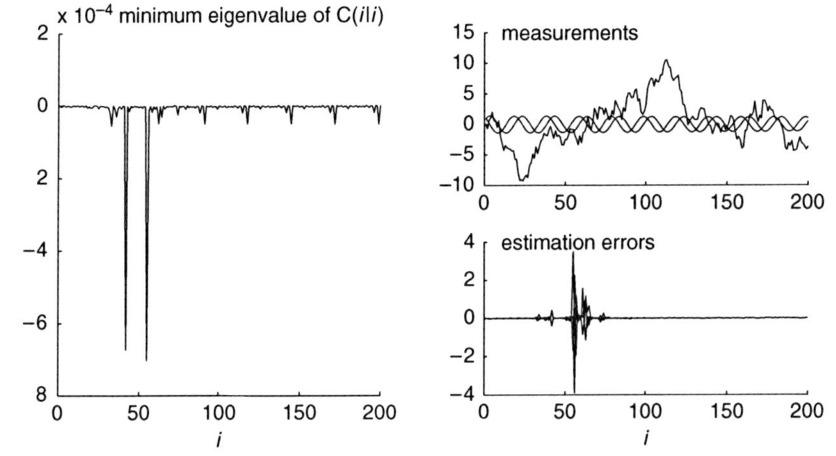

Figure 8.7 Results of a balanced implementation of the conventional Kalman filter. The filter is stable, but still an eigenvalue of $\mathbf{C}(i|i)$ is sometimes negative

As a result of all this, the Kalman filter can also be expressed in two forms. The following implementation is based on the MMSE form, i.e. equation (3.20):

update :

$$\mathbf{C}(i|i) = \left(\mathbf{C}^{-1}(i|i-1) + \mathbf{H}^T\mathbf{C}_\mathbf{v}^{-1}\mathbf{H}\right)^{-1} \qquad \text{(error covariance matrix)}$$

$$\bar{\mathbf{x}}(i|i) = \mathbf{C}(i|i)\left(\mathbf{C}^{-1}(i|i-1)\bar{\mathbf{x}}(i|i-1) + \mathbf{H}^T\mathbf{C}_\mathbf{v}^{-1}\mathbf{z}(i)\right) \qquad \text{(updated estimate)}$$

$$(8.32)$$

The Kalman form, given in (8.2), requires the inversion of $\mathbf{S}(i)$, an $N \times N$ matrix. The MMSE form requires the inversion of $\mathbf{C}(i|i-1)$ and $\mathbf{C}^{-1}(i-1|i) + \mathbf{H}^T\mathbf{C}_\mathbf{v}^{-1}\mathbf{H}$; both are $M \times M$ matrices. It also requires the inversion of $\mathbf{C}_\mathbf{v}$, but this can be done outside the Ricatti loop. Besides, $\mathbf{C}_\mathbf{v}$ is often a diagonal matrix (uncorrelated measurement noise), whose inversion is without problems. The situation where $\mathbf{C}_\mathbf{v}$ is not invertible is a degenerated case. One or more measurements are completely correlated with other measurements. Such a measurement can be removed without loss of information.

In the time-invariant case, the calculation of the term $\mathbf{H}^T\mathbf{C}_\mathbf{v}^{-1}\mathbf{H}$, appearing in (8.32), can be kept outside the loop. If so, the number of operations is $2M^3 + M^2 + M$. Thus, if there are many measurements and only a few states, i.e. $N \gg M$, the a priori form might be favourable. But in other cases, the Kalman form is often preferred.

Example 8.11 The linear-Gaussian MMSE form
Application of the MMSE form to the same system and data as in Example 8.10 yields the results as shown in Figure 8.8 (using 32-bit floating point number representations). At first sight, there seems nothing wrong with these results. However, the sudden change of the smallest eigenvalue at $i = 2$, from which point on it remains constant, is reason to become suspicious.

8.3.2 Sequential processing of the measurements

The conventional Kalman filter processes the measurement data in blocks. The data $z_n(i)$ of the sensors available at time i are collected in the measurement vector $\mathbf{z}(i)$, and processed as one unit. Another possibility is to process the individual measurements sequentially. A requirement is that the measurement noise is uncorrelated. $\mathbf{C}_v$ must be a diagonal matrix. If not, the measurement vector must be decorrelated first using techniques as described in Appendix C.3.1 (Kaminski *et al.*, 1971). In the following algorithm, the diagonal elements of $\mathbf{C}_v$ are denoted by σ_n^2. The row vector $\mathbf{h}_n$ stands for the n-th row of $\mathbf{H}$. Thus, $\mathbf{H}^T = [\,\mathbf{h}_0^T \quad \ldots \quad \mathbf{h}_{N-1}^T\,]$.

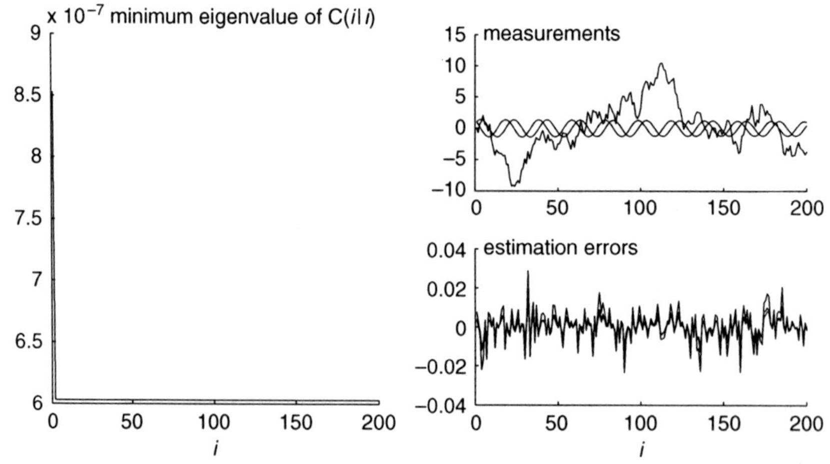

Figure 8.8 Results of the Kalman filter implemented in the MMSE form

Algorithm 8.1: Sequential update

1. Initialization:

   - $\mathbf{x}(i,0) \stackrel{def}{=} \bar{\mathbf{x}}(i|i-1)$
   - $\mathbf{C}(i,0) \stackrel{def}{=} \mathbf{C}(i|i-1)$

2. Sequential update:

   For $n = 0, 1, 2, \ldots, N-1$:

   - $s(i,n) = \mathbf{h}_n \mathbf{C}(i,n)\mathbf{h}_n^T + \sigma_n^2$ (innovation variance)
   - $\mathbf{k}_n = \dfrac{\mathbf{C}(i,n)\mathbf{h}_n^T}{s(i,n)}$ (Kalman gain vector)
   - $\mathbf{x}(i,n+1) = \mathbf{x}(i,n) + \mathbf{k}_n(z_n(i) - \mathbf{h}_n\mathbf{x}(i,n))$ (update of the estimate)
   - $\mathbf{C}(i,n+1) = \mathbf{C}(i,n) - \dfrac{\mathbf{C}(i,n)\mathbf{h}_n^T\mathbf{h}_n\mathbf{C}(i,n)}{s(i,n)}$ (update of the covariance)

3. Closure:

   - $\bar{\mathbf{x}}(i|i) = \mathbf{x}(i,n)$
   - $\mathbf{C}(i|i) = \mathbf{C}(i,n)$

The number of required operations is about $2M^2N + 2MN$. It outperforms the conventional Kalman filter. However, the subtraction that is needed to get $\mathbf{C}(i, n+1)$ can introduce negative eigenvalues. Consequently, the algorithm is still sensitive to round-off errors.

Example 8.12 Sequential processing
The results obtained by the application of sequential processing of the measurements to the problem from Example 8.10 is shown in Figure 8.9. Negative eigenvalues are not prevented, and the filter does not behave correctly.

8.3.3 The information filter

In the conventional Kalman filter it is difficult to represent a situation in which no knowledge is available about (a subspace of) the state. It would

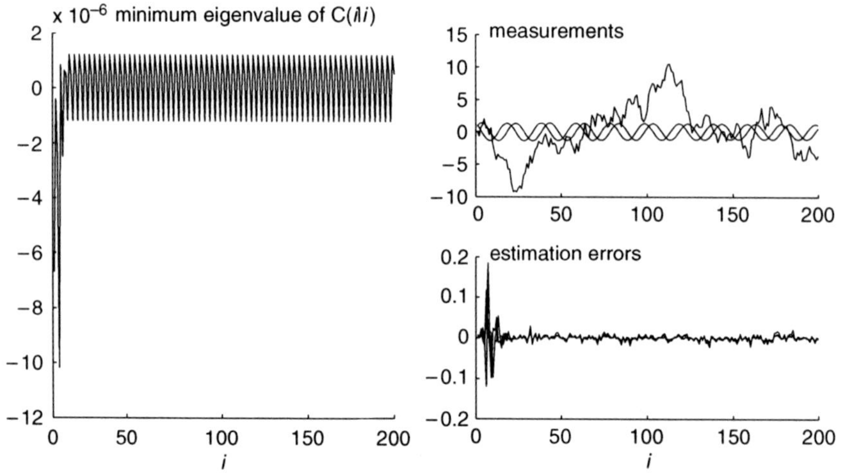

Figure 8.9 Sequential processing of the measurements

require that some eigenvalues of the corresponding error covariance matrix would be infinity. The concept of an *information matrix* circumvents this problem.

An information matrix is the inverse of a covariance matrix. If one or more eigenvalues of an information matrix are small, then a subspace exists in which the uncertainty of the random vector is large. In fact, if one or more eigenvalues are zero, then no knowledge exists about the random vector in the subspace spanned by the corresponding eigenvectors. In this situation, the covariance matrix does not exist because the information matrix is not invertible.

The information filter is an implementation of the Kalman filter in which the Ricatti loop is entirely expressed in terms of information matrices (Grewal and Andrews, 2001). Let $\mathbf{Y}(i|j)$ be the information matrix corresponding to $\mathbf{C}(i|j)$. Thus:

$$\mathbf{Y}(i|j) \stackrel{def}{=} \mathbf{C}^{-1}(i|j) \tag{8.33}$$

Using (8.32), the update in the Kalman filter is rewritten as:

update:

$$\begin{aligned} &\mathbf{Y}(i|i) = \mathbf{Y}(i|i-1) + \mathbf{H}^T\mathbf{C}_v^{-1}\mathbf{H} && \text{(information matrix)} \\ &\bar{\mathbf{x}}(i|i) = \mathbf{Y}^{-1}(i|i)(\mathbf{Y}(i|i-1)\bar{\mathbf{x}}(i|i-1) + \mathbf{H}^T\mathbf{C}_v^{-1}\mathbf{z}(i)) && \text{(updated estimate)} \end{aligned} \tag{8.34}$$

The computational cost of the update of the information matrix is only determined by the inversion of $\mathbf{Y}(i|i)$ (in the time-invariant case) because the term $\mathbf{H}^T\mathbf{C}_v^{-1}\mathbf{H}$ is constant and can be kept outside the loop. The number of required operations is about $M^3 + \frac{1}{2}M^2 + \frac{1}{2}M$.

In order to develop the expression of the predicted state information matrix, the covariance of the process noise is factored as follows $\mathbf{C}_w = \mathbf{G}\mathbf{G}^T$. As mentioned in Section 8.2.3, such a factorization is obtained by an eigenvector–eigenvalue decomposition $\mathbf{C}_w = \mathbf{V}_w\Lambda_w\mathbf{V}_w^T$. The diagonal elements of Λ_w contain the eigenvalues of $\mathbf{C}_w$. If some of the eigenvalues are zero, we remove the rows and columns in which these zero eigenvalues appear. Also, the corresponding columns in $\mathbf{V}_w$ are removed. Suppose that the number of nonzero eigenvalues is K, then Λ_w becomes a $K \times K$ matrix and $\mathbf{V}_w$ an $M \times K$ matrix. Consequently, $\mathbf{G} = \mathbf{V}_w\Lambda_w^{1/2}$ is also an $M \times K$ matrix.

Furthermore, we define the matrix $\mathbf{A}(i)$ as:

$$\mathbf{A}(i) \overset{def}{=} \left(\mathbf{F}^{-1}\right)^T \mathbf{Y}(i|i)\mathbf{F}^{-1} \tag{8.35}$$

In other words, $\mathbf{A}^{-1}(i) = \mathbf{F}\mathbf{C}(i|i)\mathbf{F}^T$ is the predicted covariance matrix *in the absence of process noise.* Here, we have silently assumed that the matrix $\mathbf{F}$ is invertible (which is the case if the time-discrete system is an approximation of a time-continuous system).

The information matrix of the predicted state follows from (8.2):

$$\begin{aligned} \mathbf{Y}(i+1|i) &= \left(\mathbf{F}\mathbf{C}(i|i)\mathbf{F}^T + \mathbf{G}\mathbf{G}^T\right)^{-1} \\ &= \left(\mathbf{A}^{-1}(i) + \mathbf{G}\mathbf{G}^T\right)^{-1} \end{aligned} \quad \text{(predicted state information)} \tag{8.36}$$

Using the matrix inversion lemma, this expression can be moulded in the easier to implement form:

$$\mathbf{Y}(i+1|i) = \mathbf{A}(i) - \mathbf{A}(i)\mathbf{G}(\mathbf{G}^T\mathbf{A}(i)\mathbf{G} + \mathbf{I})^{-1}\mathbf{G}^T\mathbf{A}(i) \tag{8.37}$$

This completes the Ricatti loop. The number of required operations of (8.36) and (8.37) is $2M^3 + 2M^2K + 2MK^2 + K^3$.

As mentioned above, the information filter can represent the situation where no information about some states is available. Typically, this occurs at the initialization of the filter, if no prior knowledge is available. However, the information matrix cannot represent a situation where states are known precisely, i.e. without uncertainty. Typically, such a situation occurs when the system $(\mathbf{F}, \mathbf{G})$ is not controllable.

The information filter also offers the possibility to define a *stochastic observability* criterion. For that purpose, consider the system without process noise. In that case, $\mathbf{Y}(i+1|i) = (\mathbf{F}^{-1})^T\mathbf{Y}(i|i)\mathbf{F}^{-1}$. Starting with complete uncertainty, i.e. $\mathbf{Y}(0|-1) = \mathbf{0}$, the prediction information at time i is found by iterative application of (8.34):

$$\mathbf{Y}(i|i-1) = \sum_{j=0}^{i} \left((\mathbf{F}^{-1})^T\right)^j \mathbf{H}^T\mathbf{C}_\mathbf{v}^{-1}\mathbf{H}(\mathbf{F}^{-1})^j \tag{8.38}$$

Note the similarity of this expression with the observability Grammian given in (8.17). The only difference is the information matrix $\mathbf{C}_\mathbf{v}^{-1}$ which weighs the importance of the measurements. Clearly, if for some i all eigenvalues of $\mathbf{Y}(i|i-1)$ are positive, then the measurements have provided information to all states.

Example 8.13 The information filter
The results obtained by the information filter are shown in Figure 8.10. The negative eigenvalues of $\mathbf{C}(i|i) = \mathbf{Y}^{-1}(i|i)$ indicate that the filter is not robust with respect to round-off errors.

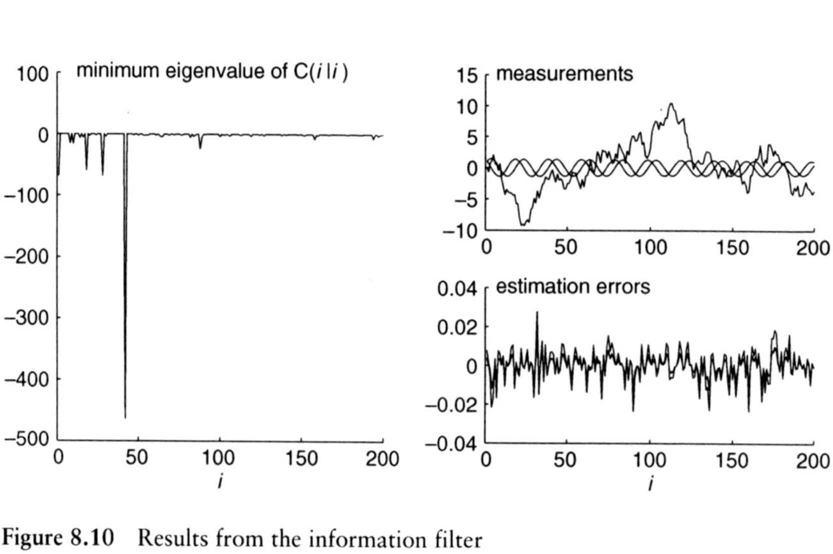

Figure 8.10 Results from the information filter

8.3.4 Square root filtering

The *square root* of a square matrix **P** is a matrix **A** such that $\mathbf{P} = \mathbf{AA}$. Sometimes the matrix **B** that satisfies $\mathbf{P} = \mathbf{B}^T\mathbf{B}$ is also called a square root, but strictly speaking such a **B** is a *Cholesky factor*, and not a square root. Anyway, square roots, Cholesky factors and other factorizations are useful matrix decomposition methods that enable stable implementations of the Kalman filter. The principal idea in *square root filtering* is to decompose a covariance matrix **P** as $\mathbf{P} = \mathbf{B}^T\mathbf{B}$ (or likewise), and to use **B** as a representation of **P**. This effectively doubles the precision of the number representation. The various factorization methods lead to various forms of square root filtering.

This section describes one particular implementation of a square root filter, the *Potter* implementation (Potter and Stern, 1963). It uses:

- *Triangular Cholesky factorization* of the error covariance matrices.
- Sequentially processing of the measurements using *symmetric elementary matrices*.
- *QR factorization* for the prediction.

The update in Potter's square root filter

We will represent the error covariance matrix $\mathbf{C}(i|j)$ by an upper triangular matrix $\mathbf{B}(i|j)$ where:

$$\mathbf{C}(i|j) = \mathbf{B}^T(i|j)\mathbf{B}(i|j) \tag{8.39}$$

An upper triangular matrix is a matrix with all elements below the diagonal equal to zero, e.g.

$$\mathbf{B} = \begin{bmatrix} b_{00} & b_{01} & b_{02} & \cdots \\ 0 & b_{11} & b_{12} & \\ 0 & 0 & b_{22} & \\ 0 & 0 & 0 & \ddots \end{bmatrix} \tag{8.40}$$

The update formula, as expressed in (8.27):

$$\mathbf{C}(i|i) = \mathbf{C}(i|i-1) - \mathbf{C}(i|i-1)\mathbf{H}^T\left(\mathbf{HC}(i|i-1)\mathbf{H}^T + \mathbf{C}_\mathbf{v}\right)^{-1}\mathbf{HC}(i|i-1)$$

turns into:

$$
\begin{aligned}
\mathbf{B}(i|i)^T\mathbf{B}(i|i) &= \mathbf{B}^T(i|i-1)\mathbf{B}(i|i-1) - \mathbf{B}^T(i|)\mathbf{B}(i|i-1)\mathbf{H}^T \\
&\qquad (\mathbf{H}\mathbf{B}^T(i|i-1)\mathbf{B}(i|i-1)\mathbf{H}^T + \mathbf{C}_v)^{-1}\mathbf{H}\mathbf{B}^T(i|i-1)\mathbf{B}(i|i-1) \\
&= \mathbf{B}^T(i|i-1)\mathbf{B}(i|i-1) - \mathbf{B}^T(i|i-1)\mathbf{M} \\
&\qquad \left(\mathbf{M}^T\mathbf{M} + \mathbf{C}_v\right)^{-1}\mathbf{M}^T\mathbf{B}(i|i-1) \\
&= \mathbf{B}^T(i|i-1)\left(\mathbf{I} - \mathbf{M}\left(\mathbf{M}^T\mathbf{M} + \mathbf{C}_v\right)^{-1}\mathbf{M}^T\right)\mathbf{B}(i|i-1)
\end{aligned}
\tag{8.41}
$$

where $\mathbf{M} \stackrel{def}{=} \mathbf{B}(i|i-1)\mathbf{H}^T$ is an $M \times N$ matrix.

If we would succeed in finding a Cholesky factor of:

$$
\mathbf{I} - \mathbf{M}\left(\mathbf{M}^T\mathbf{M} + \mathbf{C}_v\right)^{-1}\mathbf{M}^T \tag{8.42}
$$

then we would have an update formula entirely expressed in Cholesky factors. Unfortunately, in the general case it is difficult to find such a factor.

For the special case of only one measurement, $N = 1$, a factorization is within reach. If $N = 1$, then $\mathbf{H}$ becomes a row vector $\mathbf{h}$. The matrix $\mathbf{M}$ becomes an M dimensional (column) vector $\mathbf{m} = \mathbf{B}(i|i-1)\mathbf{h}^T$. Substitution in (8.42) yields:

$$
\mathbf{I} - \mathbf{m}\left(\mathbf{m}^T\mathbf{m} + \sigma_v^2\right)^{-1}\mathbf{m}^T \tag{8.43}
$$

σ_v^2 is the variance of the measurement noise.

Expression (8.43) is of the form $\mathbf{I} - \alpha\mathbf{m}\mathbf{m}^T$ with $\alpha = 1/(\|\mathbf{m}\|^2 + \sigma_v^2)$. Such a form is called a *symmetric elementary matrix*. The form can be easily factored by:

$$
\begin{aligned}
&\mathbf{I} - \alpha\mathbf{m}\mathbf{m}^T = \left(\mathbf{I} - \beta\mathbf{m}\mathbf{m}^T\right)^T\left(\mathbf{I} - \beta\mathbf{m}\mathbf{m}^T\right) \\
&\text{with } \beta = \frac{1 \pm \sqrt{1 - \alpha\|\mathbf{m}\|^2}}{\|\mathbf{m}\|^2} = \frac{1}{\|\mathbf{m}\|^2}\left(1 \pm \sqrt{\frac{\sigma_v^2}{\|\mathbf{m}\|^2 + \sigma_v^2}}\right)
\end{aligned}
\tag{8.44}
$$

Substitution of (8.44) in (8.41) yields:

$$
\mathbf{B}(i|i)^T\mathbf{B}(i|i) = \mathbf{B}^T(i|i-1)\left(\mathbf{I} - \beta\mathbf{m}\mathbf{m}^T\right)^T\left(\mathbf{I} - \beta\mathbf{m}\mathbf{m}^T\right)\mathbf{B}(i|i-1) \tag{8.45}
$$

This gives us finally the update formula in terms of Cholesky factors:

square root filtering update :

$$\mathbf{B}(i|i) = \left(\mathbf{I} - \beta\mathbf{m}\mathbf{m}^T\right)\mathbf{B}(i|i-1) \quad \text{with} \quad \beta = \frac{1}{\|\mathbf{m}\|^2}\left(1 + \sqrt{\frac{\sigma_v^2}{\|\mathbf{m}\|^2 + \sigma_v^2}}\right) \tag{8.46}$$

The general solution, when more measurements are available, is obtained by sequentially processing them in exactly the same way as discussed in Section 8.3.2.

The prediction in Potter's square root filter

The prediction step $\mathbf{C}(i+1|i) = \mathbf{F}\mathbf{C}(i|i)\mathbf{F}^T + \mathbf{C}_\mathbf{w}$ can be written as a product of Cholesky factors:

$$\mathbf{C}(i+1|i) = \mathbf{F}\mathbf{C}(i|i)\mathbf{F}^T + \mathbf{C}_\mathbf{w} = \left[\left(\mathbf{C}_\mathbf{w}^{\frac{1}{2}}\right)^T \middle| \mathbf{F}\mathbf{B}^T(i|i)\right]\left[\frac{\mathbf{C}_\mathbf{w}^{\frac{1}{2}}}{\mathbf{B}(i|i)\mathbf{F}^T}\right] \tag{8.47}$$

where $\mathbf{C}_\mathbf{w}^{\frac{1}{2}}$ is the square root of $\mathbf{C}_\mathbf{w}$. Thus, if we define the $2M \times M$ matrix $\mathbf{A} \stackrel{def}{=} [(\mathbf{C}_\mathbf{w}^{\frac{1}{2}})^T | \mathbf{F}\mathbf{B}^T(i|i)]^T$, then the prediction covariance matrix is found as $\mathbf{C}(i+1|i) = \mathbf{A}^T\mathbf{A}$.

A QR factorization of a matrix $\mathbf{A}$ (not necessarily square) produces an orthonormal matrix $\mathbf{Q}$ and an upper triangular matrix $\mathbf{R}$ such that $\mathbf{A} = \mathbf{Q}\mathbf{R}$. The matrices $\mathbf{Q}$ and $\mathbf{R}$ have compatible dimensions. Such a factorization is what we are looking for because, since $\mathbf{Q}$ is orthonormal ($\mathbf{Q}^T\mathbf{Q} = \mathbf{I}$), we have: $\mathbf{A}^T\mathbf{A} = \mathbf{R}^T\mathbf{R}$. The procedure to get $\mathbf{B}(i|i+1)$ simply boils down to constructing the matrix $\mathbf{A}$, and then performing a QR factorization.

An implementation of Potter's square root filter is given in Listing 8.4. MATLAB provides two functions for a triangular Cholesky factorization. We need such a function only at the initialization of the filter to get a factorization of $\mathbf{C}_\mathbf{x}(0)$ and $\mathbf{C}_\mathbf{w}$. The function `chol()` is applicable to positive definite matrices. In our case, both matrices can have zero eigenvalues. The `cholinc()` can also handle positive semidefinite matrices, but is only applicable to sparse matrices. The functions `sparse()` and `full()` take care of the conversions. Note that `cholinc()` returns a matrix whose number of rows, in principle, equals the number of nonzero eigenvalues. Thus, possibly, the matrix should be filled up with zeros to obtain an $M \times M$ matrix.

The QR factorization in the prediction step is taken care of by `qr()`. The size of the resulting matrix is $(M + K) \times M$ where K is the number of rows in `SqCw`. Only the first M rows carry the needed information and the remaining rows are deleted.

The update of the algorithm requires $N(M^3 + 3M^2 + M)$ operations. The number of operations of the prediction is determined partly by `qr ()`. This function requires about $\frac{1}{2}M^3 + M^2K$ operations. In full, the prediction requires $\frac{3}{2}M^3 + M^2K$ operations.

Listing 8.4
Potter's implementation in MATLAB.

```
load linsys                                   % Load a system:
[B,p]=cholinc(sparse(Cx0),'0');               % Initialize squared
B=full(B); B(p:M,:)=0;                        % prior uncertainty
x_est=x0;                                     % and prior mean
[SqCw,p]=cholinc(sparse(Cw),'0');             % Squared Cw
SqCw=full(SqCw);
while (1)                                     % Endless loop
  z=acquire_measurement_vector();
  % 1. Sequential update:
  for n=1:N                                   % For all measurements ...
    m=B*H(n,:)';                              % get row vector from H
    norm_m=m'*m;
    S=norm_m+Cv(n,n);                         % innovation variance
    K=B'*m/S;                                 % Kalman gain vector
    inno=z(n)-H(n,:)*x_est;
    x_est=x_est+K*inno;                       % update estimate
    beta=(1+sqrt(Cv(n,n)/S))/norm_m;
    B=(eye(M)-beta*m*m')*B;                   % covariance update
  end
  % 2. Prediction:
  u=get_control_vector();
  x_est=F*x_est+L*u;                          % Predict the state
  A=[SqCw; B*F'];                             % Create block matrix
  [q,B]=qr(A);                                % QR factorization
  B=B(1:M,1:M);                               % Delete irrelevant part
end
```

Example 8.14 Square root filtering
The results obtained by Potter's square root filter are shown in Figure 8.11. The double logarithmic plot of the minimum eigenvalue of $\mathbf{P} = \mathbf{C}(i|i)$ shows that the eigenvalue is of the order $O(i^{-1})$. This is exactly according to the expectations since this eigenvalue relates to a state without process noise. Thus, the variance of the

estimation error should be inversely proportional to the number of observations.

8.3.5 Comparison

In the preceding sections, five different implementations are discussed which are all mathematically equivalent. However, different implementations have different sensitivities to round-off errors and different computational cost.

Table 8.1 provides an overview of the cost expressed in the number of operations required for a single iteration. The table assumes a time invariant system so that terms like $\mathbf{H}^T\mathbf{C}_v^{-1}\mathbf{H}$ can be reused. Furthermore, the numbers are based on a straightforward MATLAB implementation without optimization with respect to computational cost. Special code that exploits the symmetry of covariance matrices can lower the number of operations a little. The computational efficiency of the square root filter can be improved by consistently maintaining the triangular structure of the matrices. (In the current implementation, the triangular structure is lost during the update, but is regained by the QR factorization.)

A quantitative, general analysis of the sensitivities of the various implementations to round-off errors is difficult. However, Table 8.1 gives an indication. The table shows the results of an experiment that relates to the system described in Example 8.10. For each implementation,

Table 8.1 Comparison of different implementation

	Computational cost update	**Computational cost prediction**	**Required no. of digits**
Conventional Kalman filter	$2M^2N + 3MN^2 + N^3$	$2M^3$	12
MMSE form	$2M^3 + M^2 + M$	$2M^3$	13
Sequential processing of measurements	$2M^2N + 2MN$	$2M^3$	12
Information filter	$M^3 + \frac{1}{2}M^2 + \frac{1}{2}M$	$2M^3 + 2M^2K + 2MK^2 + K^3$	11
Potter's square root filter	$N(M^3 + 3M^2 + M)$	$\frac{3}{2}M^3 + M^2K$	5

M = number of states.
N = number of measurements.
K = effective dimension of process noise vector.

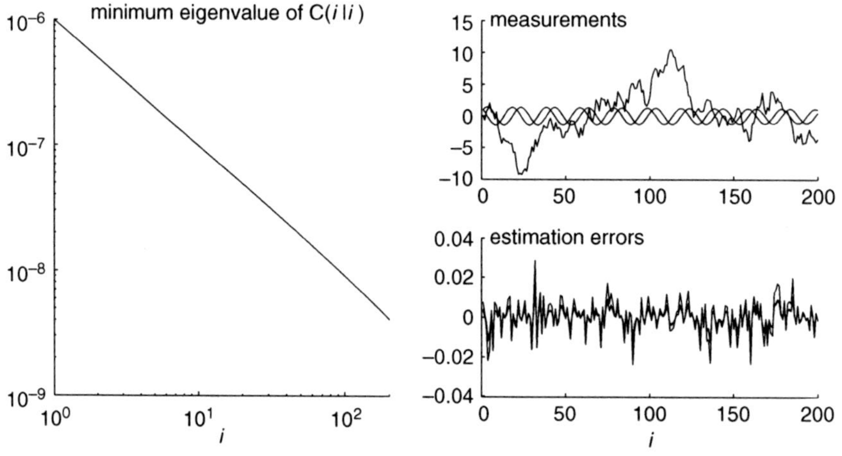

Figure 8.11 Results from Potter's square root filter

the number of digits needed for the number representations of the variables (including intermediate results) was established, in order to have a stable and consistent result.

As expected, the square root filter is the most numerical stable method, but it is also the most expensive one. Square root filtering should be considered:

- If other implementations result in covariance matrices with negative eigenvalues.
- If other implementations involve matrix inversions where the inverse condition number, i.e. $\lambda_{min}/\lambda_{max}$, of the matrix is in the same magnitude of the round-off errors.

The MMSE form is inexpensive if the number of measurements is large relative to the number of states, i.e. if $N >> M$. The sequentially processing method is inexpensive, especially when both N and M are large.

8.4 CONSISTENCY CHECKS

The purpose of this section is to provide some tools that enable the designer to check whether his design behaves consistently (Bar-Shalom and Birmiwal, 1983). As discussed in the previous sections, the two main reasons why a realized filter does not behave correctly are modelling errors and numerical instabilities of the filter.

Estimators are related with three types of error variances:

- The minimal variances that would be obtained with the most appropriate model.
- The actual variances of the estimation errors of some given estimator.
- The variances indicated by the calculated error covariance matrix of a given estimator.

Of course, the purpose is to find an estimator whose error variances equal the first one. Since we do not know whether our model approaches the most appropriate one, we do not have the guarantee that our design approaches the minimal attainable variance. However, if we have reached the optimal solution, then the actual variances of the estimation errors must coincide with the calculated variances. Such a correspondence between actual and calculated variances is a necessary condition for an optimal filter, but not a sufficient one.

Unfortunately, we need to know the real estimation errors in order to check whether the two variances coincide. This is only possible if the physical process is (temporarily) provided with additional instruments that give us reference values of the states. Usually such provisions are costly. This section discusses checks that can be accomplished without reference values.

8.4.1 Orthogonality properties

We recall that the update step of the Kalman filter is formed by the following operations (8.2):

$$\begin{aligned} \hat{\mathbf{z}}(i) &= \mathbf{H}(i)\bar{\mathbf{x}}(i|i-1) && \text{(predicted measurement)} \\ \tilde{\mathbf{z}}(i) &= \mathbf{z}(i) - \hat{\mathbf{z}}(i) && \text{(innovations)} \\ \bar{\mathbf{x}}(i|i) &= \bar{\mathbf{x}}(i|i-1) + \mathbf{K}(i)\tilde{\mathbf{z}}(i) && \text{(updated estimate)} \end{aligned} \tag{8.48}$$

The vectors $\tilde{\mathbf{z}}(i)$ are the *innovations* (*residuals*). In linear-Gaussian systems, these vectors are zero mean with the innovation matrix as covariance matrix:

$$\mathbf{S}(i) = \mathbf{H}(i)\mathbf{C}(i|i-1)\mathbf{H}^T(i) + \mathbf{C}_\mathbf{v}(i) \tag{8.49}$$

Furthermore, let $\mathbf{e}(i|j) \stackrel{def}{=} \mathbf{x}(i) - \overline{\mathbf{x}}(i|j)$ be the estimation error of the estimate $\overline{\mathbf{x}}(i|j)$. We then have the following properties:

$$\begin{aligned} &\mathrm{E}[\mathbf{e}(i|j)\mathbf{z}^T(m)] = \mathbf{0} && m \le j \\ &\mathrm{E}[\mathbf{e}(i|j)\overline{\mathbf{x}}^T(n|m)] = \mathbf{0} && m \le j \\ &\mathrm{E}[\tilde{\mathbf{z}}(i)\tilde{\mathbf{z}}^T(j)] = \delta(i,j)\mathbf{S}(i) && \text{i.e. } \tilde{\mathbf{z}}(i) \text{ is white} \end{aligned} \tag{8.50}$$

These properties follow from the *principle of orthogonality*. In the static case, any unbiased linear MMSE satisfies:

$$\mathrm{E}[\mathbf{e}(\mathbf{z}-\overline{\mathbf{z}})^T] = \mathrm{E}[(\mathbf{x} - \mathbf{K}\mathbf{z} - (\overline{\mathbf{x}} - \mathbf{K}\overline{\mathbf{z}}))(\mathbf{z}-\overline{\mathbf{z}})^T] = \mathbf{C}_{\mathbf{xz}} - \mathbf{K}\mathbf{C}_{\mathbf{z}} = \mathbf{0} \tag{8.51}$$

The last step follows from $\mathbf{K} = \mathbf{C}_{\mathbf{xz}}\mathbf{C}_{\mathbf{z}}^{-1}$. See (3.29). The principle of orthogonality, $\mathrm{E}[\mathbf{e}^T(\mathbf{z}-\overline{\mathbf{z}})] = 0$, simply states that the covariance between any component of the error and any measurement is zero. Adopting an inner product definition for two random variables e_m and z_n as $(e_m, z_m) \stackrel{def}{=} \mathrm{Cov}[e_m, z_n]$, the principle can be expressed as $\mathbf{e}\perp\mathbf{z}$.

Since $\overline{\mathbf{x}}(i|j)$ is an unbiased linear MMSE estimate, it is a linear function of the set $\{\overline{\mathbf{x}}(0), \mathbf{Z}(j)\} = \{\overline{\mathbf{x}}(0), \mathbf{z}(0), \mathbf{z}(1), \ldots, \mathbf{z}(j)\}$. According to the principle of orthogonality, we have $\mathbf{e}(i|j)\perp\mathbf{Z}(j)$. Therefore, $\mathbf{e}(i|j)$ must also be orthogonal to any $\mathbf{z}(m)$ $m \le j$, because $\mathbf{z}(m)$ is a subspace of $\mathbf{Z}(j)$. This proves the first statement in (8.50).

In addition, $\overline{\mathbf{x}}(n|m), m \le j$, is a linear combination of $\mathbf{Z}(m)$. Therefore, it is also orthogonal to $\mathbf{e}(i|j)$, which proves the second statement.

The whiteness property of the innovations follows from the following argument. Suppose $j < i$. We may write: $\mathrm{E}[\tilde{\mathbf{z}}(i)\tilde{\mathbf{z}}^T(j)] = \mathrm{E}[\mathrm{E}[\tilde{\mathbf{z}}(i)\tilde{\mathbf{z}}^T(j)|\mathbf{Z}(j)]]$. In the inner expectation, the measurements $\mathbf{Z}(j)$ are known. Since $\tilde{\mathbf{z}}(j)$ is a linear combination of $\mathbf{Z}(j), \tilde{\mathbf{z}}(j)$ is non-random. It can be taken outside the inner expectation: $\mathrm{E}[\tilde{\mathbf{z}}(i)\tilde{\mathbf{z}}^T(j)] = \mathrm{E}[\mathrm{E}[\tilde{\mathbf{z}}(i)|\mathbf{Z}(j)]\tilde{\mathbf{z}}^T(j)]$. However, $\mathrm{E}[\tilde{\mathbf{z}}(i)|\mathbf{Z}(j)]$ must be zero because the predicted measurements are unbiased estimates of the true measurements. If $\mathrm{E}[\tilde{\mathbf{z}}(i)|\mathbf{Z}(j)] = 0$, then $\mathrm{E}[\tilde{\mathbf{z}}(i)\tilde{\mathbf{z}}^T(j)] = 0$ (unless $i = j$).

8.4.2 Normalized errors

The NEES (*normalized estimation error squared*) is a test signal defined as:

$$Nees(i) = \mathbf{e}^T(i|i)\mathbf{C}^{-1}(i|i)\mathbf{e}(i) \tag{8.52}$$

In the linear-Gaussian case, the NEES has a χ^2_M distribution (chi-square with M degrees of freedom; see Appendix C.1.5).

The χ^2_M distribution of the NEES follows from the following argument. Since the state estimator is unbiased, $\mathrm{E}[\mathbf{e}(i|i)] = \mathbf{0}$. The covariance matrix of $\mathbf{e}(i|i)$ is $\mathbf{C}(i|i)$. Suppose $\mathbf{A}(i)$ is a symmetric matrix such that $\mathbf{A}(i)\mathbf{A}^T(i) = \mathbf{C}^{-1}(i|i)$. Application of $\mathbf{A}(i)$ to $\mathbf{e}(i)$ will give a random vector $\mathbf{y}(i) = \mathbf{A}(i)\mathbf{e}(i)$. The covariance matrix of $\mathbf{y}(i)$ is: $\mathbf{A}(i)\mathbf{C}(i|i)\mathbf{A}^T(i) = \mathbf{A}(i)(\mathbf{A}(i)\mathbf{A}^T(i))^{-1}\mathbf{A}^T(i) = \mathbf{I}$. Thus, the components of $\mathbf{y}(i)$ are uncorrelated, and have unit variance. If both the process noise and the measurement noise are normally distributed, then so is $\mathbf{y}(i)$. Hence, the inner product $\mathbf{y}^T(i)\mathbf{y}(i) = \mathbf{e}^T(i)\mathbf{C}^{-1}(i|i)\mathbf{e}(i)$ is the sum of M squared, independent random variables, each normally distributed with zero mean and unit variance. Such a sum has a χ^2_M distribution.

The NIS (*normalized innovation squared*) is a test signal defined as:

$$Nis(i) = \tilde{\mathbf{z}}^T(i)\mathbf{S}^{-1}(i)\tilde{\mathbf{z}}(i) \tag{8.53}$$

In the linear-Gaussian case, the NIS has a χ^2_N distribution. This follows readily from the same argument as used above.

Example 8.15 Normalized errors of second order system
Consider the following system:

$$\begin{aligned}
\mathbf{F} &= \begin{bmatrix} 0.999\cos(0.1\pi) & 0.999\sin(0.1\pi) \\ -0.999\sin(0.1\pi) & 0.999\cos(0.1\pi) \end{bmatrix} & \mathbf{H} &= [0 \quad 0.5] \\
\mathbf{C_w} &= \begin{bmatrix} 1 & 0 \\ 0 & 0 \end{bmatrix} & \mathbf{C_v} &= [1] \\
\mathbf{C_x}(0) &= \begin{bmatrix} 0.01 & 0 \\ 0 & 0.01 \end{bmatrix} & \mathrm{E}[\mathbf{x}(0)] &= \begin{bmatrix} 0 \\ 0 \end{bmatrix}
\end{aligned} \tag{8.54}$$

Figure 8.12 shows the results of a MATLAB realization of this system consisting of states and measurements. Application of the discrete Kalman filter yields estimated states and innovations. From that, the NEES and the NIS are calculated.

In this case, $M = 2$ and $N = 1$. Thus, the NEES and the NIS should obey the statistics of a χ^2_2 and a χ^2_1 distribution. The 95% percentiles of these distributions are 5.99 and 3.84, respectively. Thus, about 95% of the samples should be below these percentiles, and about 5% above. Figure 8.12 affirms this.

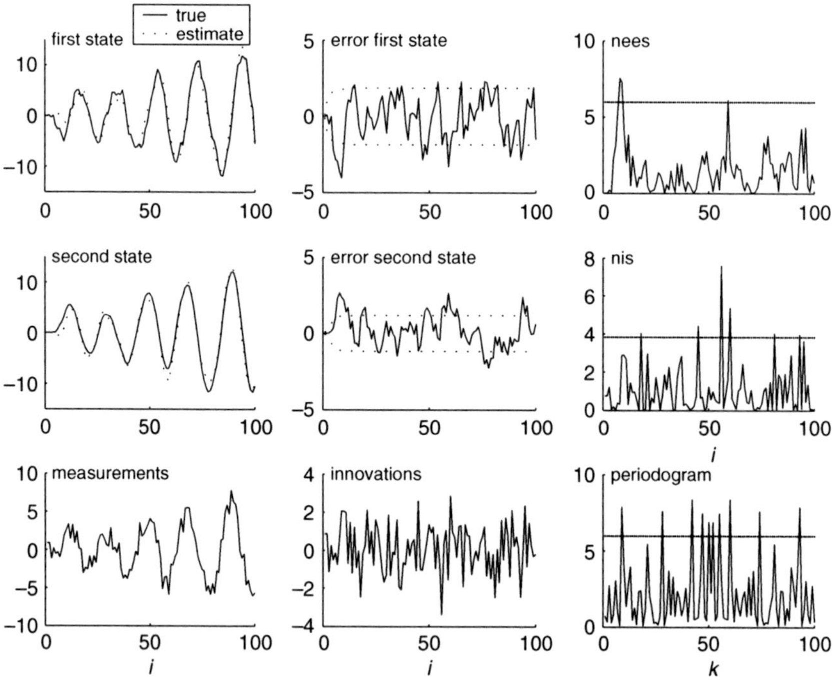

Figure 8.12 Innovations and normalized errors of a state estimator for a second order system

8.4.3 Consistency checks

From equation (8.50) and the properties of the NEES and the NIS, the following statement holds true:

If a state estimator for a linear-Gaussian system is optimal, then:

- The sequence $Nees(i)$ must be χ^2_M distributed.
- The sequence $Nis(i)$ must be χ^2_N distributed.
- The sequence $\tilde{\mathbf{z}}(i)$ (innovations) must be white.

Consistency checks of a 'state estimator-under-test' can be performed by collecting the three sequences and by applying statistical tests to see whether the three conditions are fulfilled. If one or more of these conditions are not satisfied, then we may conclude that the estimator is not optimal.

In a real design, the NEES test is only possible if the true states are known. As mentioned above, such is the case when the system is (temporarily) provided with extra measurement equipment that measures the states directly and with sufficient accuracy so that their outputs can be used as a reference. Usually, such measurement devices are too expensive and the designer has to rely on the other two tests. However, the NEES test is applicable in simulations of the physical process. It can be used to see if the design is very sensitive to changes in the parameters of the model. The other two tests use the innovations. Since the innovations are always available, even in a real design, these are of more practical significance.

In order to check the hypothesis that a data set obeys a given distribution, we have to apply a *distribution test*. Examples are the Kolmogorov–Smirnov test and the Chi-square test (Section 9.3.3). Often, instead of these rigorous tests a quick procedure suffices to give an indication. We simply determine an interval in which, say, 95% of the samples must be. Such an interval $[A,B]$ is defined by $F(B) - F(A) = 0.95$. Here, $F(\cdot)$ is the probability distribution of the random variable under test. If an appreciable part of the data set is outside this interval, then the hypothesis is rejected. For chi-square distributions there is the *one-sided 95% acceptance boundary* with $A = 0$, and B such that $F_{\chi^2_{Dof}}(B) = 0.95$. Sometimes the *two-sided 95% acceptance boundaries* are used, defined by $F_{\chi^2_{Dof}}(A) = 0.025$ and $F_{\chi^2_{Dof}}(B) = 0.975$. Table 8.2 gives values of A and B for various degrees of freedom (*Dof*).

For state estimators that have reached the steady state, $\mathrm{S}(i)$ is constant, and the whiteness property of the innovations implies that the power spectrum of any element of $\tilde{\mathbf{z}}(i)$ must be flat.[2] Suppose that $\tilde{z}_n(i)$ is an element of $\tilde{\mathbf{z}}(i)$. The variance of $\tilde{z}_n(i)$, denoted by σ_n^2, is the n-th diagonal element in S. The discrete Fourier transform of $\tilde{z}_n(i)$, calculated over $i = 0, \ldots, I-1$, is:

$$\tilde{Z}_n(k) = \sum_{i=0}^{I-1} \tilde{z}_n(i) e^{-j2\pi ki/I} \qquad k = 0, \ldots, I-1 \qquad j = \sqrt{-1} \tag{8.55}$$

The *periodogram* of $\tilde{z}_n(i)$, defined here as $P_n(k) = |\tilde{Z}_n(k)|^2/I$, is an estimate of the power spectrum (Blackman and Tukey, 1958). If the power spectrum is flat, then $\mathrm{E}[P_n(k)] = \sigma_n^2$. It can be proven that in that case the

[2] For estimators that are not in the steady state, the innovations have to be premultiplied by $\mathrm{S}^{-1/2}(i)$ so that the covariance matrix of the resulting sequence becomes constant.

Table 8.2 95% Acceptance boundaries for χ^2_{Dof} distributions

	One-sided		Two-sided	
Dof	*A*	*B*	*A*	*B*
1	0	3.841	0.0010	5.024
2	0	5.991	0.0506	7.378
3	0	7.815	0.2158	9.348
4	0	9.488	0.4844	11.14
5	0	11.07	0.8312	12.83

variables $2P_n(k)/\sigma_n^2$ is χ_2^2 distributed for all k (except for $k = 0$). Hence, the whiteness of $\tilde{z}_n(i)$ is tested by checking whether $2P_n(k)/\sigma_n^2$ is χ_2^2 distributed.

Example 8.16 Consistency checks applied to a second order system
The results of the estimator discussed in Example 8.15 and presented in Figure 8.12 pass the consistency checks successfully. Both the NEES and the NIS are about 95% of the time below the one-sided acceptance boundaries, i.e. below 5.99 and 3.84. The figure also shows the normalized periodogram calculated as $2P_n(k)/\hat{\sigma}_1^2$ with $\hat{\sigma}_1^2$ the estimated variance of the innovation. The normalized periodogram shown seems to comply with the theoretical χ_2^2 distribution.

Example 8.17 Consistency checks applied to a slightly mismatched filter
Figure 8.13 shows the results of a state estimator that is applied to the same data as used in Example 8.15. However, the model the estimator uses differs slightly. The real system matrix $\mathbf{F}$ of the generating process and the system matrix $\mathbf{F}_{filter}$ on which the design of the state estimator is based are as follows:

$$\mathbf{F} = \begin{bmatrix} 0.999\cos(0.1\pi) & 0.999\sin(0.1\pi) \\ -0.999\sin(0.1\pi) & 0.999\cos(0.1\pi) \end{bmatrix}$$

$$\mathbf{F}_{filter} = \begin{bmatrix} 0.999\cos(0.116\pi) & 0.999\sin(0.116\pi) \\ -0.999\sin(0.116\pi) & 0.999\cos(0.116\pi) \end{bmatrix}$$

Apart from that, the model used by the state estimator exactly matches the real system.

In this example, the design does not pass the whiteness test of the innovations. The peak of the periodogram at $k = 6$ is above 20.

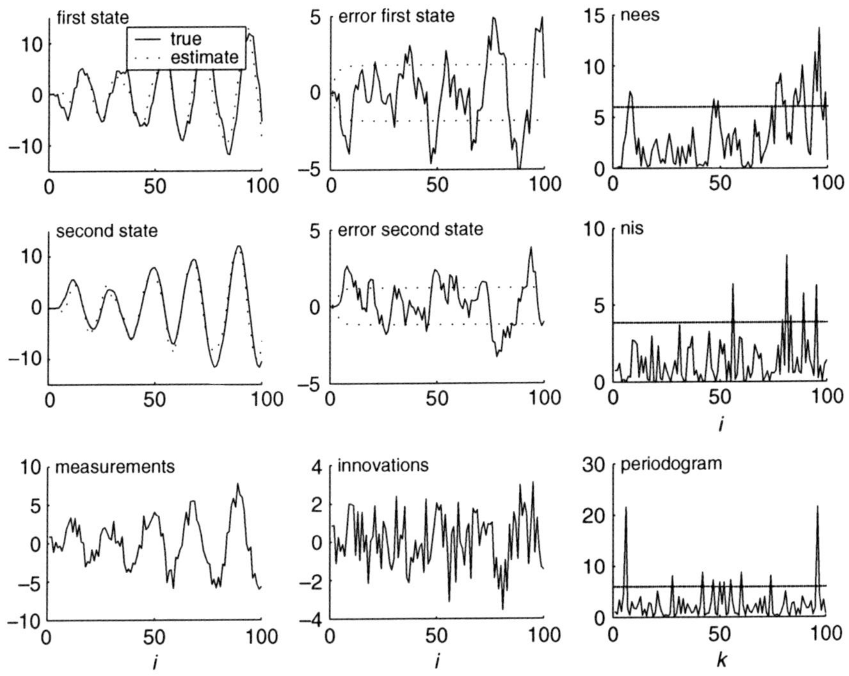

Figure 8.13 Innovations and normalized errors of a state estimator based on a slightly mismatched model

For a χ_2^2 distribution such a high value is unlikely to occur. (In fact, the chance is smaller than 1 to 20 000.)

8.4.4 Fudging

If one or more of the consistency checks fail, then somewhere a serious modelling error has occurred. The designer has to step back to an earlier stage of the design in order to identify the fault. The problem can be caused anywhere, from inaccurate modelling during the system identification to improper implementations. If the system is nonlinear and the extended Kalman filter is applied, problems may arise due to the neglect of higher order terms of the Taylor series expansion.

A heuristic method to catch the errors that arise due to approximations of the model is to deliberately increase the modelled process noise (Bar-Shalom and Li, 1993). One way to do so is by increasing the covariance matrix $\mathbf{C}_w$ of the process noise by means of a *fudge factor* γ. For instance, we simply replace $\mathbf{C}_w$ by $\mathbf{C}_w + \gamma\mathbf{I}$. Other methods to regulate a covariance matrix are discussed in Section 5.2.3. Instead of adapting $\mathbf{C}_w$ we can also increase the diagonal of the prediction covariance $\mathbf{C}(i+1|i)$ by some factor. The fudge factor should be selected such that the consistency checks now pass as successful as possible.

Fudging effectuates a decrease of faith in the model of the process, and thus causes a larger impact of the measurements. However, modelling errors give rise to deviations that are autocorrelated and not independent from the states. Thus, these deviations are not accurately modelled by white noise. The designer should maintain a critical attitude with respect to fudging.

8.5 EXTENSIONS OF THE KALMAN FILTER

The extensions considered in this section make the Kalman filter applicable to a wider class of problems. In particular, we discuss extensions to cover non-white and cross-correlated noise sequences. Also, the topic of offline estimation will be introduced.

8.5.1 Autocorrelated noise

The Kalman filter considered so far assumes white uncorrelated random sequences $\mathbf{w}(i)$ and $\mathbf{v}(i)$ for the process and measurement noise. What do we do if these assumptions do not hold in practice?

The case of autocorrelated noise is usually tackled by assuming a state space model for the noise. For instance,[3] autocorrelated process noise is represented by $\mathbf{w}(i+1) = \mathbf{F}_w\mathbf{w}(i) + \tilde{\mathbf{w}}(i)$ where $\tilde{\mathbf{w}}(i)$ is a white noise sequence with covariance matrix $\mathbf{C}_{\tilde{w}}$. *State augmentation* reduces the problem to a standard form. For that, the state vector is extended by $\mathbf{w}(i)$:

$$\begin{aligned}
\begin{bmatrix} \mathbf{x}(i+1) \\ \mathbf{w}(i+1) \end{bmatrix} &= \begin{bmatrix} \mathbf{F} & \mathbf{I} \\ 0 & \mathbf{F}_w \end{bmatrix} \begin{bmatrix} \mathbf{x}(i) \\ \mathbf{w}(i) \end{bmatrix} + \begin{bmatrix} 0 \\ \mathbf{I} \end{bmatrix} \tilde{\mathbf{w}}(i) \\
\mathbf{z}(i) &= [\mathbf{H} \quad 0] \begin{bmatrix} \mathbf{x}(i) \\ \mathbf{w}(i) \end{bmatrix} + \mathbf{v}(i)
\end{aligned} \tag{8.56}$$

[3] For convenience of notation the time index of matrices (for time variant systems) is omitted in this section.

The new, augmented state vector can be estimated using the standard Kalman filter.

The case of autocorrelated measurement noise is handled in the same way although some complications may occur (Bryson and Hendrikson, 1965). Suppose that the measurement noise is modelled by $\mathbf{v}(i+1) = \mathbf{F}_v\mathbf{v}(i) + \tilde{\mathbf{v}}(i)$ with $\tilde{\mathbf{v}}(i)$ a white noise sequence with covariance $\mathbf{C}_{\tilde{v}}$. State augmentation brings the following model:

$$\begin{aligned}\begin{bmatrix}\mathbf{x}(i+1)\\ \mathbf{v}(i+1)\end{bmatrix} &= \begin{bmatrix}\mathbf{F} & 0\\ 0 & \mathbf{F}_v\end{bmatrix}\begin{bmatrix}\mathbf{x}(i)\\ \mathbf{v}(i)\end{bmatrix} + \begin{bmatrix}\mathbf{I}\\ 0\end{bmatrix}\mathbf{w}(i) + \begin{bmatrix}0\\ \mathbf{I}\end{bmatrix}\tilde{\mathbf{v}}(i)\\ \mathbf{z}(i) &= \begin{bmatrix}\mathbf{H} & \mathbf{I}\end{bmatrix}\begin{bmatrix}\mathbf{x}(i)\\ \mathbf{v}(i)\end{bmatrix}\end{aligned} \tag{8.57}$$

The process noise of the new system consists of two terms. Taken together, the corresponding covariance matrix is $[\mathbf{I}\ \ 0]^T\mathbf{C}_w[\mathbf{I}\ \ 0]+[0\ \ \mathbf{I}]^T\mathbf{C}_{\tilde{v}}[0\ \ \mathbf{I}]$.

In the new, augmented system there is no measurement noise. The corresponding covariance matrix is zero. This is not necessarily a problem because the standard Kalman filter does not use the inverse of the measurement covariance matrix. However, the computation of the Kalman gain matrix does require the inverse of $\mathbf{HC}(i|i-1)\mathbf{H}^T + \mathbf{C}_v$. If in this expression $\mathbf{C}_v = 0$, the feasibility of calculating $\mathbf{K}$ depends on the invertibility of $\mathbf{HC}(i|i-1)\mathbf{H}^T$. The statement $\mathbf{C}_v = 0$ implies that some linear combinations of the state vector is known without any uncertainty. If in this subspace of the state space no process noise is active, ultimately $\mathbf{HC}(i|i-1)\mathbf{H}^T$ becomes near singular, and the filter becomes unstable. The solution for this potential problem is to apply *differencing*. Instead of using $\mathbf{z}(i)$ directly, we use the differenced measurements $\mathbf{y}(i) \overset{def}{=} \mathbf{z}(i) - \mathbf{F}_v\mathbf{z}(i-1) = \mathbf{Hx}(i) - \mathbf{F}_v\mathbf{Hx}(i-1) + \tilde{\mathbf{v}}(i-1)$.

Example 8.18 Suppression of 50 Hz emf interference
Consider a signal $x(t)$ that is disturbed by a 50 (Hz) emf interference caused, for instance, by an inductive crosstalk of an electric power line. The bandwidth of the signal is $B = 10$ (rad/s), and the sampling period is $\Delta = 1$ (ms). We model the signal by a first order system $\dot{x} = -Bx + w$ which in discrete time becomes $x(i+1) = (1 - B\Delta)x(i) + w(i)$.

An inductive coupling with a power line induces an interfering periodic waveform with a ground harmonic of $f_0 = 50$ (Hz). Usually, such a waveform also contains a component with double frequency (and possibly higher harmonics, but these will be neglected here). The disturbance can be modelled by two second order equations, that is:

$$\mathbf{v}(i+1) = d\begin{bmatrix} \begin{matrix} \cos(2\pi f_0\Delta) & \sin(2\pi f_0\Delta) \\ -\sin(2\pi f_0\Delta) & \cos(2\pi f_0\Delta) \end{matrix} & 0 \\ 0 & \begin{matrix} \cos(4\pi f_0\Delta) & \sin(4\pi f_0\Delta) \\ -\sin(4\pi f_0\Delta) & \cos(4\pi f_0\Delta) \end{matrix} \end{bmatrix}\mathbf{v}(i) + \tilde{\mathbf{v}}(i) \tag{8.58}$$

The factor d is selected close to one, modelling the fact that the magnitudes of each component vary in time only slowly.

Application of the augmented state estimator to a simulation of the process shows results as depicted in Figure 8.14. The Bode diagram clearly shows that the state estimator acts as a double-notch filter. The width of the notch depends on the choice of d. The Bode diagram, valid for the steady state Kalman filter, is obtained with the

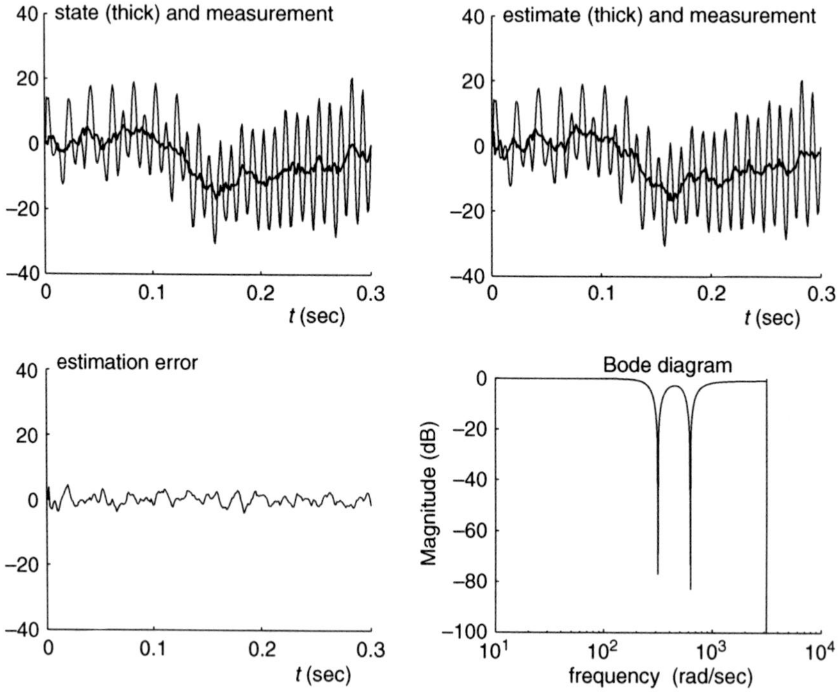

Figure 8.14 Suppression of 50 Hz emf interference based on Kalman filtering

following MATLAB fragment (making use of the Control System Toolbox):

```
. . .
sys=ss(Fn,eye(5),H,0,Ts);               % Create a state space model
[Kest,L,P,M,Z]=kalman(sys,Wn,0);        % Find the steady state KF
bodemag(Kest(2,:),'k');                 % Plot the Bode diagram
```

8.5.2 Cross-correlated noise

Another situation occurs when the process and measurement noise are cross-correlated, that is, $\mathbf{C}_{\mathbf{wv}} = E[\mathbf{w}(i)\mathbf{v}^T(i)] \neq \mathbf{0}$. Such might happen if both the physical process and measurement system is affected by the same source of disturbance. Examples are changes of temperature and electrical inference due to induction.

The strategy to bring this situation to the standard estimation problem is to introduce a modified state equation by including a term $\mathbf{T}(\mathbf{z}(i) - \mathbf{H}\mathbf{x}(i) - \mathbf{v}(i)) \equiv \mathbf{0}$:

$$\begin{aligned}
\mathbf{x}(i+1) &= \mathbf{F}\mathbf{x}(i) + \mathbf{w}(i) \\
&= \mathbf{F}\mathbf{x}(i) + \mathbf{w}(i) + \mathbf{T}(\mathbf{z}(i) - \mathbf{H}\mathbf{x}(i) - \mathbf{v}(i)) \\
&= (\mathbf{F} - \mathbf{T}\mathbf{H})\mathbf{x}(i) + \mathbf{w}(i) - \mathbf{T}\mathbf{v}(i) + \mathbf{T}\mathbf{z}(i)
\end{aligned} \tag{8.59}$$

The factor $\mathbf{F} - \mathbf{T}\mathbf{H}$ is the modified transition matrix. The terms $\mathbf{w}(i) - \mathbf{T}\mathbf{v}(i)$ are regarded as process noise. The term $\mathbf{T}\mathbf{z}(i)$ is known and can be regarded as a control input.

Note that $\mathbf{T}$ can be selected arbitrarily because $\mathbf{T}(\mathbf{z}(i) - \mathbf{H}\mathbf{x}(i) - \mathbf{v}(i)) \equiv \mathbf{0}$. Therefore, we can select $\mathbf{T}$ such that the new process noise becomes uncorrelated with respect to the measurement noise. That is, $E[(\mathbf{w}(i) - \mathbf{T}\mathbf{v}(i))\mathbf{v}^T(i)] = \mathbf{0}$. Or: $\mathbf{C}_{\mathbf{wv}} - \mathbf{T}\mathbf{C}_{\mathbf{v}} = \mathbf{0}$. In other words, if $\mathbf{T} = \mathbf{C}_{\mathbf{wv}}\mathbf{C}_{\mathbf{v}}^{-1}$, the modified state equation is in the standard form without cross-correlation.

8.5.3 Smoothing

Up to now only online estimation of continuous states has been considered. The topic of prediction has been touched on in Section 4.2.1. Here, we introduce the subject of offline estimation, generally referred to as *smoothing*. Assuming a linear-Gaussian model of the process and measurements, we have recorded a set of measurements $\mathbf{Z}_K = \{\mathbf{z}(0), \mathbf{z}(1), \ldots, \mathbf{z}(K)\}$. Using the prior knowledge $E[\mathbf{x}(0)]$ and $\mathbf{C_x}(0)$ we want estimates for some points in time $0 \leq k \leq K$. We emphasize once again that this section is only introductory. For details we refer to the pertinent literature (Gelb *et al.*, 1974).

The problem is often divided into three types of problems:

- Fixed interval smoothing: K is fixed. k is variable between 0 and K.
- Fixed point smoothing: k is fixed. K increases with time, $K = i$.
- Fixed lag smoothing: both k and K increase with time, but with $K - k$ fixed to the so-called $lag \stackrel{def}{=} K - k$. Thus, $K = i$ and $k = i - lag$.

Fixed interval smoothing is needed most often. The problem occurs when an experiment has been done, the data has been acquired and stored, and the data is analysed afterwards.

The general approach to smoothing is the same as for discrete states. See Section 4.3.3. The estimation occurs in two passes. In the first pass, the data is processed forward in time to yield estimates $\hat{\mathbf{x}}_f(k) = \bar{\mathbf{x}}(k|k)$. In the second pass, the data is processed backward in time. Starting with $k = K$ we recursively estimate the previous states using only data from the 'future'. Thus, the backward estimate $\hat{\mathbf{x}}_b(k)$ only uses information from $\mathbf{Z}_K - \mathbf{Z}_{k+1} = \{\mathbf{z}(k+1), \ldots, \mathbf{z}(K)\}$. For that, a 'reversed time' Kalman filter can be used. The final estimate $\bar{\mathbf{x}}(k|K)$ is obtained by optimally combining $\hat{\mathbf{x}}_f(k)$ and $\hat{\mathbf{x}}_b(k)$.

Although this approach yields the desired optimally smoothed states, it is not computationally efficient. For each of the three different types of smoothing problems more efficient algorithms have been proposed. One of them is the well-known *Rauch–Tung–Striebel smoother* (Rauch *et al.*, 1965). The algorithm implements a fixed interval smoother. It does not explicitly use $\hat{\mathbf{x}}_b(k)$. Instead, it uses recursively $\bar{\mathbf{x}}(k|K)$. The algorithm is as follows:

Algorithm 8.2: Rauch–Tung–Striebel smoother

1. Apply the standard discrete Kalman filter to find the offline estimates and store the results, that is, the estimates $\bar{\mathbf{x}}(k|k), \mathbf{C}(k|k)$, along with the one-step-ahead predictions $\bar{\mathbf{x}}(k+1|k), \mathbf{C}(k+1|k)$.
2. For $k = K - 1$ looping back to $k = 0$ with step $= -1$:
   2.1. $\mathbf{A} = \mathbf{C}(k|k)\mathbf{F}^T\mathbf{C}^{-1}(k+1|k)$
   2.2. $\bar{\mathbf{x}}(k|K) = \bar{\mathbf{x}}(k|k) + \mathbf{A}(\bar{\mathbf{x}}(k+1|K) - \bar{\mathbf{x}}(k+1|k))$
   2.3. $\mathbf{C}(k|K) = \mathbf{C}(k|k) + \mathbf{A}(\mathbf{C}(k+1|K) - \mathbf{C}(k+1|k))\mathbf{A}^T$

Step 2.3 is only required if we are interested in the smoothing covariance matrix.

Example 8.19 Estimation of a transient of an electrical RC circuit Figure 8.15 shows an electric circuit consisting of a capacitor connected by means of a switch to a resistor. The resistor represents the input impedance of an AD converter that measures the voltage z. The voltage across the capacitor is x. At $t = 0$ the switch closes, giving rise to a measured voltage $z = x + v$ (v is regarded as sensor noise). The system obeys the following state equation $\dot{x} = -x/(RC)$ with $RC = 10\,(\mu s)$.

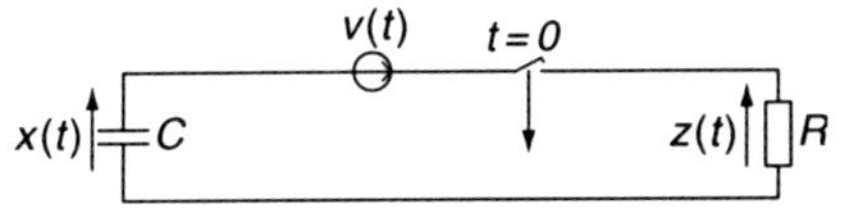

Figure 8.15 The measurement of a transient in an electrical RC network

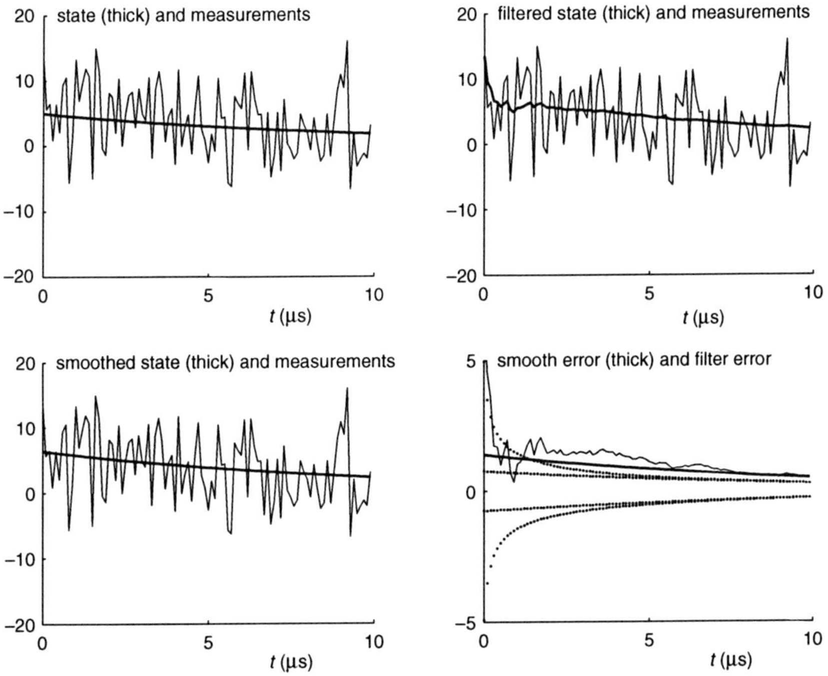

Figure 8.16 Estimation of a transient by means of filtering and by smoothing

In discrete time this becomes $x(i+1) = (1 - \Delta/(RC))x(i)$. The sampling period is $\Delta = 0.1$ (μs). $K = 100$. No prior knowledge about $x(0)$ is available.

Figure 8.16 shows observed measurements along with the corresponding estimated states and the result from the Rauch–Tung–Striebel smoother. Clearly, the uncertainty of the offline obtained estimate is much smaller than the uncertainty of the Kalman filtered result. This holds true especially in the beginning where the online filter has only a few measurements at its disposal. The offline estimator can take advantage of all measurements.

8.6 REFERENCES

Åmström, K.J. and Wittenmark, B., *Computer-Controlled Systems – Theory and Design*, second edition, Prentice Hall, Englewood Cliffs, NJ, 1990.

Bar-Shalom, Y. and Birmiwal, K., Consistency and robustness of PDAF for target tracking in cluttered environments, *Automatica*, **19**, 431–7, July 1983.

Bar-Shalom, Y. and Li, X.R., *Estimation and Tracking – Principles, Techniques, and Software*, Artech House, Boston, MA, 1993.

Blackman, R.B. and Tukey, J.W., *The Measurement of Power Spectra*, Dover, New York, 1958.

Box, G.E.P. and Jenkins, G.M., *Time Series Analysis: Forecasting and Control*, Holden-Day, San Francisco, 1976.

Bryson, A.E. and Hendrikson, L.J., Estimation using sampled data containing sequentially correlated noise, *Journal of Spacecraft and Rockets*, **5**(6), 662–5, June 1968.

Eykhoff, P., *System Identification, Parameter and State Estimation*, Wiley, London, 1974.

Gelb, A., Kasper, J.F., Nash, R.A., Price, C.F. and Sutherland, A.A., *Applied Optimal Estimation*, MIT Press, Cambridge, MA, 1974.

Grewal, M.S. and Andrews, A.P., *Kalman Filtering – Theory and Practice Using MATLAB*, second edition, Wiley, New York, 2001.

Kaminski, P.G., Bryson, A.E. and Schmidt, J.F., Discrete square root filtering: a survey of current techniques, *IEEE Transactions on Automatic Control*, **16**(6), 727–36, December 1971.

Ljung, L., *System Identification – Theory for the User*, 2nd edition, Prentice Hall, Upper Saddle River, NJ, 1999.

Ljung, L. and Glad, T., *Modeling of Dynamic Systems*, Prentice Hall, Englewood Cliffs, NJ, 1994.

Potter, J.E. and Stern, R.G., Statistical filtering of space navigation measurements, *Proceedings of AIAA Guidance and Control Conference*, AIAA, New York, 1963.

Rauch, H.E., Tung, F. and Striebel, C.T., Maximum likelihood estimates of linear dynamic systems, *AIAA Journal*, **3**(8), 1445–50, August 1965.

Söderström, T. and Stoica, P., *System Identification*, Prentice Hall, International, London, 1989.

8.7 EXERCISES

1. The observed sequence of data shown in Figure 8.17 is available in the data file `C8exercise1.mat`. Use MATLAB to determine the smallest order of an autoregressive model that is still able to describe the data well. Determine the parameters of that model. (*).
2. Given the system:

$$\mathbf{F} = \begin{bmatrix} 0.32857 & -0.085714 \\ 1.1429 & 1.0714 \end{bmatrix} \quad \mathbf{H} = [10 \quad 5]$$

Determine the observability Grammian and the observability matrix. What are the eigenvalues of these matrices? What can be said about the observability? (0)

3. Given the system $\mathbf{x}(i+1) = \mathbf{F}\mathbf{x}(i) + \mathbf{G}w(i)$ and $\mathbf{z}(i) = \mathbf{H}\mathbf{x}(i) + v(i)$ with

$$\mathbf{F} = \begin{bmatrix} 0.65 & -0.06 \\ -0.375 & 0.65 \end{bmatrix} \quad \mathbf{G} = \begin{bmatrix} 2 \\ 5 \end{bmatrix} \quad \mathbf{H} = [1 \quad 1]$$

$w(i)$ and $v(i)$ are white noise sequences with unit variances. Examine the observability and the controllability of this system. Does the steady state Kalman filter exist? If so, determine the Kalman gain, the innovation matrix, the prediction covariance matrix and the error covariance matrix. (*)

4. Repeat exercise 3, but this time with:

$$\mathbf{F} = \begin{bmatrix} 0.85 & -0.14 \\ -0.875 & 0.85 \end{bmatrix} \quad (*)$$

5. Repeat exercise 3, but now with:

$$\mathbf{F} = \begin{bmatrix} 0.75 & -0.1 \\ -0.625 & 0.75 \end{bmatrix} \quad (*)$$

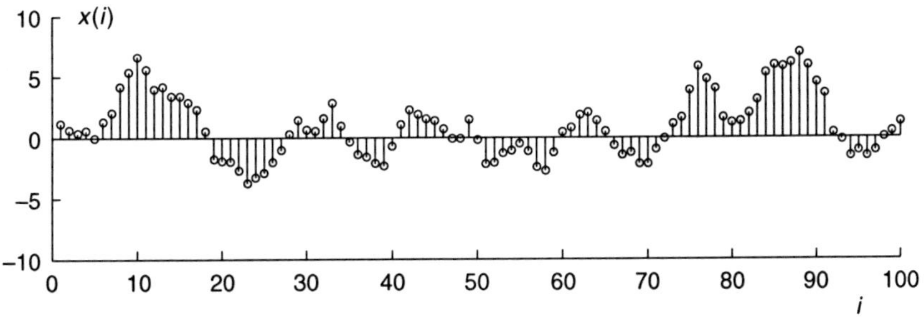

Figure 8.17 Observed random sequence in exercise 1

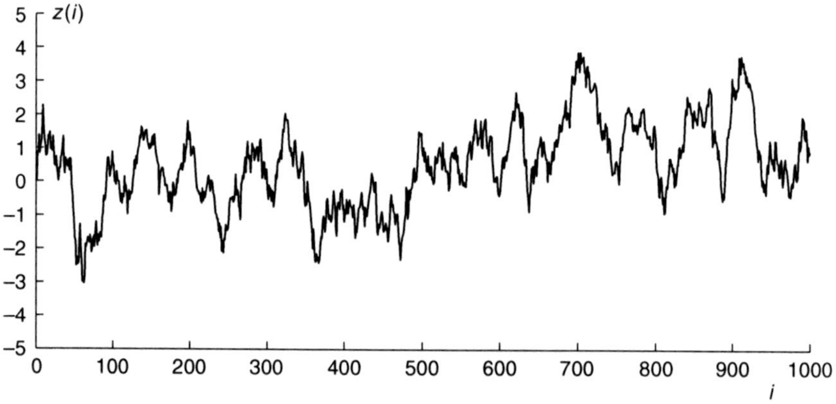

Figure 8.18 Observed measurements from a drifting sensor

6. Explain the different results obtained in exercises 3, 4 and 5, by examining the eigenvalues of F in the different cases. (**)
7. Determine the computational complexity of the information filter. (*)
8. Drift in the measurements.

   We consider a physical quantity $x(i)$ that is sensed by a drifting sensor whose output is modelled by $z(i) = x(i) + v(i)$ and $v(i+1) = \beta v(i) + \tilde{v}(i)$ with $\beta = 0.999$. $\tilde{v}(i)$ is a white noise sequence with zero mean and variance $\sigma_{\tilde{v}}^2 = 0.002$. The physical quantity has a limited bandwidth modelled by $x(i+1) = \alpha x(i) + w(i)$ with $\alpha = 0.95$. The process noise is white and has a variance $\sigma_w^2 = 0.0975$. A record of the measurements is shown in Figure 8.18. The data is available in the file `C8exercise8.mat`.

   - Give a state space model of this system. (0)
   - Examine the observability and controllability of this system. (0)
   - Give the solution of the discrete Lyapunov equation (0)
   - Realize the discrete Kalman filter. Calculate and plot the estimates, its σ boundary, and the innovations and the periodogram of the innovations. (*)
   - Compare the signal-to-noise ratios before and after filtering. (0)
   - Perform the consistency checks. (*)

9

Worked Out Examples

In this final chapter, three worked out examples will be given of the topics discussed in this book: classification, parameter estimation and state estimation. They will take the form of a step-by-step analysis of data sets obtained from real-world applications. The examples demonstrate the techniques treated in the previous chapters. Furthermore, they are meant to illustrate the standard approach to solving these types of problems. Obviously, the MATLAB and PRTools algorithms as they were presented in the previous chapters will be used. The data sets used here are available through the website accompanying this book.

9.1 BOSTON HOUSING CLASSIFICATION PROBLEM

9.1.1 Data set description

The Boston Housing data set is often used to benchmark data analysis tools. It was first described in Harrison and Rubinfield (1978). This paper investigates which features are linked to the air pollution in several areas in Boston. The data set can be downloaded from the UCI Machine Learning repository at http://www.ics.uci.edu/~mlearn/MLRepository.html.

Each feature vector from the set contains 13 elements. Each feature element provides specific information on an aspect of an area of a suburb. Table 9.1 gives a short description of each feature element.

Classification, Parameter Estimation and State Estimation: An Engineering Approach using MATLAB
F. van der Heijden, R.P.W. Duin, D. de Ridder and D.M.J. Tax
 ISBN: 0-470-09013-8

Table 9.1 The features of the Boston Housing data set

	Feature name	Description	Feature range and type
1	CRIME	Crime rate per capita	0–89, real valued
2	LARGE	Proportion of area dedicated to lots larger than 25 000 square feet	0–100, real valued, but many have value 0
3	INDUSTRY	Proportion of area dedicated to industry	0.46–27.7, real valued
4	RIVER	1 = borders the Charles river, 0 = does not border the Charles river	0–1, nominal values
5	NOX	Nitric oxides concentration (in parts per 10 million)	0.38–0.87, real valued
6	ROOMS	Average number of rooms per house	3.5–8.8, real valued
7	AGE	Proportion of houses built before 1940	2.9–100, real valued
8	WORK	Average distance to employment centres in Boston	1.1–12.1, real valued
9	HIGHWAY	Highway accessibility index	1–24, discrete values
10	TAX	Property tax rate (per $10 000)	187–711, real valued
11	EDUCATION	Pupil–teacher ratio	12.6–22.0, real valued
12	AA	$1000(A - 0.63)^2$ where A is the proportion of African-Americans	0.32–396.9, real valued
13	STATUS	Percentage of the population which has lower status	1.7–38.0, real valued

The goal we set for this section is to predict whether the median price of a house in each area is larger than or smaller than $20 000. That is, we formulate a two-class classification problem. In total, the data set consists of 506 areas, where for 215 areas the price is lower than $20 000 and for 291 areas it is higher. This means that when a new object has to be classified using only the class prior probabilities, assuming the data set is a fair representation of the true classification problem, it can be expected that in $(215/506) \cdot 100\% = 42.5\%$ of the cases we will misclassify the area.

After the very first inspection of the data, by just looking what values the different features might take, it appears that the individual features differ significantly in range. For instance, the values for the feature TAX varies between 187 and 711 (a range of more than 500), while the feature values of NOX are between 0.38 and 0.87 (a range of less than 0.5). This suggests that some scaling might be necessary. A second important observation is that some of the features can only have discrete values, like RIVER and HIGHWAY. How to combine real valued features with discrete features is already a problem in itself. In this analysis we will ignore this problem, and we will assume that all features are real valued. (Obviously, we will lose some performance using this assumption.)

9.1.2 Simple classification methods

Given the varying nature of the different features, and the fact that further expert knowledge is not given, it will be difficult to construct a good model for this data. The scatter diagram of Figure 9.1 shows that an assumption of Gaussian distributed data is clearly wrong (if only by the presence of the discrete features), but when just classification performance is considered, the decision boundary might still be good enough. Perhaps more flexible methods such as the Parzen density or the κ-nearest neighbour method will perform better; after a suitable feature selection and feature scaling.

Let us start with some baseline methods and train a linear and quadratic Bayes classifier, `ldc` and `qdc`:

Listing 9.1

```
% Load the housing dataset, and set the baseline performance
load housing.mat;
z                                   % Show what dataset we have
w=ldc;                              % Define an untrained linear
                                      classifier
err_ldc_baseline=crossval(z,w,5)    % Perform 5-fold
                                      cross-validation
err_qdc_baseline=crossval(z,qdc,5)  % idem for the quadratic
                                      classifier
```

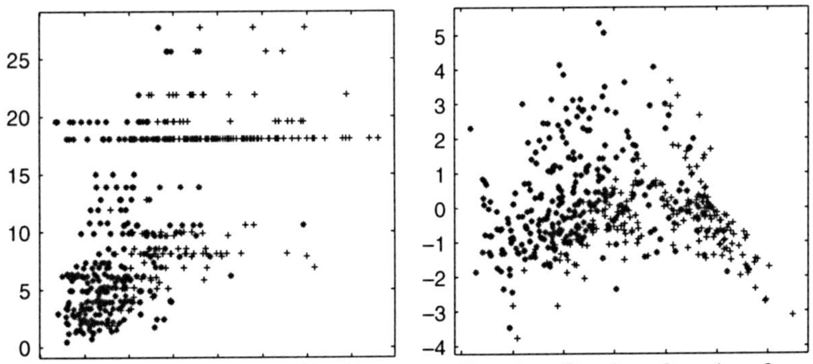

Figure 9.1 Scatter plots of the Boston Housing data set. The left subplot shows features STATUS and INDUSTRY, where the discrete nature of INDUSTRY can be spotted. In the right subplot, the data set is first scaled to unit variance, after which it is projected onto its first two principal components

The five-fold cross-validation errors of `ldc` and `qdc` are 13.0% and 17.4%, respectively. Note that by using the function `crossval`, we avoided having to use the training set for testing. If we had done that, the errors would be 11.9% and 16.2%, but this estimate would be biased and too optimistic. The results also depend on how the data is randomly split into batches. When this experiment is repeated, you will probably find slightly different numbers. Therefore, it is advisable to repeat the entire experiment a number of times (say, 5 or 10) to get an idea of the variation in the cross-validation results. However, for many experiments (such as the feature selection and neural network training below), this may lead to unacceptable training times; therefore, the code given does not contain any repetitions. For all the results given, the standard deviation is about 0.005, indicating that the difference between `ldc` and `qdc` is indeed significant. Notice that we use the word 'significant' here in a slightly loose sense. From a statistical perspective it would mean that for the comparison of the two methods, we should state a null-hypothesis that both methods perform the same, and define a statistical test (for instance a t-test) to decide if this hypothesis holds or not. In this discussion we use the simple approach in which we just look at the standard deviation of the classifier performances, and call the performance difference significant when the averages differ by more than two times their standard deviations.

Even with these simple models on the raw, not preprocessed data, a relative good test error of 13.0% can be achieved (with respect to the simplest approach by looking at the class probabilities). Note that `qdc`, which is a more powerful classifier, gives the worst performance. This is due to the relatively low sample size: two covariance matrices, with $\frac{1}{2}13(13+1) = 91$ free parameters each, have to be estimated on 80% of the data available for the classes (i.e. 172 and 233 samples, respectively). Clearly, this leads to poor estimates.

9.1.3 Feature extraction

It might be expected that more flexible methods, like the κ-nearest neighbour classifier (`knnc`, with κ optimized using the leave-one-out method) or the Parzen density estimator (`parzenc`) give better results. Surprisingly, a quick check shows that then the errors become 19.6% and 21.9% (with a standard deviation of about 1.1% and 0.6%, respectively), clearly indicating that some overtraining occurs and/or that some feature selection and feature scaling might be required. The next step, therefore, is to rescale the data to have zero mean and unit variance in all feature directions. This can be performed using the PRTools function `scalem([],'variance')`:

Listing 9.2

```
load housing.mat;
% Define an untrained linear classifier w/scaled input data
w_sc=scalem([],'variance');
w=w_sc*ldc;
% Perform 5-fold cross-validation
err_ldc_sc=crossval(z,w,5)
% Do the same for some other classifiers
err_qdc_sc=crossval(z,w_sc*qdc,5)
err_knnc_sc=crossval(z,w_sc*knnc,5)
err_parzenc_sc=crossval(z,w_sc*parzenc,5)
```

First note, that when we introduce a preprocessing step, this step should be defined *inside* the mapping **w**. The obvious approach, to map the whole data set `z_sc = z*scalem(a,'variance')` and then to apply the cross-validation to estimate the classifier performance, is *incorrect*. In that case, some of the testing data is already used in the scaling of the data, resulting in an overtrained classifier, and thus in an unfair estimate of the error. To avoid this, the mapping should be extended from `ldc` to `w_sc*ldc`. The routine `crossval` then takes care of fitting both the scaling and the classifier.

By scaling the features, the performance of the first two classifiers, `ldc` and `qdc`, should not change. The normal density based classifiers are insensitive to the scaling of the individual features, because they already use their variance estimation. The performance of `knnc` and `parzenc` on the other hand improve significantly, to 14.1% and 13.1%, respectively (with a standard deviation of about 0.4%). Although `parzenc` approaches the performance of the linear classifier, it is still slightly worse. Perhaps feature extraction or feature selection will improve the results.

As discussed in Chapter 7, principal component analysis (PCA) is one of the most often used feature extraction methods. It focuses on the high-variance directions in the data, and removes low-variance directions. In this data set we have seen that the feature values have very different scales. Applying PCA directly to this data will put high emphasis on the feature TAX and will probably ignore the feature NOX. Indeed, when

PCA preprocessing is applied such that 90% of the variance is retained, the performance of all the methods significantly decreases. To avoid this clearly undesired effect, we will first rescale the data to have unit variance and apply PCA on the resulting data. The basic training procedure now becomes:

Listing 9.3

```
load housing.mat;
% Define a preprocessing
w_pca=scalem([],'variance')*pca([],0.9);
% Define the classifier
w=w_sc*ldc;
% Perform 5-fold cross-validation
err_ldc_pca=crossval(z,w,5)
```

It appears that, compared with normal scaling, the application of `pca([],0.9)` does not significantly improve the performances. For some methods, the performance increases slightly (16.6% (±0.6%) error for `qdc`, 13.6% (±0.9%) for `knnc`), but for other methods, it decreases. This indicates that the high-variance features are not much more informative than the low-variance directions.

9.1.4 Feature selection

The use of a simple supervised feature extraction method, such as the Bhattacharrya mapping (implemented by replacing the call to `pca` by `bhatm([])`), also decreases the performance. We will therefore have to use better feature selection methods to reduce the influence of noisy features and to gain some performance.

We will first try branch-and-bound feature selection to find five features, with the simple inter–intra class distance measure as a criterion, finding the optimal number of features. Admittedly, the number of features selected, five, is arbitrary, but the branch-and-bound method does not allow for finding the optimal subset size.

Listing 9.4

```
% Load the housing dataset
load housing.mat;
% Construct scaling and feature selection mapping
w_fsf=featselo([],'in-in',5)*scalem([],'variance');
```

```
% Calculate cross-validation error for classifiers
% trained on the optimal 5-feature set
err_ldc_fsf=crossval(z,w_fsf*ldc,5)
err_qdc_fsf=crossval(z,w_fsf*qdc,5)
err_knnc_fsf=crossval(z,w_fsf*knnc,5)
err_parzenc_fsf=crossval(z,w_fsf*parzenc,5)
```

The feature selection routine often selects features STATUS, AGE and WORK, plus some others. The results are not very good: the performance decreases for all classifiers. Perhaps we can do better if we take the performance of the actual classifier to be used as a criterion, rather than the general inter–intra class distance. To this end, we can just pass the classifier to the feature selection algorithm. Furthermore, we can also let the algorithm find the optimal number of features by itself. This means that branch-and-bound feature selection can now no longer be used, as the criterion is not monotonically increasing. Therefore, we will use forward feature selection, `featself`.

Listing 9.5

```
% Load the housing dataset
load housing.mat;
% Optimize feature set for ldc
w_fsf=featself([],ldc,0)*scalem([],'variance');
err_ldc_fsf=crossval(z,w_fsf*ldc,5)
% Optimize feature set for qdc
w_fsf=featself([],qdc,0)*scalem([],'variance');
err_qdc_fsf=crossval(z,w_fsf*qdc,5)
% Optimize feature set for knnc
w_fsf=featself([],knnc,0)*scalem([],'variance');
err_knnc_fsf=crossval(z,w_fsf*knnc,5)
% Optimize feature set for parzenc
w_fsf=featself([],parzenc,0)*scalem([],'variance');
err_parzenc_fsf=crossval(z,w_fsf*parzenc,5)
```

This type of feature selection turns out to be useful only for `ldc` and `qdc`, whose performances improve to 12.8% (±0.6%) and 14.9% (±0.5%), respectively. `knnc` and `parzenc`, on the other hand, give 15.9% (±1.0%) and 13.9% (±1.7%), respectively. These results do not differ significantly from the previous ones. The `featself` routine often selects the same rather large set of features (from most to least significant): STATUS, AGE, WORK, INDUSTRY, AA, CRIME, LARGE, HIGHWAY, TAX. But these features are highly correlated, and the set used can be reduced to the first three with just a small increase in error of about 0.005. Nevertheless, feature selection in general does not seem to help much for this data set.

9.1.5 Complex classifiers

The fact that `ldc` already performs so well on the original data indicates that the data is almost linearly separable. A visual inspection of the scatter plots in Figure 9.1 seems to strengthen this hypothesis. It becomes even more apparent after the training of a linear support vector classifier (`svc([],'p',1)`) and a Fisher classifier (`fisherc`), both with a cross-validation error of 13.0%, the same as for `ldc`.

Given that `ldc` and `parzenc` thus far performed best, we might try to train a number of classifiers based on these concepts, for which we are able to tune classifier complexity. Two obvious choices for this are neural networks and support vector classifiers (SVCs). Starting with the latter, we can train SVCs with polynomial kernels of degree close to 1, and with radial basis kernels of radius close to 1. By varying the degree or radius, we can vary the resulting classifier's nonlinearity:

Listing 9.6

```
load housing.mat;                          % Load the housing dataset
w_pre = scalem([], 'variance');            % Scaling mapping
degree = 1:3;                              % Set range of parameters
radius = 1:0.25:3;
for i = 1:length(degree)
  err_svc_p(i) = ...                       % Train polynomial SVC
    crossval(z,w_pre*svc([],'p',degree(i)),5);
end;
for i = 1:length(radius)
  err_svc_r(i) = ...                       % Train radial basis SVC
    crossval(z,w_pre*svc([], 'r',radius(i)),5);
end;
figure; clf; plot(degree,err_svc_p);
figure; clf; plot(radius,err_svc_r);
```

The results of a single repetition are shown in Figure 9.2: the optimal polynomial kernel SVC is a quadratic one (a degree of 2), with an average error of 12.5%, and the optimal radial basis kernel SVC (a radius of around 2) is slightly better, with an average error of 11.9%. Again, note that we should really repeat the experiment a number of times, to get an impression of the variance in the results. The standard deviation is

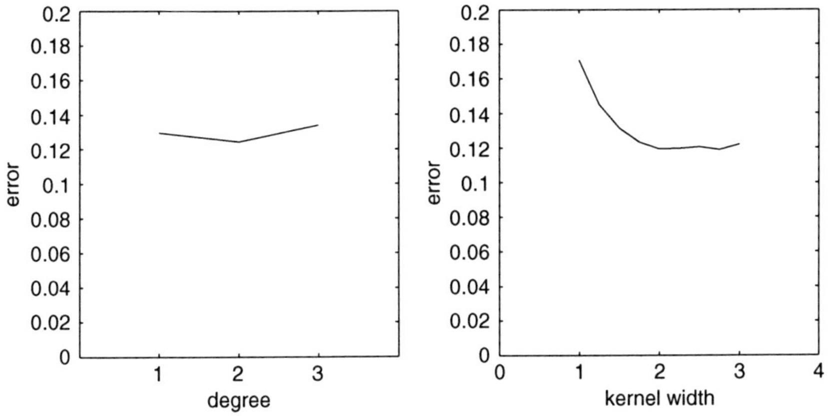

Figure 9.2 Performance of a polynomial kernel SVC (left, as a function of the degree of the polynomial) and a radial basis function kernel SVC (right, as a function of the basis function width)

between 0.2% to 1.0%. This means that the minimum in the right subfigure is indeed significant, and that the radius parameter should indeed be around 2.0. On the other hand, in the left subplot the graph is basically flat and a linear SVC is therefore probably to be preferred.

For the sake of completeness, we can also train feed-forward neural networks with varying numbers of hidden layers and units. In PRTools, there are three routines for training feed-forward neural networks. The `bpxnc` function trains a network using the back-propagation algorithm, which is slow but does not often overtrain it. The `lmnc` function uses a second-order optimization routine, Levenberg–Marquardt, which speeds up training significantly but often results in overtrained networks. Finally, the `neurc` routine attempts to counteract the overtraining problem of `lmnc` by creating an artificial tuning set of 1000 samples by perturbing the training samples (see `gendatk`) and stops training when the error on this tuning set increases. It applies `lmnc` three times and returns the neural network giving the best result on the tuning set. Here, we apply both `bpxnc` and `neurc`:

Listing 9.7

```
load housing.mat;                        % Load the housing dataset
w_pre = scalem([], 'variance');          % Scaling mapping
networks = {bpxnc, neurc};               % Set range of parameters
nlayers = 1:2;
nunits = [4 8 12 16 20 30 40];
for i = 1:length(networks)
```

```
  for j = 1:length(nlayers)
    for k = 1:length(nunits)
      % Train a neural network with nlayers(j) hidden layers
      % of nunits(k) units each, using algorithm network{i}
      err_nn(i,j,k) = crossval(z, ...
        w_pre*networks{i}([],ones(1,nlayers(j))*nunits(k)),5);
    end;
  end;
  figure; clear all;                      % Plot the errors
  plot(nunits,err_nn(i,1,:), '-'); hold on;
  plot(nunits,err_nn(i,2,:), '--');
  legend('1 hidden layer', '2 hidden layers');
end;
```

Training neural networks is a computationally intensive process; and here they are trained for a large range of parameters, using cross-validation. The algorithm above takes more than a day to finish on a modern workstation, although per setting just a single neural network is trained.

The results, shown in Figure 9.3, seem to be quite noisy. After repeating the algorithm several times, it appears that the standard deviation is in the order of 1%. Ideally, we would expect the error as a function of the number of hidden layers and units per hidden layer to have a clear global optimum. For bpxnc, this is roughly the case, with a minimal cross-validation error of 10.5% for a network with one hidden layer of 30 units, and 10.7% for a network with two hidden layers of 16 units. Normally, we would prefer to choose the network with the lowest

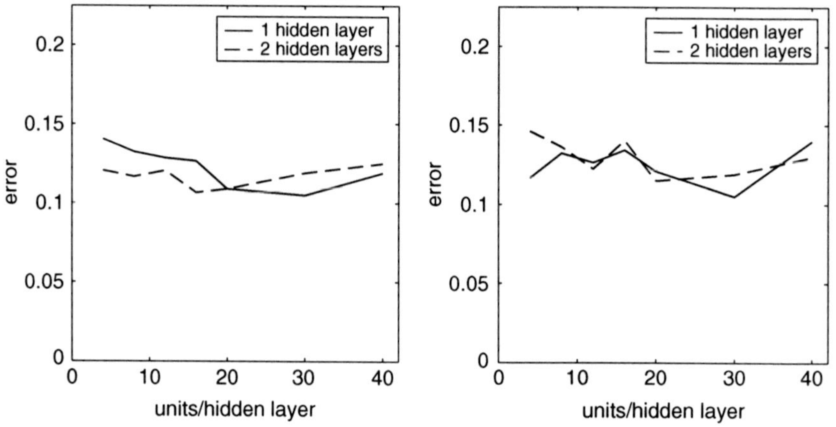

Figure 9.3 Performance of neural networks with one or two hidden layers as a function of the number of units per hidden layer, trained using *bpxnc* (left) and *neurc* (right)

complexity as the variance in the error estimate would be lowest. However, the two optimal networks here have roughly the same number of parameters. So, there is no clear best choice between the two.

The cross-validation errors of networks trained with `neurc` show more variation (Figure 9.3). The minimal cross-validation error is again 10.5% for a network with a single hidden layer of 30 units. Given that the graph for `bpxnc` is much smoother than that of `neurc`, we would prefer to use a `bpxnc`-trained network.

9.1.6 Conclusions

The best overall result on the housing data set was obtained using a `bpxnc`-trained neural network (10.5% cross-validation error), slightly better than the best SVC (11.9%) or a simple linear classifier (13.0%). However, remember that neural network training is a more noisy process than training an SVC or linear classifier: the latter two will find the same solution when run twice on the same data set, whereas a neural network may give different results due to the random initialization. Therefore, using an SVC in the end may be preferable.

Of course, the analysis above is not exhaustive. We could still have tried more exotic classifiers, performed feature selection using different criteria and search methods, searched through a wider range of parameters for the SVCs and neural networks, investigated the influence of possible outlier objects and so on. However, this will take a lot of computation, and for this application there seems to be no reason to believe we might obtain a significantly better classifier than those found above.

9.2 TIME-OF-FLIGHT ESTIMATION OF AN ACOUSTIC TONE BURST

The determination of the time of flight (ToF) of a tone burst is a key issue in acoustic position and distance measurement systems. The length of the acoustic path between a transmitter and receiver is proportional to the ToF, that is, the time evolved between the departure of the waveform from the transmitter and the arrival at the receiver (Figure 9.4). The position of an object is obtained, for instance, by measuring the distances from the object to a number of acoustic beacons. See also Figure 1.2.

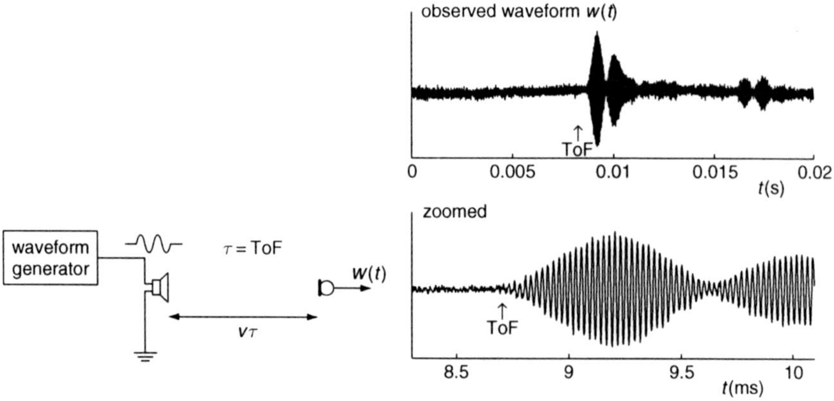

Figure 9.4 Set-up of a sensory system for acoustic distance measurements

The quality of an acoustic distance measurement is directly related to the quality of the ToF determination. Electronic noise, acoustic noise, atmospheric turbulence and temperature variations are all factors that influence the quality of the ToF measurement. In indoor situations, objects in the environment (wall, floor, furniture, etc.) may cause echoes that disturb the nominal response. These unwanted echoes can cause hard-to-predict waveforms, thus making the measurement of the ToF a difficult task.

The transmitted waveform can take various forms, for instance, a frequency modulated (chirped) continuous waveform (CWFM), a frequency or phase shift-keyed signal or a tone burst. The latter is a pulse consisting of a number of periods of a sine wave. An advantage of a tone burst is that the bandwidth can be kept moderate by adapting the length of the burst. Therefore, this type of signal is suitable for use in combination with piezoelectric transducers, which are cheap and robust, but have a narrow bandwidth.

In this section, we design an estimator for the determination of the ToFs of tone bursts that are acquired in indoor situations using a set-up as shown in Figure 9.4. The purpose is to determine the time delay between sending and receiving a tone burst. A learning and evaluation data set is available that contains 150 records of waveforms acquired in different rooms, different locations in the rooms, different distances and different heights above the floor. Figure 9.4 shows an example of one of the waveforms. Each record is accompanied by a reference ToF indicating the true value of the ToF. The standard deviation of the reference ToF is estimated at 10 (μs). The applied sampling period is $\Delta = 2$ (μs).

The literature roughly mentions three concepts to determine the ToF, i.e. thresholding, data fitting (regression) and ML (maximum likelihood) estimation (Heijden van der *et al.*, 2003). Many variants of these concepts have been proposed. This section only considers the main representatives of each concept:

- Comparing the envelope of the wave against a threshold that is proportional to the magnitude of the waveform.
- Fitting a one-sided parabola to the foot of the envelope of the waveform.
- Conventional matched filtering.
- Extended matched filtering based on a covariance model of the signal.

The first one is a heuristic method that does not optimize any criterion function. The second one is a regression method, and as such a representative of data fitting. The last two methods are ML estimators. The difference between these is that the latter uses an explicit model of multiple echoes. In the former case, such a model is missing.

The section first describes the methods. Next, the optimization of the design parameters using the data set is explained, and the evaluation is reported. The data set and the MATLAB listings of the various methods can be found on the accompanying website.

9.2.1 Models of the observed waveform

The moment of time at which a transmission begins is well defined since it is triggered under full control of the sensory system. The measurement of the moment of arrival is much more involved. Due to the narrow bandwidth of the transducers the received waveform starts slowly. A low SNR makes the moment of arrival indeterminate. Therefore, the design of a ToF estimator requires the availability of a model describing the arriving waveform. This waveform $w(t)$ consists of three parts: the nominal response $a{\cdot}h(t-\tau)$ to the transmitted wave; the interfering echoes $a{\cdot}r(t-\tau)$; and the noise $v(t)$:

$$w(t) = a{\cdot}h(t-\tau) + a{\cdot}r(t-\tau) + v(t) \tag{9.1}$$

We assume that the waveform is transmitted at time $t = 0$, so that τ is the ToF. (9.1) simply states that the observed waveform equals the nominal response $h(t)$, but now time-shifted by τ and attenuated by a. Such an assumption is correct for a medium like air because, within the bandwidth of interest, the propagation of a waveform through air does not show a significant dispersion. The attenuation coefficient a depends on many factors, but also on the distance, and thus also on τ. However, for the moment we will ignore this fact. The possible echoes are represented by $a \cdot r(t - \tau)$. They share the same time shift τ because no echo can occur before the arrival of the nominal response. The additional time delays of the echoes are implicitly modelled within $r(t)$. The echoes and the nominal response also share a common attenuation factor. The noise $v(t)$ is considered white.

The actual shape of the nominal response $h(t)$ depends on the choice of the tone burst and on the dynamic properties of the transducers. Sometimes, a parametric empirical model is used, for instance:

$$h(t) = t^m \exp(-t/T) \cos(2\pi f t + \varphi) \quad t \geq 0 \tag{9.2}$$

f is the frequency of the tone burst; $\cos(2\pi f t + \varphi)$ is the carrier; and $t^m \exp(-t/T)$ is the envelope. The factor t^m describes the rise of the waveform (m is empirically determined; usually between 1 and 3). The factor $\exp(-t/T)$ describes the decay. Another possibility is to model $h(t)$ non-parametrically. In that case, a sampled version of $h(t)$, obtained in an anechoic room where echoes and noise are negligible, is recorded. The data set contains such a record. See Figure 9.5.

Often, the existence of echoes is simply ignored, $r(t) = 0$. Sometimes, a single echo is modelled $r(t) = d_1 \cdot h(t - \tau_1)$ where τ_1 is the delay of the echo with respect to $t = \tau$. The most extensive model is when multiple echoes are considered $r(t) = \sum_k d_k h(t - \tau_k)$. The sequences d_k and τ_k are hardly predictable and therefore regarded as random. In that case, $r(t)$

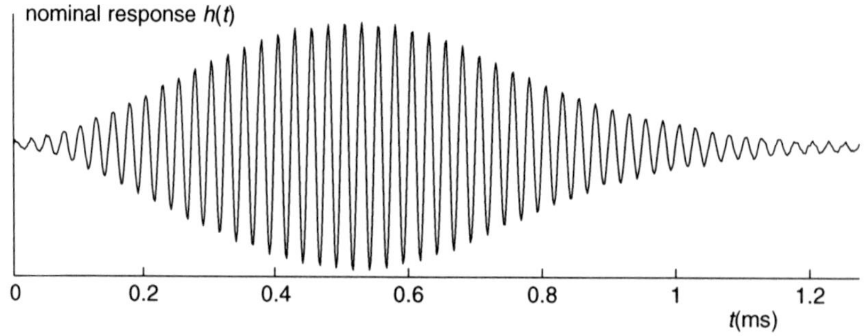

Figure 9.5 A record of the nominal response $h(t)$

becomes random too, and the echoes are seen as disturbing noise with non-stationary properties.

The observed waveform $\mathbf{z} = [\, z_0 \;\; \ldots \;\; z_{N-1} \,]^T$ is a sampled version of $w(t)$:

$$z_n = w(n\Delta) \tag{9.3}$$

Δ is the sampling period. N is the number of samples. Hence, $N\Delta$ is the registration period. With that, the noise $v(t)$ manifests itself as a random vector $\mathbf{v}$, with zero mean and covariance matrix $\mathbf{C_v} = \sigma_v^2\mathbf{I}$.

9.2.2 Heuristic methods for determining the ToF

Some applications require cheap solutions that are suitable for direct implementation using dedicated hardware, such as instrumental electronics. For that reason, a popular method to determine the ToF is simply thresholding the observed waveform at a level T. The estimated ToF $\hat{\tau}_{thres}$ is the moment at which the waveform crosses a threshold level T.

Due to the slow rising of the nominal response, the moment $\hat{\tau}_{thres}$ of level crossing appears just after the true τ, thus causing a bias. Such a bias can be compensated afterwards. The threshold level T should be chosen above the noise level. The threshold operation is simple to realize, but a disadvantage is that the bias depends on the magnitude of the waveform. Therefore, an improvement is to define the threshold level relative to the maximum of the waveform, that is $T = \alpha \max(w(t))$. α is a constant set to, for instance, 30%.

The observed waveform can be written as $g(t)\cos(2\pi ft + \varphi) + n(t)$ where $g(t)$ is the envelope. The carrier $\cos(2\pi ft + \varphi)$ of the waveform causes a resolution error equal to $1/f$. Therefore, rather than applying the threshold operation directly, it is better to apply it to the envelope. A simple, but inaccurate method to get the envelope is to rectify the waveform and to apply a low-pass filter to the result. The optimal method, however, is much more involved and uses *quadrature filtering*. A simpler approximation is as follows. First, the waveform is band-filtered to reduce the noise. Next, the filtered signal is phase-shifted over 90° to obtain the quadrature component $q(t) = g(t)\sin(2\pi ft + \varphi) + n_q(t)$. Finally, the envelope is estimated using $\hat{g}(t) = \sqrt{w^2_{\text{band-filtered}}(t) + q^2(t)}$. Figure 9.6 provides an example.

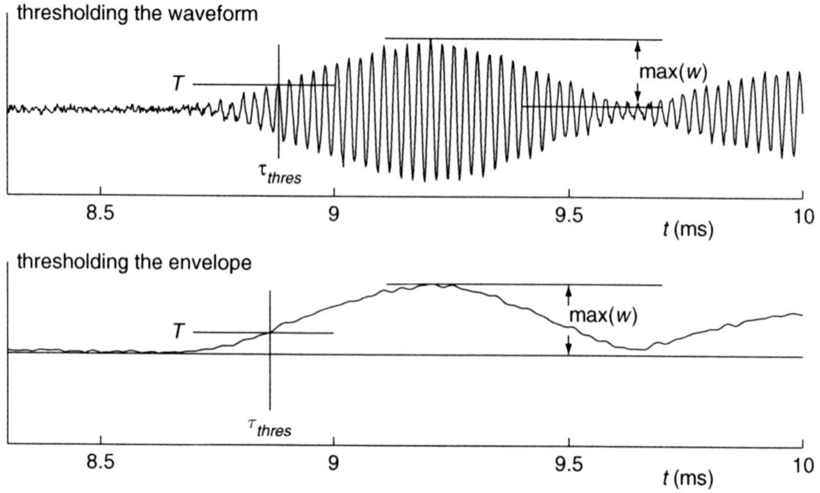

Figure 9.6 ToF measurements based on thresholding operations

The design parameters are the relative threshold α, the bias compensation b, and the cut-off frequencies of the band-filter.

9.2.3 Curve fitting

In the curve-fitting approach, a functional form is used to model the envelope. The model is fully known except for some parameters, one of which is the ToF. As such, the method is based on the regression techniques introduced in Section 3.3.3. On adoption of an error criterion between the observed waveform and the model, the problem boils down to finding the parameters that minimize the criterion. We will use the SSD criterion discussed in Section 3.3.1. The particular function that will be fitted is the one-sided parabola defined by:

$$f(t, \mathbf{x}) = \begin{cases} x_0 + x_1(t - x_2)^2 & \text{if } t > x_2 \\ x_0 & \text{elsewhere} \end{cases} \tag{9.4}$$

The final estimate of the ToF is $\hat{\tau}_{curve} = x_2$.

The function must be fitted to the foot of the envelope. Therefore, an important task is to determine the interval $t_b < t < t_e$ that makes up the foot. The choice of t_b and t_e is critical. If the interval is short, then the noise sensitivity is large. If the interval is too large, modelling errors become too influential. The strategy is to find two anchor points t_1 and t_2 that are stable enough under the various conditions.

The first anchor point t_1 is obtained by thresholding a low-pass filtered version $g_{\text{filtered}}(t)$ of the envelope just above the noise level. If σ_v is the standard deviation of the noise in $w(t)$, then $g_{\text{filtered}}(t)$ is thresholded at a level $3\frac{1}{2}\sigma_v$, thus yielding t_1. The standard deviation σ_v is estimated during a period preceding the arrival of the tone. A second suitable anchor point t_2 is the first location just after t_1 where $g_{\text{filtered}}(t)$ is maximal, i.e. the location just after t_1 where $dg_{\text{filtered}}(t)/dt = 0$. t_e is defined midway between t_1 and t_2 by thresholding $g_{\text{filtered}}(t)$ at a level $3\frac{1}{2}\sigma_v + \alpha(g_{\text{filtered}}(t_2) - 3\frac{1}{2}\sigma_v)$. Finally, t_b is calculated as $t_b = t_1 - \beta(t_e - t_1)$. Figure 9.7 provides an example.

Once the interval of the foot has been established, the curve can be fitted to the data (which in this case is the original envelope $\hat{g}(t)$). Since the curve is nonlinear in **x**, an analytical solution, such as in the polynomial regression in Section 3.3.3, cannot be applied. Instead a numerical procedure, for instance using MATLAB's `fminsearch`, should be applied.

The design parameters are α, β and the cut-off frequency of the low-pass filter.

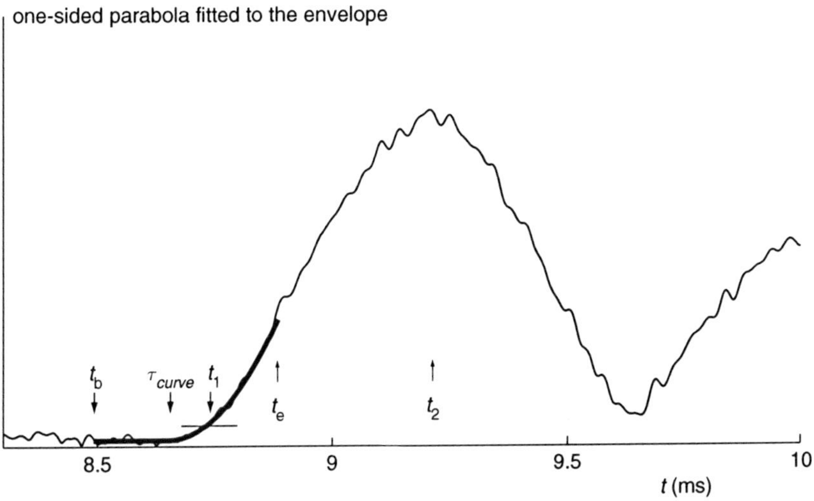

Figure 9.7 ToF estimation by fitting a one-sided parabola to the foot of the envelope

9.2.4 Matched filtering

This conventional solution is achieved by neglecting the reflections. The measurements are modelled by a vector **z** with N elements:

$$z_n = a{\cdot}h(n\Delta - \tau) + v(n\Delta) \tag{9.5}$$

The noise is represented by a random vector **v** with zero mean and covariance matrix $\mathbf{C}_\mathbf{v} = \sigma_v^2\mathbf{I}$. Upon introduction of a vector $\mathbf{h}(\tau)$ with elements $h(n\Delta - \tau)$ the conditional probability density of **z** is:

$$p(\mathbf{z}|\tau) = \frac{1}{\sqrt{(2\pi\sigma_n^2)^N}}\exp\left(-\frac{1}{2\sigma_n^2}(\mathbf{z} - a\mathbf{h}(\tau))^T(\mathbf{z} - a\mathbf{h}(\tau))\right) \tag{9.6}$$

Maximization of this expression yields the maximum likelihood estimate for τ. In order to do so, we only need to minimize the L_2 norm of $\mathbf{z} - a\mathbf{h}(\tau)$:

$$(\mathbf{z} - a\mathbf{h}(\tau))^T(\mathbf{z} - a\mathbf{h}(\tau)) = \mathbf{z}^T\mathbf{z} + a^2\mathbf{h}(\tau)^T\mathbf{h}(\tau) - 2a\mathbf{z}^T\mathbf{h}(\tau) \tag{9.7}$$

The term $\mathbf{z}^T\mathbf{z}$ does not depend on τ and can be ignored. The second term is the signal energy of the direct response. A change of τ only causes a shift of it. But, if the registration period is long enough, the signal energy is not affected by such a shift. Thus, the second term can be ignored as well. The maximum likelihood estimate boils down to finding the τ that maximizes $a\mathbf{z}^T\mathbf{h}(\tau)$. A further simplification occurs if the extent of $h(t)$ is limited to, say, $K\Delta$ with $K << N$. In that case $a\mathbf{z}^T\mathbf{h}(\tau)$ is obtained by cross-correlating z_n by $a{\cdot}h(n\Delta + \tau)$:

$$y(\tau) = a\sum_{k=0}^{K-1} h(k\Delta - \tau)z_k \tag{9.8}$$

The value of τ which maximizes $y(\tau)$ is the best estimate. The operator expressed by (9.8) is called a matched filter or a correlator. Figure 9.8 shows a result of the matched filter. Note that apart from its sign, the amplitude a does not affect the outcome of the estimate. Hence, the fact that a is usually unknown doesn't matter much. Actually, if the nominal response is given in a non-parametric way, the matched filter doesn't have any design parameters.

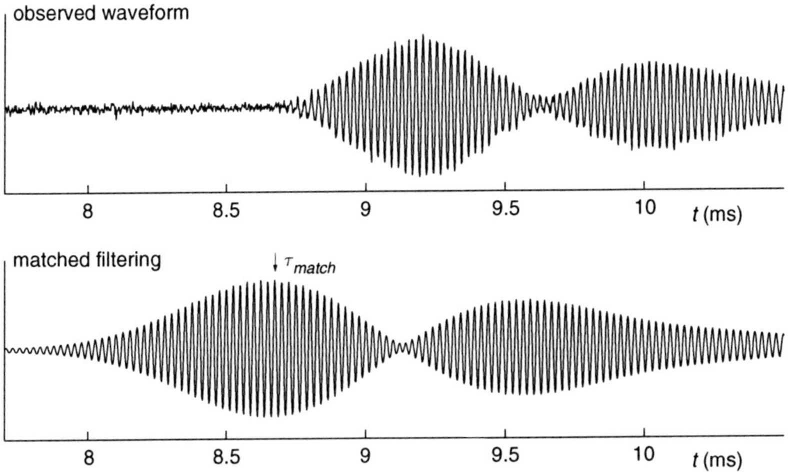

Figure 9.8 Matched filtering

9.2.5 ML estimation using covariance models for the reflections

The matched filtered is not designed to cope with interfering reflections. Especially, if an echo partly overlaps the nominal response, the results are inaccurate. In order to encompass situations with complex interference patterns the matched filter must be extended. A possibility is to model the echoes explicitly. A tractable model arises if the echoes are described by a non-stationary autocovariance function.

Covariance models

The echoes are given by $r(t) = \sum_k d_k h(t - \tau_k)$. The points in time, τ_k, are a random sequence. Furthermore we have $\tau_k > 0$ since all echoes appear after the arrival of the direct response. The attenuation factors d_k have a range of values. We will model them as independent Gaussian random variables with zero mean and variance σ_d^2. Negative values of d_k are allowed because of the possible phase reversal of an echo. We limit the occurrence of an echo to an interval $0 < \tau_k < \mathrm{T}$, and assume a uniform distribution. Then the autocovariance function of a single echo is:

$$\begin{aligned} C_k(t_1, t_2) &= \mathrm{E}\left[d_k^2 h(t_1 - \tau_k) h(t_2 - \tau_k)\right] \\ &= \frac{\sigma_d^2}{\mathrm{T}} \int_{\tau_k=0}^{\mathrm{T}} h(t_1 - \tau_k) h(t_2 - \tau_k) d\tau_k \end{aligned} \tag{9.9}$$

If there are K echoes, the autocovariance function of $r(t)$ is

$$C_r(t_1, t_2) = KC_k(t_1, t_2) = \frac{K\sigma_d^2}{\mathrm{T}} \int_{\tau_k=0}^{\mathrm{T}} h(t_1 - \tau_k) h(t_2 - \tau_k) d\tau_k \tag{9.10}$$

because the factors d_k and random points τ_k are independent.

For arbitrary τ, the reflections are shifted accordingly. The sampled version of the reflections is $r(n\Delta - \tau)$ which can be brought in a vector $\mathbf{r}(\tau)$. The elements $C_{\mathbf{r}|\tau}(n,m)$ of the covariance matrix $\mathbf{C}_{\mathbf{r}|\tau}$ of $\mathbf{r}(\tau)$, conditioned on τ, become:

$$C_{\mathbf{r}|\tau}(n, m) = C_r(n\Delta - \tau,\ m\Delta - \tau) \tag{9.11}$$

If the registration period is sufficiently large, the determinant $|\mathbf{C}_{\mathbf{r}|\tau}|$ does not depend on τ.

The observed waveform $w(t) = a(h(t - \tau) + r(t - \tau)) + v(t)$ involves two unknown factors, the amplitude a and the ToF τ. The prior probability density of the latter is not important because the maximum likelihood estimator that we will apply does not require it. However, the first factor a is a nuisance parameter. We deal with it by regarding a as a random variable with its own density $p(a)$. The influence of a is integrated in the likelihood function by means of Bayes' theorem for conditional probabilities, i.e. $p(\mathbf{z}|\tau) = \int p(\mathbf{z}|\tau,a)p(a)da$.

Preferably, the density $p(a)$ reflects our state of knowledge that we have about a. Unfortunately, taking this path is not easy, for two reasons. It would be difficult to assess this state of knowledge quantitatively. Moreover, the result will not be very tractable. A more practical choice is to assume a zero mean Gaussian density for a. With that, conditioned on τ, the vector $a\mathbf{h}(\tau)$ with elements $a{\cdot}h(n\Delta - \tau)$ becomes zero mean and Gaussian with covariance matrix:

$$\mathbf{C}_{\mathbf{h}|\tau} = \sigma_a^2 \mathbf{h}(\tau)\mathbf{h}^T(\tau) \tag{9.12}$$

where σ_a^2 is the variance of the amplitude a.

At first sight it seems counterintuitive to model a as a zero mean random variable since small and negative values of a are not very likely. The only reason for doing so is that it paves the way to a mathematically tractable model. In Section 9.2.4 we noticed already that the actual value of a does not influence the solution. We simply hope that in the extended matched filter a does not have any influence either. The advantage is that the dependence of τ on $\mathbf{z}$ is now captured in a concise model, i.e. a single covariance matrix:

$$\mathbf{C}_{\mathbf{z}|\tau} = \sigma_a^2(\mathbf{h}(\tau)\mathbf{h}^T(\tau) + \mathbf{C}_{\mathbf{r}|\tau}) + \sigma_v^2\mathbf{I} \tag{9.13}$$

This matrix completes the covariance model of the measurements. In the sequel, we assume a Gaussian conditional density for $\mathbf{z}$. Strictly speaking, this holds true only if sufficient echoes are present since in that case the central limit theorem applies.

Maximum likelihood estimation of the time-of-flight

With the measurements modelled as a zero mean, Gaussian random vector with the covariance matrix given in (9.13), the likelihood function for τ becomes:

$$p(\mathbf{z}|\tau) = \frac{1}{\sqrt{(2\pi)^K|\mathbf{C}_{\mathbf{z}|\tau}|}}\exp\left(-\frac{1}{2}\mathbf{z}^T\mathbf{C}_{\mathbf{z}|\tau}^{-1}\mathbf{z}\right) \tag{9.14}$$

The maximization of this probability with respect to τ yields the maximum likelihood estimate; see Section 3.1.4. Unfortunately, this solution is not practical because it involves the inversion of the matrix $\mathbf{C}_{\mathbf{z}|\tau}$. The size of $\mathbf{C}_{\mathbf{z}|\tau}$ is $N \times N$ where N is the number of samples of the registration (which can easily be in the order of 10^4).

Principal component analysis

Economical solutions are attainable by using PCA techniques (Section 7.1.1). If the registration period is sufficiently large, the determinant $|\mathbf{C}_{\mathbf{z}|\tau}|$ will not depend on τ. With that, we can safely ignore this factor. What remains is the maximization of the argument of the exponential:

$$\Lambda(\mathbf{z}|\tau) \stackrel{def}{=} -\mathbf{z}^T\mathbf{C}_{\mathbf{z}|\tau}^{-1}\mathbf{z} \tag{9.15}$$

The functional $\Lambda(\mathbf{z}|\tau)$ is a scaled version of the log-likelihood function.

The first computational savings can be achieved if we apply a principal component analysis to $\mathbf{C}_{\mathbf{z}|\tau}$. This matrix can be decomposed as follows:

$$\mathbf{C}_{\mathbf{z}|\tau} = \sum_{n=0}^{N-1} \lambda_n(\tau)\mathbf{u}_n(\tau)\mathbf{u}_n^T(\tau) \tag{9.16}$$

$\lambda_n(\tau)$ and $\mathbf{u}_n(\tau)$ are eigenvalues and eigenvectors of $\mathbf{C}_{\mathbf{z}|\tau}$. Using (9.16) the expression for $\Lambda(\mathbf{z}|\tau)$ can be moulded into the following equivalent form:

$$\Lambda(\mathbf{z}|\tau) = -\mathbf{z}^T \left(\sum_{n=0}^{N-1} \frac{\mathbf{u}_n(\tau)\mathbf{u}_n^T(\tau)}{\lambda_n(\tau)} \right) \mathbf{z} = -\sum_{n=0}^{N-1} \frac{(\mathbf{z}^T\mathbf{u}_n(\tau))^2}{\lambda_n(\tau)} \tag{9.17}$$

The computational savings are obtained by discarding all terms in (9.17) that do not capture much information about the true value of τ. Suppose that λ_n and $\mathbf{u}_n$ are arranged according to their importance with respect to the estimation, and that above some value of n, say J, the importance is negligible. With that, the number of terms in (9.17) reduces from N to J. Experiments show that J is in the order of 10. A speed up by a factor of 1000 is feasible.

Selection of good components

The problem addressed now is how to order the eigenvectors in (9.17) such that the most useful components come first, and thus will be selected. The eigenvectors $\mathbf{u}_n(\tau)$ are orthonormal and span the whole space. Therefore:

$$\sum_{n=0}^{N-1} (\mathbf{z}^T\mathbf{u}_n(\tau))^2 = \|\mathbf{z}\|^2 \tag{9.18}$$

Substitution in (9.17) yields:

$$\begin{aligned} \Lambda(\mathbf{z}|\tau) &= -\sum_{n=0}^{N-1} \frac{(\mathbf{z}^T\mathbf{u}_n(\tau))^2}{\lambda_n(\tau)} + \sum_{n=0}^{N-1} \frac{(\mathbf{z}^T\mathbf{u}_n(\tau))^2}{\sigma_v^2} - \frac{\|\mathbf{z}\|^2}{\sigma_v^2} \\ &= \sum_{n=0}^{N-1} \frac{\lambda_n(\tau) - \sigma_v^2}{\lambda_n(\tau)\sigma_v^2} (\mathbf{z}^T\mathbf{u}_n(\tau))^2 - \frac{\|\mathbf{z}\|^2}{\sigma_v^2} \end{aligned} \tag{9.19}$$

The term containing $\|\mathbf{z}\|$ does not depend on τ and can be omitted. The maximum likelihood estimate for τ appears to be equivalent to the one that maximizes:

$$\sum_{n=0}^{N-1} \gamma_n(\tau)(\mathbf{z}^T\mathbf{u}_n(\tau))^2 \quad \text{with} \quad \gamma_n(\tau) = \frac{\lambda_n(\tau) - \sigma_v^2}{\lambda_n(\tau)\sigma_v^2} \tag{9.20}$$

The weight $\gamma_n(\tau)$ is a good criterion to measure the importance of an eigenvector. Hence, a plot of the γ_n versus n is helpful to find a reasonable value of J such that $J << N$. Hopefully, γ_n is large for the first few n, and then drops down rapidly to zero.

The computational structure of the estimator based on a covariance model

A straightforward implementation of (9.20) is not very practical. The expression must be evaluated for varying values of τ. Since the dimension of $\mathbf{C}_{\mathbf{z}|\tau}$ is large, this is not computationally feasible.

The problem will be tackled as follows. First, we define a moving window for the measurements z_n. The window starts at $n = i$ and ends at $n = i + I - 1$. Thus, it comprises I samples. We stack these samples into a vector $\mathbf{x}(i)$ with elements $x_n(i) = z_{n+i}$. Each value of i corresponds to a hypothesized value $\tau = i\Delta$. Thus, under this hypothesis, the vector $\mathbf{x}(i)$ contains the direct response with $\tau = 0$, i.e. $x_n(i) = a{\cdot}h(n\Delta) + a{\cdot}r(n\Delta) + v(n\Delta)$. Instead of applying operation (9.20) for varying τ, τ is fixed to zero and $\mathbf{z}$ is replaced by the moving window $\mathbf{x}(i)$:

$$y(i) = \sum_{n=0}^{J-1} \gamma_n(0)(\mathbf{x}(i)^T\mathbf{u}_n(0))^2 \tag{9.21}$$

If $\hat{i}$ is the index that maximizes $y(i)$, then the estimate for τ is found as $\hat{\tau}_{cvm} = \hat{i}\Delta$.

The computational structure of the estimator is shown in Figure 9.9. It consists of a parallel bank of J filters/correlators, one for each eigenvector $\mathbf{u}_n(0)$. The results of that are squared, multiplied by weight factors $\gamma_n(0)$ and then accumulated to yield the signal $y(i)$. It can be proven that if we set $\sigma_d^2 = 0$, i.e. a model without reflection, the estimator degenerates to the classical matched filter.

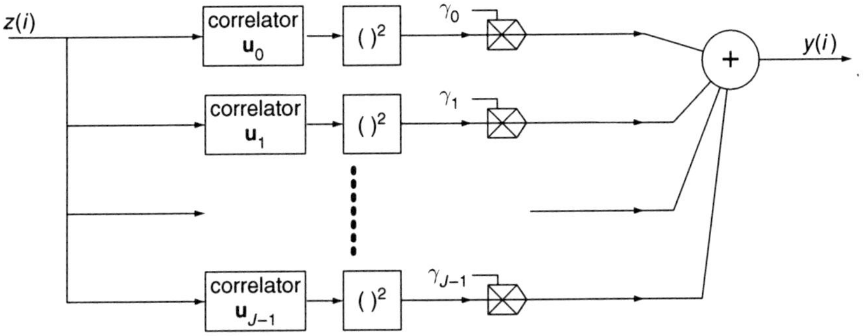

Figure 9.9 ML estimation based on covariance models

The design parameters of the estimators are the $SNR \stackrel{def}{=} \sigma_a^2/\sigma_v^2$, the duration of echo generation T, the echo strength $S_r \stackrel{def}{=} K\sigma_d^2$, the number of correlators J and the window size I.

Example

In the following example, the selected design parameters are $SNR = 100$, $T = 0.8$ (ms), $S_r = 0.2$ and $I = 1000$. Using (9.11) and (9.13), we calculate $C_{z|\tau}$, and from that the eigenvectors and corresponding eigenvalues and weights are obtained. Figure 9.10 shows the result. As expected the response of the first filter/correlator is similar to the direct response, and this part just implements the conventional matched filter. From the seventh filter on, the weights decrease, and from the fifteenth filter on the weights are near zero. Thus, the useful number of filters is between 7 and 15. Figure 9.11 shows the results of application of the filters to an observed waveform.

9.2.6 Optimization and evaluation

In order to find the estimator with the best performance the best parameters of the estimators must be determined. The next step then is to assess their performances.

Cross-validation

In order to prevent overfitting we apply a three-fold cross-validation procedure to the data set consisting of 150 records of waveforms. The corresponding MATLAB code is given in Listing 9.8. Here, it is assumed that the estimator under test is realized in a MATLAB function called `ToF_estimator()`.

The optimization of the operator using a training set occurs according to the procedure depicted in Figure 9.12. In Listing 9.8 this is implemented by calling the function `opt_ToF_estimator()`. The actual code for the optimization, given in Listing 9.9, uses the MATLAB function `fminsearch()`.

We assume here that the operator has a bias that can be compensated for. The value of the compensation is just another parameter of the estimator. However, for its optimization the use of the function `fminsearch()` is not needed. This would unnecessarily increase the

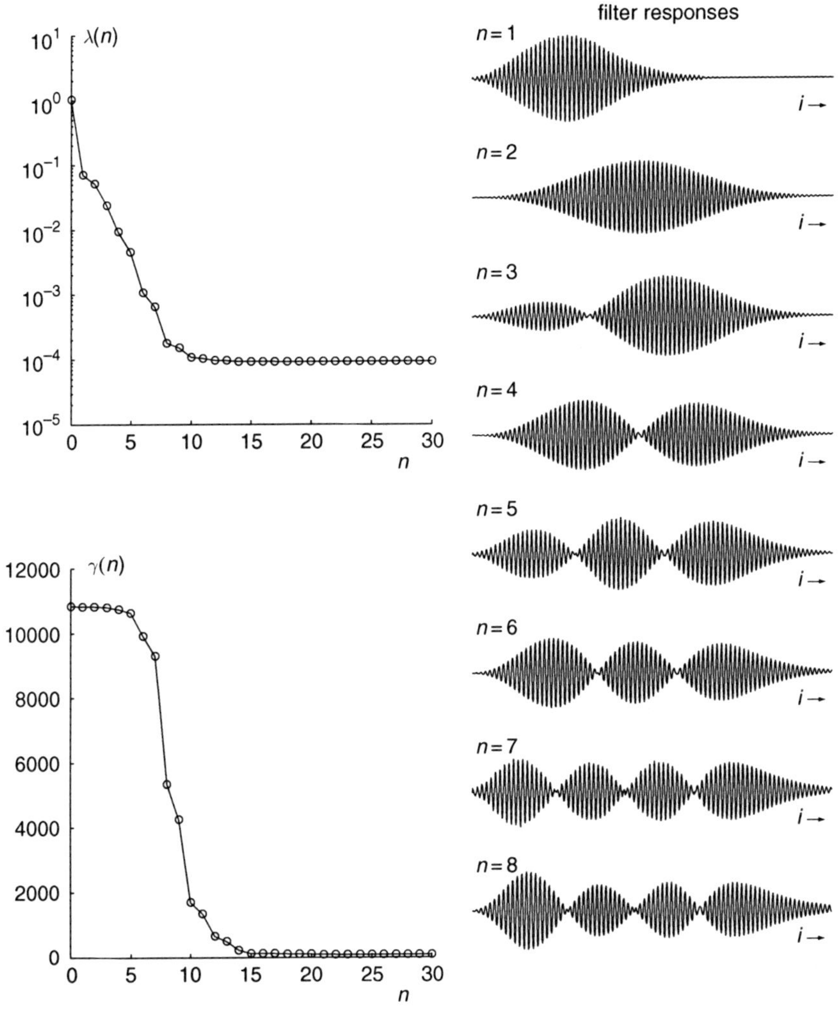

Figure 9.10 Eigenvalues, weights and filter responses of the covariance model based estimator

search space of the parameters. Instead, we simply use the (estimated) variance as the criterion to optimize, thereby ignoring a possible bias for a moment. As soon as the optimal set of parameters has been found, the corresponding bias is estimated afterwards by applying the optimized estimator once again to the learning set.

Note, however, that the uncertainty in the estimated bias causes a residual bias in the compensated ToF estimate. Thus, the compensation of the bias does not imply that the estimator is necessarily unbiased.

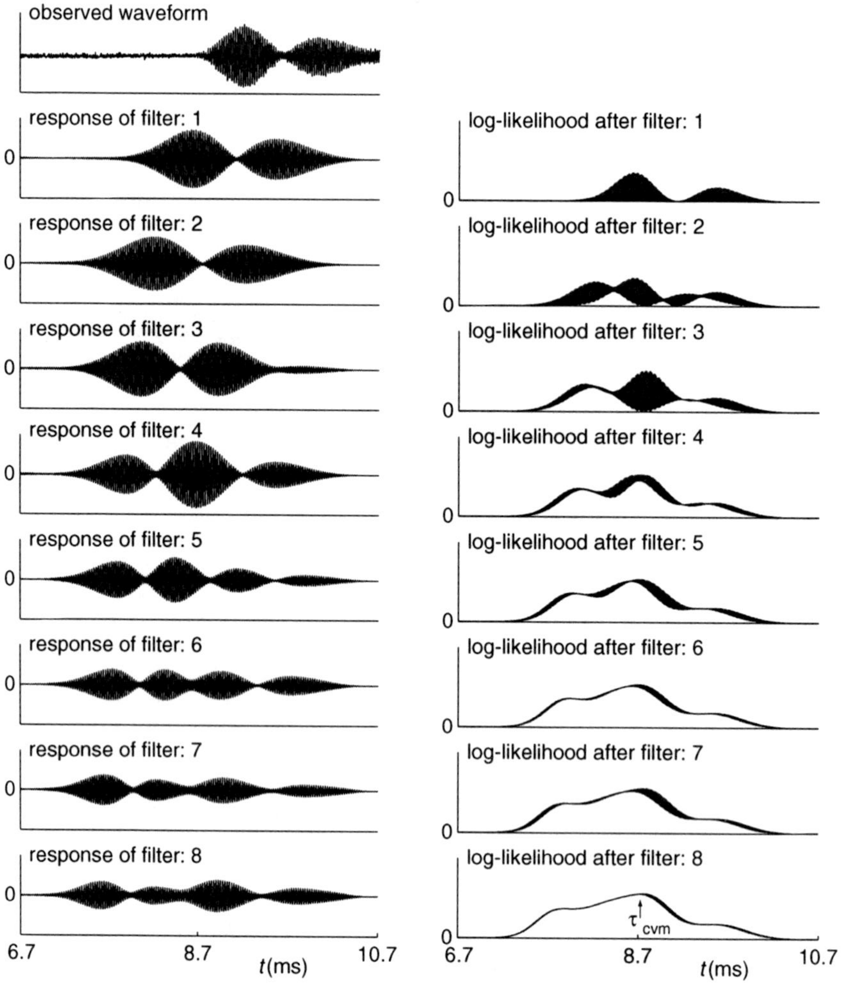

Figure 9.11 Results of the estimator based on covariance models

Therefore, the evaluation of the estimator should not be restricted to assessment of the variance alone.

Listing 9.8
MATLAB listing for cross-validation.

```
load tofdata.mat;          % Load tof dataset containing 150 waveforms
Npart = 3;                 % Number of partitions
Nchunk = 50;               % Number of waveforms in one partition

% Create 3 random partitions of the data set
```

```
p=randperm(Npart*Nchunk); % Find random permutation of 1:150
for n=1:Npart
    for i=1:Nchunk
        Zp{n,i}=Zraw{p((n-1)*Nchunk+i)};
        Tp(n,i)=TOFindex(p((n-1)*Nchunk+i));
    end
end

% Cross-validation
for n=1:Npart
    % Create a learn set and an evaluation set
    Zlearn=Zp;
    Tlearn=Tp;
    for i=1:Npart
        if (i==n)
            Zlearn(i,:)=[]; Tlearn(i,:)=[];
            Zeval=Zp(i,:); Teval=Tp(i,:);
        end
    end
    Zlearn=reshape(Zlearn,(Npart-1)*Nchunk,1);
    Tlearn=reshape(Tlearn,1,(Npart-1)*Nchunk);
    Zeval=reshape(Zeval ,1, Nchunk);

    % Optimize a ToF estimator
    [parm,learn_variance,learn_bias]= ...
    opt_ToF_estimator(Zlearn,Tlearn);

    % Evaluate the estimator
    for i=1:Nchunk
        index(i)=ToF_estimator(Zeval{i},parm);
    end
    variance(n)=var(Teval-index);
    bias(n)=mean(Teval-index-learn_bias);
end
```

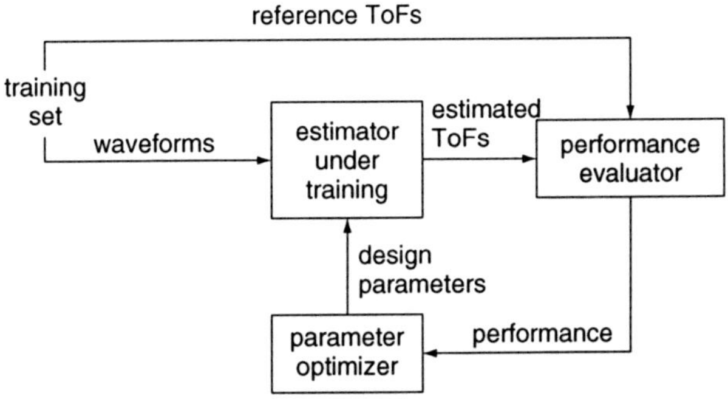

Figure 9.12 Training the estimators

Listing 9.9
MATLAB listing for the optimization of a ToF estimator.

```
function [parm,variance,bias] = . . .
  opt_ToF_estimator(Zlearn,Tlearn)

  % Optimize the parameters of a ToF estimator
  parm= [0.15, 543, 1032]; % Initial parameters
  parm= fminsearch(@objective,parm, [],Zlearn,Tlearn);
  [variance,bias] =objective(parm,Zlearn,Tlearn);
return;

% Objective function:
% estimates the variance (and bias) of the estimator
function [variance,bias] =objective(parm,Z,TOFindex)
  for i =1:length(Z)
    index(i) =ToF_estimator(Z{i},parm);
  end
  variance=var(TOFindex - index);
  bias=mean(TOFindex - index);
return
```

Results

Table 9.2 shows the results obtained from the cross-validation. The first row of the table gives the variances directly obtained from the training data during optimization. They are obtained by averaging over the three variances of the three partitions. The second row tabulates the variances obtained from the evaluation data (also obtained by averaging over the three partitions). Inspection of these results reveals that the threshold method is overfitted. The reason for this is that some of the records have a very low signal-to-noise ratio. If – by chance – these records do not

Table 9.2 Results of three-fold cross-validation of the four ToF estimators

	Envelope thresholding	Curve fitting	Matched filtering	CVM based estimator
Variance (learn data) (μs²)	356	378	14339	572
Variance (test data) (μs²)	181010	384	14379	591
corrected variance (μs²)	180910	284	14279	491
Bias (μs)	35	2	10	2
RMS (μs)	427 ± 25	17 ± 1.2	119 ± 7	22 ± 1.4

occur in the training set, then the threshold level will be set too low for these noisy waveforms.

The reference values of the ToFs in the data set have an uncertainty of 10 (μs). Therefore, the variance estimated from the evaluation sets have a bias of 100 (μs^2). The third row in the table shows the variances after correction for this effect.

Another error source to account for is the residual bias. This error can be assessed as follows. The statistical fluctuations due to the finite data set cause uncertainty in the estimated bias. Suppose that σ^2 is the variance of a ToF estimator. Then, according to (5.11), the bias estimated from a training set of N_S samples has a variance of σ^2/N_S. The residual bias has an order of magnitude of $\sigma/\sqrt{N_S}$. In the present case, σ^2 includes the variance due to the uncertainty in the reference values of the ToFs. The calculated residual biases are given in the fourth row in Table 9.2.

The final evaluation criterion must include both the variance and the bias. A suitable criterion is the mean square error, or equivalently the root mean square error (RMS) defined as $\text{RMS} = \sqrt{\text{variance} + \text{bias}^2}$. The calculated RMSs are given in the fifth row.

Discussion

Examination of Table 9.2 reveals that the four methods are ranked according to: curve fitting, CVM[1] based, matched filtering and finally thresholding. The performance of the threshold method is far behind the other methods. The reason of the failure of the threshold method is a lack of robustness. The method fails for a few samples in the training set. The robustness is simply improved by increasing the relative threshold, but at the cost of a larger variance. For instance, if the threshold is raised to 50%, the variance becomes around 900 (μs^2).

The poor performance of the matched filter is due to the fact that it is not able to cope with the echoes. The covariance model clearly helps to overcome this problem. The performance of the CVM method is just below that of curve fitting. Apparently, the covariance model is an important improvement over the simple white noise model of the matched filter, but still too inaccurate to beat the curve-fitting method.

---

[1] CVM = covariance model

One might argue that the difference between the performances of curve fitting and CVM is not a statistically significant one. The dominant factor in the uncertainty of the RMSs is brought forth by the estimated variance. Assuming Gaussian distributions for the randomness, the variance of the estimation error of the variance is given by (5.13):

$$\mathrm{Var}[\widehat{\sigma^2}] = \frac{2}{N_S}\sigma^4 \tag{9.22}$$

where $N_S = 150$ is the number of samples in the data set. This error propagates into the RMS with a standard deviation of $\sigma/\sqrt{2N_S}$. The corresponding margins are shown in Table 9.2. It can be seen that the difference between the performances is larger than the margins, thus invalidating the argument.

Another aspect of the design is the computational complexity of the methods. The threshold method and the curve-fitting method both depend on the availability of the envelope of the waveform. The quadrature filtering that is applied to calculate the envelope requires the application of the Fourier transform. The waveforms in the data set are windowed to $N = 8192$ samples. Since N is a power of two, MATLAB's implementations of the fast Fourier transforms, `fft()` and `ifft()`, have a complexity of $2N^2 \log_2 N$. The threshold method does not need further substantial computational effort. The curve-fitting method needs an additional numerical optimization, but since the number of points of the curve is not large, such an optimization is not very expensive.

A Fourier-based implementation of a correlator also has a complexity of $2N^2 \log_2 N$. Thus, the computational cost of the matched filter is in the same order as the envelope detector. The CVM based method uses J correlators. Its complexity is $(J + 1)N^2 \log_2 N$. Since typically $J = 10$, the CVM method is about 10 times more expensive than the other methods.

In conclusion, the most accurate method for ToF estimation appears to be the curve-fitting method with a computational complexity that is in the order of $2N^2 \log_2 N$. With this method, ToF estimation with an uncertainty of about 17 (μs) is feasible. The maximum likelihood estimator based on a covariance model follows the curve-fitting method closely with an uncertainty of 22 (μs). Additional modelling of the occurrence of echoes is needed to improve its performance.

9.3 ONLINE LEVEL ESTIMATION IN AN HYDRAULIC SYSTEM

This third example considers the hydraulic system already introduced in Section 8.1, where it illustrated some techniques for system identification. Figure 8.2 shows an overview of the system. The goal in the present section is to design an online state estimator for the two levels $h_1(t)$ and $h_2(t)$, and for the input flow q_0.

The model that appeared to be best is Torricelli's model. In discrete time, with $x_1(i) \stackrel{def}{=} h_1(i\Delta)$ and $x_2(i) \stackrel{def}{=} h_2(i\Delta)$, the model of (8.3) becomes:

$$\begin{bmatrix} x_1(i+1) \\ x_2(i+1) \end{bmatrix} = \begin{bmatrix} x_1(i) \\ x_2(i) \end{bmatrix} + \frac{\Delta}{C} \begin{bmatrix} -\sqrt{R_1(x_1(i) - x_2(i))} + q_0(i) \\ \sqrt{R_1(x_1(i) - x_2(i))} - \sqrt{R_2 x_2(i)} \end{bmatrix} \tag{9.23}$$

R_1 and R_2 are two constants that depend on the areas of the cross-sections of the pipelines of the two tanks. C is the capacity of the two tanks. These parameters are determined during the system identification. $\Delta = 5$ (s) is the sampling period.

The system in (9.23) is observable with only one level sensor. However, in order to allow consistency checks (which are needed to optimize and to evaluate the design) the system is provided with two level sensors. The redundancy of sensors is only needed during the design phase, because once a consistently working estimator has been found, one level sensor suffices.

The experimental data that is available for the design is a record of the measured levels as shown in Figure 9.13. The levels are obtained by means of pressure measurements using the Motorola MPX2010 pressure sensor. Some of the specifications of this sensor are given in Table 9.3. The full-scale span V_{FSS} corresponds to $P_{max} = 10$ kPa. In turn, this maximum pressure corresponds to a level of $h_{max} = P_{max}/\rho g \approx 1000$ (cm). Therefore, the linearity is specified between -10 (cm) and $+10$ (cm). This specification is for the full range. In our case, the swing of the levels is limited to 20 (cm), and the measurement system was calibrated at 0 (cm) and 25 (cm). Therefore, the linearity will be much less. The pressure hysteresis is an error that depends on whether the pressure is increasing or decreasing. The pressure hysteresis can induce a maximal level error between -1 (cm) and $+1$ (cm). Besides these sensor errors, the measurements are also contaminated by electronic noise

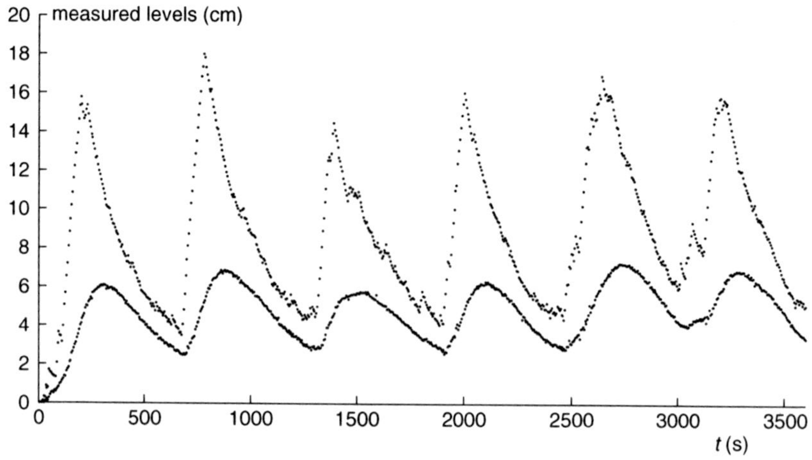

Figure 9.13 Measured levels of two interconnected tanks

Table 9.3 Specifications of the MPX2010 pressure sensor

Characteristic	Symbol	Min	Typ	Max	Unit
Pressure range	P	0	—	10	kPa
Full-scale span	V_{FSS}	24	25	26	mV
Sensitivity	$\Delta V/\Delta P$	—	2.5	—	mV/kPa
Linearity error	—	−1.0	—	1.0	$\%V_{FSS}$
Pressure hysteresis	—	—	±0.1	—	$\%V_{FSS}$

modelled by white noise $v(i)$ with a standard deviation of $\sigma_v = 0.04$ (cm). With that, the model of the sensory system becomes:

$$\begin{bmatrix} z_1(i) \\ z_2(i) \end{bmatrix} = \begin{bmatrix} x_1(i) \\ x_2(i) \end{bmatrix} + \begin{bmatrix} e_1(i) \\ e_2(i) \end{bmatrix} + \begin{bmatrix} v_1(i) \\ v_2(i) \end{bmatrix} \tag{9.24}$$

or more concisely: $\mathbf{z}(i) = \mathbf{H}\mathbf{x}(i) + \mathbf{e}(i) + \mathbf{v}(i)$ with $\mathbf{H} = \mathbf{I}$. The error $\mathbf{e}(i)$ represents the linearity and hysteresis error.

In the next sections, the linearized Kalman filter, the extended Kalman filter and the particle filter will be examined. In all three cases a model is needed for the input flow $q_0(i)$. The linearized Kalman filter can only handle linear models. The extended Kalman filter can handle nonlinear models, but only if the nonlinearities are smooth. Particle filtering offers the largest freedom of modelling. All three models need parameter estimation in order to adapt the model to the data. Consistency checks must indicate which estimator is most suitable.

The prior knowledge that we assume is that the tanks at $i = 0$ are empty.

9.3.1 Linearized Kalman filtering

The simplest dynamic model for $q_0(i)$ is the first order AR model:

$$q_0(i+1) = \bar{\bar{q}}_0 + \alpha\left(q_0(i) - \bar{\bar{q}}_0\right) + w(i) \tag{9.25}$$

$\bar{\bar{q}}_0$ is a constant input flow that maintains the equilibrium in the tanks. The Gaussian process noise $w(i)$ causes random perturbations around this equilibrium. The factor α regulates the bandwidth of the fluctuations of the input flow. The design parameters are $\bar{\bar{q}}_0$, σ_w and α.

The augmented state vectors are $\mathbf{x}(i) = [\, x_1(i)\ x_2(i)\ q_0(i)\,]$. Equations (9.23) and (9.25) make up the augmented state equation $\mathbf{x}(i+1) = \mathbf{f}(\mathbf{x}(i), w(i))$, which is clearly nonlinear. The equilibrium follows from equating $\bar{\bar{\mathbf{x}}} = \mathbf{f}(\bar{\bar{\mathbf{x}}}, 0)$:

$$\begin{bmatrix} \bar{\bar{x}}_1 \\ \bar{\bar{x}}_2 \end{bmatrix} = \begin{bmatrix} \bar{\bar{x}}_1 \\ \bar{\bar{x}}_2 \end{bmatrix} + \frac{\Delta}{C}\begin{bmatrix} -\sqrt{R_1\left(\bar{\bar{x}}_1 - \bar{\bar{x}}_2\right)} + \bar{\bar{q}}_0 \\ \sqrt{R_1\left(\bar{\bar{x}}_1 - \bar{\bar{x}}_2\right)} - \sqrt{R_2\bar{\bar{x}}_2} \end{bmatrix} \quad \Rightarrow \quad \begin{aligned} \bar{\bar{x}}_2 &= \bar{\bar{q}}_0^2 / R_2 \\ \bar{\bar{x}}_1 &= \frac{R_1 + R_2}{R_1}\bar{\bar{x}}_2 \end{aligned} \tag{9.26}$$

The next step is to calculate the Jacobian matrix of $\mathbf{f}()$:

$$\mathbf{F}(\mathbf{x}) = \frac{\partial \mathbf{f}()}{\partial \mathbf{x}} = \begin{bmatrix} 1 - \frac{\Delta R_1}{2C\sqrt{R_1(x_1 - x_2)}} & \frac{\Delta R_1}{2C\sqrt{R_1(x_1 - x_2)}} & \frac{\Delta}{C} \\ \frac{\Delta R_1}{2C\sqrt{R_1(x_1 - x_2)}} & 1 - \frac{\Delta R_1}{2C\sqrt{R_1(x_1 - x_2)}} - \frac{\Delta R_2}{2C\sqrt{R_2 x_2)}} & 0 \\ 0 & 0 & \alpha \end{bmatrix} \tag{9.27}$$

The linearized model arises by application of a truncated Taylor series expansion around the equilibrium

$$\mathbf{x}(i+1) = \bar{\bar{\mathbf{x}}} + \mathbf{F}(\bar{\bar{\mathbf{x}}})\left(\mathbf{x}(i) - \bar{\bar{\mathbf{x}}}\right) + \mathbf{G}w(i) \tag{9.28}$$

with $\mathbf{G} = [\, 0 \quad 0 \quad 1\,]^T$.

The three unknown parameters $\overline{\overline{q}}_0$, α and σ_w must be estimated from the data. Various criteria can be used to find these parameters, but in most of these criteria the innovations play an important role. For a proper design the innovations are a white Gaussian sequence of random vectors with zero mean and known covariance matrix, i.e. the innovation matrix. See Section 8.4. In the sequel we will use the NIS (normalized innovations squared; Section 8.4.2). For a consistent design, the NIS must have a χ^2_N distribution. In the present case $N = 2$. The expectation of a χ^2_N-distributed variable is N. Thus, here, ideally the mean of the NIS is 2. A simple criterion is one which drives the average of the calculated NIS to 2. A possibility is:

$$J(\overline{\overline{q}}_0, \alpha, \sigma_w) = \left| \frac{1}{I} \sum_{i=0}^{I-1} Nis(i) - 2 \right| \tag{9.29}$$

The values of $\overline{\overline{q}}_0$, α and σ_w that minimize J are considered optimal.

The strategy is to realize a MATLAB function `J = flinDKF(y)` that implements the linearized Kalman filter, applies it to the data, calculates $Nis(i)$ and returns J according to eq. (9.29). The input argument `y` is an array containing the variables $\overline{\overline{q}}_0$, α and σ_w. Their optimal values are found by one of MATLAB's optimization functions, e.g. `[parm,J] = fminsearch(@flinDKF, [20 0.95 5])`.

Application of this strategy revealed an unexpected phenomenon. The solution of $(\overline{\overline{q}}_0, \alpha, \sigma_w)$ that minimizes J is not unique. The values of α and σ_w do not influence J as long as they are both sufficiently large. For instance, Figure 9.14 shows the results for $\alpha = 0.98$ and $\sigma_w = 1000$ (cm^3/s). But any value of α and σ_w above this limit gives virtually the same results and the same minimum. The minimum obtained is $J = 25.6$. This is far too large for a consistent solution. Also the NIS, shown in Figure 9.14, does not obey the statistics of a χ^2_2 distribution. During the transient, in the first 200 seconds, the NIS reaches extreme levels indicating that some unmodelled phenomena occur there. But also during the remaining part of the process the NIS shows some unwanted high peaks. Clearly the estimator is not consistent.

Three modelling errors might explain the anomalous behaviour of the linearized Kalman filter. First, the linearization of the system equation might fail. Second, the AR model of the input flow might be inappropriate. Third, the ignorance of possible linearization errors of the sensors might not be allowed. In the next two sections, the first two possible explanations will be examined.

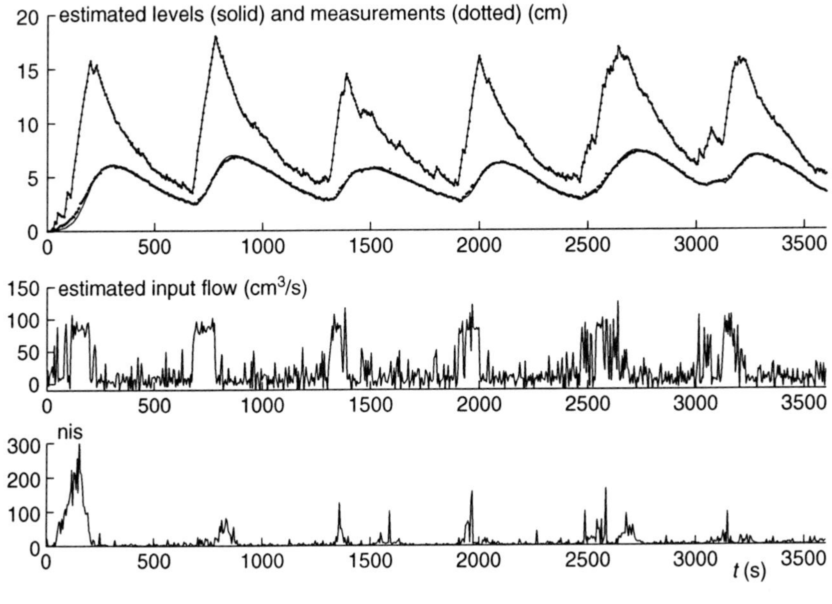

Figure 9.14 Results from the linearized Kalman filter

9.3.2 Extended Kalman filtering

Linearization errors of the system equation are expected to be influential when the levels deviate largely from the equilibrium state. The abnormality during the transient in Figure 9.14 might be caused by this kind of error because there the nonlinearity is strongest. The extended Kalman filter is able to cope with smooth linearity errors. Therefore, this method might give an improvement.

In order to examine this option, a MATLAB function `J = fextDKF(y)` was realized that implements the extended Kalman filter. Again, `y` is an array containing the variables $\bar{\bar{q}}_0$, α and σ_w. Application to the data, calculation of *J* and minimization with `fminsearch()` yields estimates as shown in Figure 9.15.

The NIS of the extended Kalman filter, compared with that of the linearized Kalman filter, is much better now. The optimization criterion *J* attains a minimal value of 3.6 instead of 25.6; a significant improvement has been reached. However, the NIS still doesn't obey a χ^2_2 distribution. Also, the optimization of the parameters $\bar{\bar{q}}_0$, α and σ_w is not without troubles. Again, the minimum is not unique. Any solution of α and σ_w satisfies as long as both parameters are sufficiently

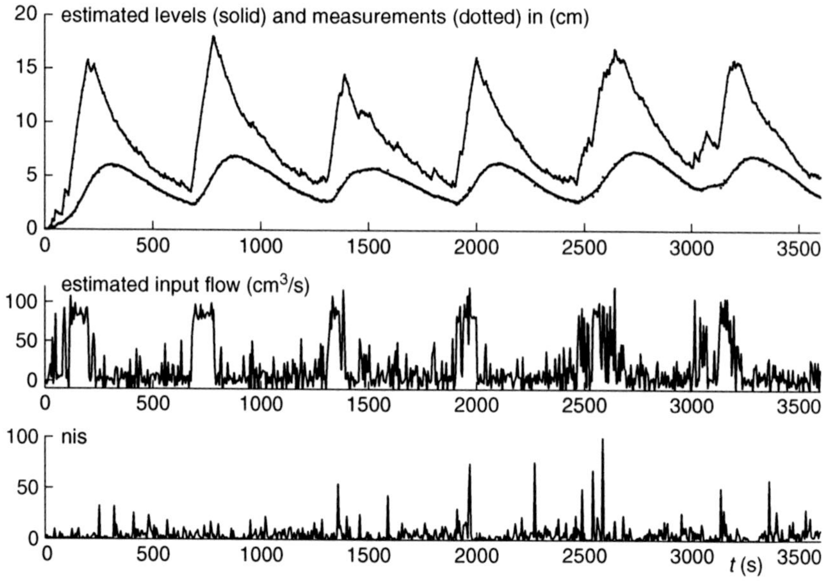

Figure 9.15 Results from the extended Kalman filter

large. Moreover, it now appears that the choice of $\bar{\bar{q}}_0$ does not have any influence at all.

The explanation of this behaviour is as follows. First of all, we observe that in the prediction step of the Kalman filter a large value of σ_w induces a large uncertainty in the predicted input flow. As a result, the prediction of the level in the first tank is also very uncertain. In the next update step, the corresponding Kalman gain will be close to one, and the estimated level in the first tank will closely follow the measurement, $\bar{x}_1(i|i) \approx z_1(i)$. If α is sufficiently large, so that the autocorrelation in the sequence $q_0(i)$ is large, the estimate $\bar{q}_0(i|i)$ is derived from the differences of succeeding samples $\bar{x}_1(i|i)$ and $\bar{x}_1(i-1|i-1)$. Since $\bar{x}_1(i|i) \approx z_1(i)$ and $\bar{x}_1(i-1|i-1) \approx z_1(i-1)$, the estimate $\bar{q}_0(i|i)$ only depends on measurements. The value of $\bar{\bar{q}}_0$ does not influence that. The conclusion is that the AR model does not provide useful prior knowledge for the input flow.

9.3.3 Particle filtering

A visual inspection of the estimated input flow in Figure 9.15 indeed reveals that an AR model does not fit well. The characteristic of the input flow is more like that of a random binary signal modelled by two discrete states $\{q_{\min}, q_{\max}\}$, and a transition probability $P_t(q_0(i)|q_0(i-1))$; see Section 4.3.1. The estimated flow in Figure 9.15 also suggests that $q_{\min} = 0$. This actually models an on/off control mechanism of the input flow. With this assumption, the transition probability is fully defined by two probabilities:

$$\begin{aligned} P_{up} &= P_t(q_0(i) = q_{\max}|q_0(i-1) = 0) \\ P_{down} &= P_t(q_0(i) = 0|q_0(i-1) = q_{\max}) \end{aligned} \tag{9.30}$$

The unknown parameters of the new flow model are $q_{\max}$, P_{up} and P_{down}. These parameters must be retrieved from the data by optimizing some consistency criterion of an estimator. Kalman filters cannot cope with binary signals. Therefore, we now focus on particle filtering, because this method is able to handle discrete variables (see Section 4.4).

Again, the strategy is to realize a MATLAB function `J = fpf(y)` that implements a particle filter, applies it to the data and returns a criterion J. The input argument `y` contains the design parameters $q_{\max}$, P_{up} and P_{down}. The output J must express how well the result of the particle filter is consistent. Minimization of this criterion gives the best attainable result.

An important issue is how to define the criterion J. The NIS, which was previously used, exists within the framework of Kalman filters, i.e. for linear-Gaussian systems. It is not trivial to find a concept within the framework of particle filters that is equivalent to the NIS. The general idea is as follows. Suppose that, using all previous measurement $\mathbf{Z}(i-1)$ up to time $i-1$, the probability density of the state $\mathbf{x}(i)$ is $p(\mathbf{x}(i)|\mathbf{Z}(i-1))$. Then, the probability of $\mathbf{z}(i)$ is:

$$\begin{aligned} p(\mathbf{z}(i)|\mathbf{Z}(i-1)) &= \int_{\mathbf{x}} p(\mathbf{z}(i), \mathbf{x}(i)|\mathbf{Z}(i-1))d\mathbf{x} \\ &= \int_{\mathbf{x}} p(\mathbf{z}(i)|\mathbf{x}(i))p(\mathbf{x}(i)|\mathbf{Z}(i-1))d\mathbf{x} \end{aligned} \tag{9.31}$$

The density $p(\mathbf{z}(i)|\mathbf{x}(i))$ is simply the model of the sensory system and as such known. The probability $p(\mathbf{x}(i)|\mathbf{Z}(i-1))$ is represented by the predicted samples. Therefore, using (9.31) the probability $p(\mathbf{z}(i)|\mathbf{Z}(i-1))$ can be calculated. The filter is consistent only if the sequence of observed measurement $\mathbf{z}(i)$ obeys the statistics prescribed by the sequence of densities $p(\mathbf{z}(i)|\mathbf{Z}(i-1))$.

A test of whether all $\mathbf{z}(i)$ comply with $p(\mathbf{z}(i)|\mathbf{Z}(i-1))$ is not easy, because $p(\mathbf{z}(i)|\mathbf{Z}(i-1))$ depends on i. The problem will be tackled by treating each scalar measurement separately. We consider the n-th element $z_n(i)$ of the measurement vector, and assume that $p_n(z,i)$ is its hypothesized marginal probability density. Suppose that the cumulative distribution of $z_n(i)$ is $F_n(z,i) = \int_{-\infty}^{z} p_n(\zeta,i)d\zeta$. Then the random variable $u_n(i) = F_n(z_n(i),i)$ has a uniform distribution between 0 and 1. The consistency check boils down to testing whether the set $\{u_n(i)|i = 0,\ldots,I-1 \text{ and } n = 1,\ldots,N\}$ indeed has such a uniform distribution.

In the literature, the statistical test of whether a set of variables has a given distribution function is called a *goodness-of-fit* test. There are various methods for performing such a test. A particular one is the chi-square test and is as follows (Kreyszig, 1970):

Algorithm 9.1: Goodness of fit (Chi-square test for the uniform distribution)
Input: a set of variables that are within the interval [0, 1]. The size of the set is B.

1. Divide the interval [0, 1] into a number of L equally spaced containers and count the number of times that the random variable falls in the ℓ-th container. Denote this count by b_ℓ. Thus, $\sum_{\ell=1}^{L} b_\ell = B$. (Note: L must be such that $b_\ell > 5$ for each ℓ.)
2. Set $e = B/L$. This is the expected number of variables in one container.
3. Calculate the test variable $\mathrm{J} = \sum_{\ell=1}^{L} \frac{(b_\ell - e)^2}{e}$.
4. If the set is uniformly distributed, then J has a χ^2_{L-1} distribution. Thus, if J is too large to be compatible with the χ^2_{L-1} distribution, the hypothesis of a uniform distribution must be rejected. For instance, if $L = 20$, the probability that $\mathrm{J} > 43.8$ equals 0.001.

The particular particle filter that was implemented is based on the condensation algorithm described in Section 4.4.3. The MATLAB code is given in Listing 9.10. Some details of the implementation follow next. See Algorithm 4.4.

- The prior knowledge that we assume is that at $i = 0$ both tanks are empty, that is $x_1(0) = x_2(0) = 0$. The probability that the input flow is 'on' or 'off' is 50/50.
- During the update, the importance weights should be set to $w^{(k)} = p(\mathbf{z}(i)|\mathbf{x}^{(k)})$. Assuming a Gaussian distribution of the measurement errors, the weights are calculated as $w^{(k)} = \exp\left(-\frac{1}{2}(\mathbf{z}(i) - \mathbf{x}^{(k)})^T \mathbf{C}_v^{-1}(\mathbf{z}(i) - \mathbf{x}^{(k)})\right)$. The normalizing constant of the Gaussian

can be ignored because the weights are going to be normalized anyway.

- 'Finding the smallest j such that $w_{cum}^{(j)} \geq r^{(k)}$' is implemented by the bisection method of finding a root using the golden rule.
- In the prediction step, 'finding the samples $\mathbf{x}^{(k)}$ drawn from the density $p(\mathbf{x}(i)|\mathbf{x}(i-1) = \mathbf{x}_{selected}^{(k)})$' is done by generating random input flow transitions from 'on' to 'off' and from 'off' to 'on' according to the transition probabilities P_{down} and P_{up}, and to apply these to the state equation (implemented by `f(Ys,R1,R2)`). However, the time-discrete model only allows the transitions to occur at exactly the sampling periods $t_i = i\Delta$. In the real time-continuous world, the transitions can take place anywhere in the time interval between two sampling periods. In order to account for this effect, the level of the first tank is randomized with a correction term randomly selected between 0 and $q_{\max}\Delta/C$. Such a correction is only needed for samples where a transition takes place.
- The conditional mean, approximated by $\hat{\mathbf{x}}(i) = \sum_k w^{(k)}\mathbf{x}^{(k)} / \sum_k w^{(k)}$, is used as the final estimate.
- The probability density $p(\mathbf{z}(i)|\mathbf{Z}(i-1))$ is represented by samples $\mathbf{z}^{(k)}$, which are derived from the predicted samples $\mathbf{x}^{(k)}$ according to the model $\mathbf{z}^{(k)} = \mathbf{H}\mathbf{x}^{(k)} + \mathbf{v}^{(k)}$ where $\mathbf{v}^{(k)}$ is Gaussian noise with covariance matrix $\mathbf{C_v}$. The marginal probabilities $p_n(z, i)$ are then represented by the scalar samples $z_n^{(k)}$. The test variables $u_n(i)$ are simply obtained by counting the number of samples for which $z_n^{(k)} < z_n(i)$ and dividing it by K, the total number of samples.

Listing 9.10

MATLAB listing of a function implementing the condensation algorithm.

```
function J=fpf (y)
load hyddata.mat;     % Load the dataset (measurements in Z)
I=length (Z);         % Length of the sequence
R1=105.78;            % Friction constant (cm^5/s^2)
R2=84.532;            % Friction constant (cm^5/s^2)
qmin=0;               % Minimal input flow
delta=5;              % Sampling period (s)
C=420;                % Capacity of tank (cm^2)
sigma_v=0.04;         % Standard deviation sensor noise (cm)
Ncond=1000;           % Number of particles
M=3;                  % Dimension of state vector
N=2;                  % Dimension of measurement vector
```

```
% Set the design parameters
qmax=y(1);                 % Maximum input flow
Pup=y(2);                  % Transition probability up
Pdn=y(3);                  % Transition probability down

% Initialisation
hmax=0; hmin=0;                        % Margins of levels at i=0
H=eye(N,M);                            % Measurement matrix
invCv=inv(sigma_v^2 * eye(N));         % Inv. cov. of sensor noise

% Generate the samples
Xs(1,:)=hmin+(hmax-hmin)*rand(1,Ncond);
Xs(2,:)=hmin+(hmax-hmin)*rand(1,Ncond);
Xs(3,:)=qmin+(qmax-qmin)*(rand(1,Ncond)>0.5);

for i=1:I
  % Generate predicted meas. representing p(z(i)|Z(i-1))
  Zs=H*Xs+sigma_v*randn(2,Ncond);

  % Get uniform distributed rv
  u(1,i)=sum((Zs(1,:) < Z(1,i)))/Ncond;
  u(2,i)=sum((Zs(2,:) < Z(2,i)))/Ncond;

  % Update
  res=H*Xs - Z(:,i)*ones(1,Ncond);          % Residuals
  W=exp(-0.5*sum(res.*(invCv*res)))';       % Weights
  if (sum(W)==0), error('process did not converge'); end
  W=W/sum(W); CumW=cumsum(W);

  xest(:,i)=Xs(:,:)*W;                      % Sample mean

  % Find an index permutation using golden rule root finding
  for j=1:Ncond
    R=rand; ja=1; jb=Ncond;
    while (ja < jb-1)
      jx=floor(jb-0.382*(jb-ja));
      fa=R-CumW(ja); fb=R-CumW(jb); fxx=R-CumW(jx);
      if (fb*fxx < 0), ja=jx; else, jb=jx; end
    end
    ind(j)=jb;
  end

  % Resample
  for j=1:Ncond, Ys(:,j)=Xs(:,ind(j)); end

  % Predict
  Tdn=(rand(1,Ncond)<Pdn);                  % Random transitions
```

```
  Tup=(rand(1,Ncond)<Pup);                  % idem
  kdn=find((Ys(3,:)==qmax) & Tdn);          % Samples going down
  kup=find((Ys(3,:)==qmin) & Tup);          % Samples going up
  Ys(3,kdn)=qmin;                           % Turn input flow off
  Ys(1,kdn)=Ys(1,kdn)+...                   % Randomize level 1
     (qmax-qmin)*delta*rand(1,length(kdn))/C;
   Ys(3,kup)=qmax;                          % Turn input flow on
   Ys(1,kup)=Ys(1,kup)-...                  % Randomize level 1
     (qmax-qmin)*delta*rand(1,length(kup))/C;
   Xs=f(Ys,R1,R2);                          % Update samples
  end

  e=I/10;                                   % Expected number of rv in
                                              one bin
  % Get histograms (10 bins) and calculate test variables
  for i=1:2, b=hist(u(i,:)); c(i)=sum((b-e).^2/e); end
  J=sum(c);                                 % Full test (chi-square
                                              with 19 DoF)
return
```

Figure 9.16 shows the results obtained using particle filtering. With the number of samples set to 1000, minimization of J with respect to $q_{\max}$, P_{up} and P_{down} yielded a value of about $J = 1400$; a clear indication

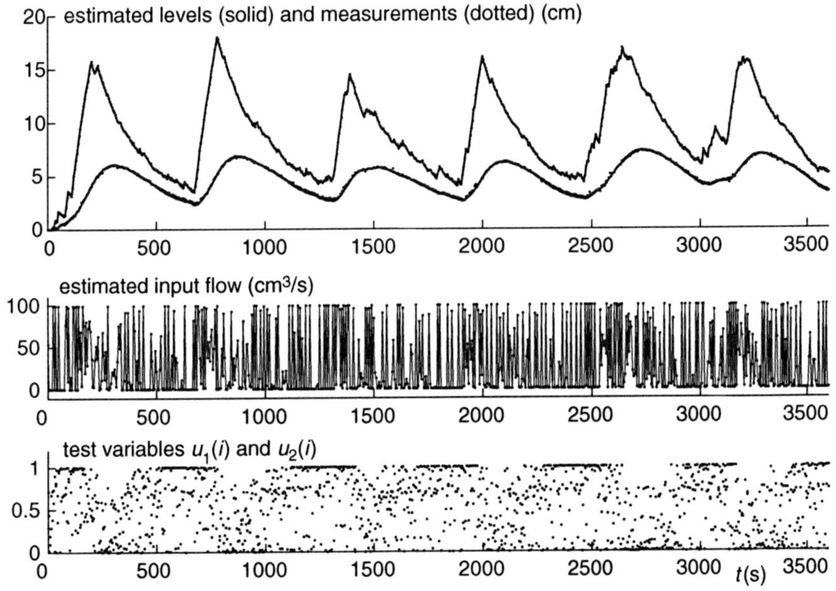

Figure 9.16 Results from particle filtering

that the test variables $u_1(i)$ and $u_2(i)$ are not uniformly distributed because J should have a χ^2_{19}-distribution. The graph in Figure 9.16, showing these test variables, confirms this statement. The values of P_{up} and P_{down} that provided the minimum were about 0.3. This minimum was very flat though.

The large transition probabilities that are needed to minimize the criterion indicate that we still have not found an appropriate model for the input flow. After all, large transition probabilities mean that the particle filter has much freedom of selecting an input flow that fits with the data.

9.3.4 Discussion

Up to now we have not succeeded in constructing a consistent estimator. The linearized Kalman filter failed because of the nonlinearity of the system, which is simply too severe. The extended Kalman filter was an improvement, but its behaviour was still not regular. The filter could only be put to action if the model for the input flow was unrestrictive, but the NIS of the filter was still too large. The particle filter that we tested used quite another model for the input flow than the extended Kalman filter, but yet this filter also needed an unrestrictive model for the input flow.

At this point, we must face another possibility. We have adopted a linear model for the measurements, but the specifications of the sensors mention linearity and hysteresis errors $\mathbf{e}(i)$ which can – when measured over the full-scale span – become quite large. Until now we have ignored the error $\mathbf{e}(i)$ in (9.24), but without really demonstrating that doing so is allowed. If it is not, that would explain why both the extended Kalman filter and the particle filter only work out with unrestrictive models for the input signal. In trying to get estimates that fit the data from both sensors the estimators cannot handle any further restrictions on the input signal.

The best way to cope with the existence of linearity and hysteresis errors is to model them properly. A linearity error – if reproducible – can be compensated by a calibration curve. A hysteresis error is more difficult to compensate because it depends on the dynamics of the states. In the present case, a calibration curve can be deduced for the second sensor using the previous results. In the particle filter of Section 9.3.3 the estimate of the first level is obtained by closely following the measurements in the first tank. The estimates of the second level are obtained merely by using the model of the hydraulic system. Therefore, these estimates can be used as a reference for the measurements in the second tank.

Figure 9.17 shows a scatter diagram of the errors versus estimated levels of the second tank. Of course, the data is contaminated by measurement noise, but nevertheless the trend is appreciable. Using MATLAB's `polyfit()` and `polyval()` polynomial regression has been applied to find a curve that fits the data. The resulting polynomial model is used to compensate the linearity errors of the second sensor. The results, illustrated in Figure 9.18, show test variables that are much more uniformly distributed than the same variables in Figure 9.16. The chi-square test gives a value of $J = 120$. This is a significant improvement, but nevertheless still too much for a consistent filter. However, Figure 9.18 only goes to show that the compensation of the linearity errors of the sensors do have a large impact.

For the final design, the errors of the first sensor should also be compensated. Additional measurements are needed to obtain the calibration curve for that sensor. Once this curve is available, the design procedure as described above starts over again, but this time with compensation of linearity errors included. There is no need to reconsider the linearized Kalman filter because we have already seen that its linearization errors are too severe.

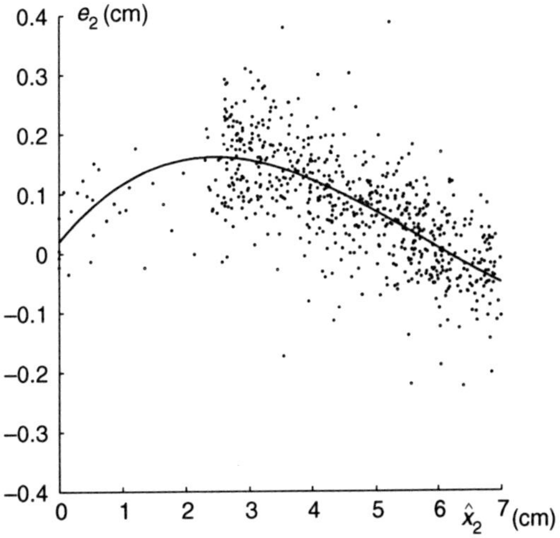

Figure 9.17 Calibration curve for the second sensor

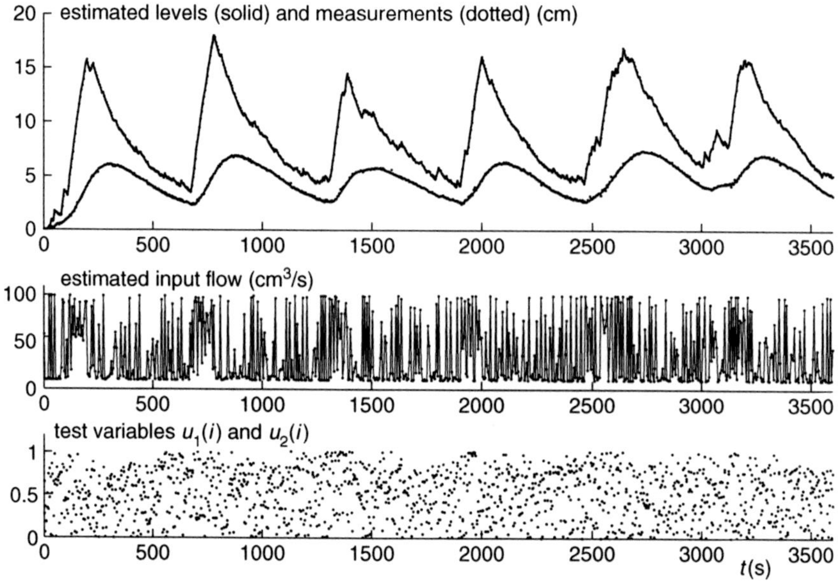

Figure 9.18 Results from particle filtering after applying a linearity correction

9.4 REFERENCES

Harrison, D. and Rubinfeld, D.L., Hedonic prices and the demand for clean air, *Journal of Environmental Economics and Management*, 5, 81–102, 1978.

Heijden, F. van der, Túquerres, G. and Regtien, P.P.L., Time-of-flight estimation based on covariance models, *Measurement Science and Technology*, **14**, 1295–304, 2003.

Kreyszig, E., *Introductory Mathematical Statistics – Principles and Methods*, Wiley, New York, 1970.

Appendix A

Topics Selected from Functional Analysis

This appendix summarizes some concepts from functional analysis. The concepts are part of the mathematical background required for understanding this book. Mathematical peculiarities not relevant in this context are omitted. Instead, at the end of the appendix references to more detailed treatments are given.

A.1 LINEAR SPACES

A *linear space* (or *vector space*) over a field F is a set R with elements (*vectors*) $\mathbf{f}, \mathbf{g}, \mathbf{h}, \ldots$ equipped with two operations:

- *addition* ($\mathbf{f} + \mathbf{g}$): $R \times R \to R$
- *scalar multiplication* ($\alpha\mathbf{f}$ with $\alpha \in F$): $F \times R \to R$

Usually, the field F is the set $\mathbb{R}$ of real numbers, or the set $\mathbb{C}$ of complex numbers. The addition and the multiplication operation must satisfy the following axioms:

(a) $\mathbf{f} + \mathbf{g} = \mathbf{g} + \mathbf{f}$
(b) $(\mathbf{f} + \mathbf{g}) + \mathbf{h} = \mathbf{f} + (\mathbf{g} + \mathbf{h})$

Classification, Parameter Estimation and State Estimation: An Engineering Approach using MATLAB
F. van der Heijden, R.P.W. Duin, D. de Ridder and D.M.J. Tax
 ISBN: 0-470-09013-8

(c) a so-called *zero element* $\mathbf{0} \in R$ exists such that $\mathbf{f} + \mathbf{0} = \mathbf{f}$
(d) a *negative element* $-\mathbf{f}$ exists for each $\mathbf{f}$ such that $\mathbf{f} + (-\mathbf{f}) = \mathbf{0}$
(e) $\alpha(\mathbf{f} + \mathbf{g}) = \alpha\mathbf{f} + \alpha\mathbf{g}$
(f) $(\alpha + \beta)\mathbf{f} = \alpha\mathbf{f} + \beta\mathbf{f}$
(g) $(\alpha\beta)\mathbf{f} = \alpha(\beta\mathbf{f})$
(h) $1\mathbf{f} = \mathbf{f}$

A *linear subspace* S of a linear space R is a subset of R which itself is linear. A condition sufficient and necessary for a subset $S \subset R$ to be linear is that $\alpha\mathbf{f} + \beta\mathbf{g} \in S$ for all $\mathbf{f}$, $\mathbf{g} \in S$ and for all α, $\beta \in F$.

Examples

$\mathbb{C}[a,b]$ is the set of all complex functions $f(x)$ continuous in the interval $[a,b]$. With the usual definition of addition and scalar multiplication this set is a linear space.[1] The set of polynomials of degree N:

$$f(x) = c_0 + c_1 x + c_2 x^2 + \cdots + c_N x^N \quad \text{with} \quad c_n \in \mathbb{C}$$

is a linear subspace of $\mathbb{C}[a,b]$.

The set $\mathbb{R}^\infty$ consisting of an infinite, countable series of real numbers $\mathbf{f} = (f_0, f_1, \ldots)$ is a linear space provided that the addition and multiplication takes place element by element.[1] The subset of $\mathbb{R}^\infty$ that satisfies the convergence criterion:

$$\sum_{n=0}^{\infty} |f_n|^2 < \infty$$

is a linear subspace of $\mathbb{R}^\infty$.

The set $\mathbb{R}^N$ consisting of N real numbers $\mathbf{f} = (f_0, f_1, \ldots f_{N-1})$ is a linear space provided that the addition and multiplication takes place element by element. Any linear hyperplane containing the null vector (zero element) is a linear subspace.

[1] Throughout this appendix the examples relate to vectors which are either real or complex. However, these examples can be converted easily from real to complex or vice versa. The set of all real functions continuous in the interval $[a,b]$ is denoted by $\mathbb{R}[a,b]$. The set of infinite and finite countable complex numbers is denoted by $\mathbb{C}^\infty$ and $\mathbb{C}^N$, respectively.

Any vector that can be written as:

$$\mathbf{f} = \sum_{i=0}^{n-1} \alpha_i \mathbf{f}_i \qquad \alpha_i \in F \tag{a.1}$$

is called a *linear combination* of $\mathbf{f}_0, \mathbf{f}_1, \ldots, \mathbf{f}_{n-1}$. The vectors $\mathbf{f}_0, \mathbf{f}_1, \ldots, \mathbf{f}_{m-1}$ are *linear dependent* if a set of numbers β_i exists, not all zero, for which the following equation holds:

$$\sum_{i=0}^{m-1} \beta_i \mathbf{f}_i = \mathbf{0} \tag{a.2}$$

If no such set exists, the vectors $\mathbf{f}_0, \mathbf{f}_1, \ldots, \mathbf{f}_{m-1}$ are said to be *linear independent*.

The *dimension* of a linear space R is defined as the non-negative integer number N for which N independent linear vectors in R exist, while any set of $N + 1$ vectors in R is linear dependent. If for each number N there exist N linear independent vectors in R, the dimension of R is ∞.

Example
$\mathbb{C}[a,b]$ and $\mathbb{R}^\infty$ have dimension ∞. $\mathbb{R}^N$ has dimension N.

A.1.1 Normed linear spaces

A *norm* $\|\mathbf{f}\|$ of a linear space is a mapping $R \to \mathbb{R}$ (i.e. a *real function* or a *functional*) such that:

(a) $\|\mathbf{f}\| \geq 0$, where $\|\mathbf{f}\| = 0$ if and only if $\mathbf{f} = \mathbf{0}$
(b) $\|\alpha \mathbf{f}\| = |\alpha| \|\mathbf{f}\|$
(c) $\|\mathbf{f} + \mathbf{g}\| \leq \|\mathbf{f}\| + \|\mathbf{g}\|$

A linear space equipped with a norm is called a *normed linear space*.

Examples
The following real functions satisfy the axioms of a norm:
In $\mathbb{C}[a,b]$:

$$\|f(x)\|_p = \left(\int_{x=a}^{b} |f(x)|^p dx \right)^{\frac{1}{p}} \quad \text{with: } p \geq 1 \tag{a.3}$$

In $\mathbb{R}^\infty$:

$$\|\mathbf{f}\|_p = \left(\sum_{n=0}^{\infty} |f_n|^p\right)^{\frac{1}{p}} \quad \text{with: } p \geq 1 \tag{a.4}$$

In $\mathbb{R}^N$:

$$\|\mathbf{f}\|_p = \left(\sum_{n=0}^{N-1} |f_n|^p\right)^{\frac{1}{p}} \quad \text{with: } p \geq 1 \tag{a.5}$$

These norms are called the L_p *norm* (continuous case, e.g. $\mathbb{C}[a,b]$) and the l_p *norm* (discrete case, e.g. $\mathbb{C}^\infty$). Graphical representations of the norm are depicted in Figure A.1. Special cases occur for particular choices of the parameter p. If $p = 1$, the norm is the sum of the absolute magnitudes, e.g.

$$\|f(x)\|_1 = \int_{x=a}^{b} |f(x)|dx \quad \text{and} \quad \|\mathbf{f}\|_1 = \sum_n |f_n| \tag{a.6}$$

If $p = 2$, we have the *Euclidean norm*:

$$\|f(x)\|_2 = \sqrt{\int_{x=a}^{b} |f(x)|^2 dx} \quad \text{and} \quad \|\mathbf{f}\|_2 = \sqrt{\sum_n |f_n|^2} \tag{a.7}$$

In $\mathbb{R}^3$ and $\mathbb{R}^2$ the norm $\|\mathbf{f}\|_2$ is the length of the vector as defined in geometry. If $p \to \infty$, the L_p and l_p norms are the maxima of the absolute magnitudes, e.g.

$$\|f(x)\|_\infty = \max_{x \in [a,b]} |f(x)| \quad \text{and} \quad \|\mathbf{f}\|_\infty = \max_n |f_n| \tag{a.8}$$

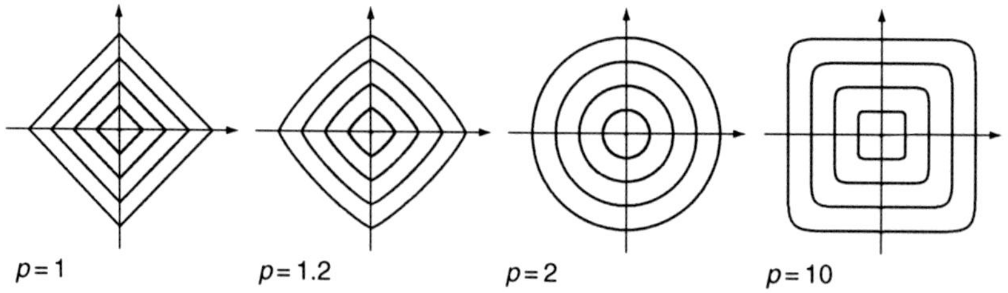

Figure A.1 'Circles' in $\mathbb{R}^2$ equipped with the l_p norm

A.1.2 Euclidean spaces or inner product spaces

A *Euclidean space* (also called *inner product space*) R is a linear space for which the *inner product* is defined. The inner product $(\mathbf{f},\mathbf{g})$ between two vectors $\mathbf{f},\mathbf{g} \in R$ over a field F is a mapping $R \times R \rightarrow F$ that satisfies the following axioms:

(a) $(\mathbf{f}+\mathbf{g},\mathbf{h}) = (\mathbf{f},\mathbf{h}) + (\mathbf{g},\mathbf{h})$
(b) $(\alpha\mathbf{f},\mathbf{g}) = \alpha(\mathbf{f},\mathbf{g})$
(c) $(\mathbf{g},\mathbf{f}) = \overline{(\mathbf{f},\mathbf{g})}$
(d) $(\mathbf{f},\mathbf{f}) \geq 0$, real
(e) $(\mathbf{f},\mathbf{f}) = 0 \Rightarrow \mathbf{f} = 0$

In (c) the number $\overline{(\mathbf{f},\mathbf{g})}$ is the complex conjugated of $(\mathbf{f},\mathbf{g})$. Of course, this makes sense only if F is complex. If F is the set of real numbers, then $(\mathbf{g},\mathbf{f}) = (\mathbf{f},\mathbf{g})$.

Examples
$\mathbb{C}[a,b]$ is a Euclidean space if the inner product is defined as:

$$(f(x), g(x)) = \int_{x=a}^{b} f(x)\overline{g(x)}dx$$

$\mathbb{R}^\infty$ is a Euclidean space if with $\mathbf{f} = (f_0, f_1, \ldots)$ and $\mathbf{g} = (g_0, g_1, \ldots)$:

$$(\mathbf{f},\mathbf{g}) = \sum_{n=0}^{\infty} f_i g_i$$

$\mathbb{R}^N$ is a Euclidean space if with $\mathbf{f} = (f_0, f_1, \ldots, f_{N-1})$ and $\mathbf{g} = (g_0, g_1, \ldots, g_{N-1})$:

$$(\mathbf{f},\mathbf{g}) = \sum_{n=0}^{N-1} f_i g_i$$

In accordance with (a.7), any Euclidean space becomes a normed linear space as soon as it is equipped with the *Euclidean norm*:

$$\|\mathbf{f}\| = +\sqrt{(\mathbf{f},\mathbf{f})} \tag{a.9}$$

Given this norm, any two vectors **f** and **g** satisfy the *Schwarz inequality*:

$$|(\mathbf{f}, \mathbf{g})| \leq \|\mathbf{f}\|\|\mathbf{g}\| \tag{a.10}$$

Two vectors **f** and **g** are said to be *orthogonal* (notation: $\mathbf{f}\perp\mathbf{g}$) whenever $(\mathbf{f},\mathbf{g}) = 0$. If two vectors **f** and **g** are orthogonal then (*Pythagoras*):

$$\|\mathbf{f}+\mathbf{g}\|^2 = \|\mathbf{f}\|^2 + \|\mathbf{g}\|^2 \tag{a.11}$$

Given a vector $\mathbf{g} \neq \mathbf{0}$, an arbitrary vector **f** can be decomposed into a component that coincides with **g** and an orthogonal component: $\mathbf{f} = \alpha\mathbf{g} + \mathbf{h}$ with $\mathbf{g}\perp\mathbf{h}$. The scalar α follows from:

$$\alpha = \frac{(\mathbf{f}, \mathbf{g})}{(\mathbf{g}, \mathbf{g})} = \frac{(\mathbf{f}, \mathbf{g})}{\|\mathbf{g}\|^2} \tag{a.12}$$

The term $\alpha\mathbf{g}$ is called the *projection* of **f** on **g**. The *angle* φ between two vectors **f** and **g** is defined such that it satisfies:

$$\cos(\varphi) = \frac{(\mathbf{f}, \mathbf{g})}{\|\mathbf{f}\|\|\mathbf{g}\|} \tag{a.13}$$

A.2 METRIC SPACES

A *metric space* R is a set equipped with a *distance measure* $\rho(\mathbf{f}, \mathbf{g})$ that maps any couple of elements $\mathbf{f},\mathbf{g} \in R$ into a non-negative real number: $R \times R \to \mathbb{R}^+$. The distance measure must satisfy the following axioms:

(a) $\rho(\mathbf{f},\mathbf{g}) = 0$ if and only if $\mathbf{f} = \mathbf{g}$
(b) $\rho(\mathbf{f},\mathbf{g}) = \rho(\mathbf{g},\mathbf{f})$
(c) $\rho(\mathbf{f},\mathbf{h}) \leq \rho(\mathbf{f},\mathbf{g}) + \rho(\mathbf{g},\mathbf{h})$

All normed linear spaces become a metric space, if we set:

$$\rho(\mathbf{f}, \mathbf{g}) = \|\mathbf{f} - \mathbf{g}\| \tag{a.14}$$

Consequently, the following mappings satisfy the axioms of a distance measure. In $\mathbb{C}[a,b]$:

$$\rho(f(x), g(x)) = \left(\int_{x=a}^{b} |f(x) - g(x)|^p dx \right)^{\frac{1}{p}} \quad \text{with:} \quad p \geq 1 \tag{a.15}$$

In $\mathbb{R}^\infty$:

$$\rho(\mathbf{f}, \mathbf{g}) = \left(\sum_{n=0}^{\infty} |f_n - g_n|^p \right)^{\frac{1}{p}} \quad \text{with:} \quad p \geq 1 \tag{a.16}$$

In $\mathbb{R}^N$:

$$\rho(\mathbf{f}, \mathbf{g}) = \left(\sum_{n=0}^{N-1} |f_n - g_n|^p \right)^{\frac{1}{p}} \quad \text{with:} \quad p \geq 1 \tag{a.17}$$

These are the *Minkowski distances*. A theorem related to these measures is *Minkowski's inequality*. In $\mathbb{R}^N$ and $\mathbb{C}^\infty$ this equality states that:[2]

$$\left(\sum_n |f_n + g_n|^p \right)^{\frac{1}{p}} \leq \left(\sum_n |f_n|^p \right)^{\frac{1}{p}} + \left(\sum_n |g_n|^p \right)^{\frac{1}{p}} \tag{a.18}$$

Special cases occur for particular choices of the parameter p. If $p = 1$, the distance measure equals the sum of absolute differences between the various coefficients, e.g.

$$\rho(f(x), g(x)) = \int_{x=a}^{b} |f(x) - g(x)| dx \quad \text{and} \quad \rho(\mathbf{f}, \mathbf{g}) = \sum_n |f_n - g_n| \tag{a.19}$$

This measure is called the *city-block distance* (also known as *Manhattan, magnitude, box-car* or *absolute value distance*). If $p = 2$, we have the ordinary *Euclidean distance measure*. If $p \to \infty$ the maximum of the absolute differences between the various coefficients is determined, e.g.

[2] In $\mathbb{C}[a,b]$ the inequality is similar.

$$\rho(f(x), g(x)) = \max_{x \in [a,b]} |f(x) - g(x)| \quad \text{and} \quad \rho(\mathbf{f}, \mathbf{g}) = \max_n |f_n - g_n| \tag{a.20}$$

This measure is the *chessboard distance* (also known as *Chebyshev* or *maximum value distance*).

The *quadratic measure* is defined as:

$$\rho(\mathbf{f}, \mathbf{g}) = +\sqrt{(\mathbf{f} - \mathbf{g}, \mathbf{A}\mathbf{f} - \mathbf{A}\mathbf{g})} \tag{a.21}$$

where $\mathbf{A}$ is a self-adjoint operator with non-negative eigenvalues. This topic will be discussed in Section A.4 and Section B.5. This measure finds its application in, for instance, pattern classification (where it is called the *Mahalanobis distance*).

Another distance measure is:

$$\rho(\mathbf{f}, \mathbf{g}) = \begin{cases} 1 & \text{if } \mathbf{f} = \mathbf{g} \\ 0 & \text{if } \mathbf{f} \neq \mathbf{g} \end{cases} \tag{a.22}$$

An application of this measure is in Bayesian estimation and classification theory where it is used to express a particular cost function. Note that in contrast with the preceding examples this measure cannot be derived from a norm. For every metric derived from a norm we have $\rho(\alpha\mathbf{f},\mathbf{0}) = |\alpha|\rho(\mathbf{f},\mathbf{0})$. However, this equality does not hold for (a.22).

A.3 ORTHONORMAL SYSTEMS AND FOURIER SERIES

In a Euclidean space R with the norm given by (a.9), a subset $S \subset R$ is an *orthogonal system* if every couple $\mathbf{b}_i,\mathbf{b}_j$ in S is orthogonal; i.e. $(\mathbf{b}_i,\mathbf{b}_j) = 0$ whenever $i \neq j$. If in addition each vector in S has unit length, i.e. $\|\mathbf{b}_i\| = 1$, then S is called an *orthonormal system.*

Examples

In $\mathbb{C}[a,b]$ the following harmonic functions form an orthonormal system:

$$w_n(x) = \frac{1}{\sqrt{b-a}} \exp\left(\frac{2\pi \mathrm{j} n x}{b-a}\right) \quad \text{with: } \mathrm{j} = \sqrt{-1} \tag{a.23}$$

In $\mathbb{C}^N$ the following vectors are an orthonormal system:

$$\mathbf{w}_n = \left[\frac{1}{\sqrt{N}} \frac{1}{\sqrt{N}}\exp\left(\frac{2\pi \mathrm{j} n}{N}\right) \cdots \frac{1}{\sqrt{N}}\exp\left(\frac{2\pi \mathrm{j}(N-1)n}{N}\right)\right] \tag{a.24}$$

Let $S = \{\mathbf{w}_0 \quad \mathbf{w}_1 \ldots \mathbf{w}_{N-1}\}$ be an orthonormal system in a Euclidean space R, and let $\mathbf{f}$ be an arbitrary vector in R. Then the *Fourier coefficients* of $\mathbf{f}$ with respect to S are the inner products:

$$\phi_k = (\mathbf{f}, \mathbf{w}_k) \qquad k = 0, 1, \ldots, N-1 \tag{a.25}$$

Furthermore, the series:

$$\sum_{k=0}^{N-1} \phi_k \mathbf{w}_k \tag{a.26}$$

is called the *Fourier series* of $\mathbf{f}$ with respect to the system S. Suppose we wish to approximate $\mathbf{f}$ by a suitably chosen linear combination of the system S. The best approximation (according to the norm in R) is given by (a.26). This follows from the following inequality:

$$\left\| \mathbf{f} - \sum_{k=0}^{N-1} \phi_k \mathbf{w}_k \right\| \le \left\| \mathbf{f} - \sum_{k=0}^{N-1} \beta_k \mathbf{w}_k \right\| \quad \text{for arbitrary } \beta_k \tag{a.27}$$

The approximation improves as the number N of vectors increases. This follows readily from *Bessel's inequality*:

$$\sum_{k=0}^{N-1} |\varphi_k|^2 \le \|\mathbf{f}\|^2 \tag{a.28}$$

Let $S = \{\mathbf{w}_0 \ \mathbf{w}_1 \ \ldots\}$ be an orthonormal system in a Euclidean space R. The number of vectors in S may be infinite. Suppose that no vector $\tilde{\mathbf{w}}$ exists for which the system S augmented by $\tilde{\mathbf{w}}$, i.e. $\tilde{S} = \{\tilde{\mathbf{w}} \ \mathbf{w}_0 \ \mathbf{w}_1 \ \ldots\}$ is also an orthonormal system. Then, S is called an *orthonormal basis*. In that case, the smallest linear subspace containing S is the whole space R.

The number of vectors in an orthonormal basis S may be finite (as in $\mathbb{R}^N$ and $\mathbb{C}^N$), countable infinite (as in $\mathbb{R}^\infty, \mathbb{C}^\infty, \mathbb{R}[a,b]$, and $\mathbb{C}[a,b]$ with $-\infty < a < b < \infty$), or uncountable infinite (as in $\mathbb{R}[-\infty,\infty]$, and $\mathbb{C}[-\infty,\infty]$).

Examples
In $\mathbb{C}[a,b]$ with $-\infty < a < b < \infty$ an orthonormal basis is given by the harmonic functions in (a.23). The number of such functions is countable infinite. The Fourier series defined in (a.26) essentially corresponds to the Fourier series expansion of a periodic signal. In $\mathbb{C}[-\infty,\infty]$ this expansion evolves into the Fourier integral.[3]

In $\mathbb{C}^N$ an orthonormal basis is given by the vectors in (a.24). The Fourier series in (a.26) is equivalent to the discrete Fourier transform.

The examples given above are certainly not the only orthonormal bases. In fact, even in $\mathbb{R}^N$ (with $N > 1$) infinitely many orthonormal bases exist.

If S is an orthonormal basis, and $\mathbf{f}$ an arbitrary vector with Fourier coefficients ϕ_k with respect to S, then the following theorems hold:

$$\mathbf{f} = \sum_k \phi_k \mathbf{w}_k \tag{a.29}$$

$$(\mathbf{f},\mathbf{g}) = \sum_k (\mathbf{f},\mathbf{w}_k)(\mathbf{g},\mathbf{w}_k) \qquad \text{(Parseval)} \tag{a.30}$$

$$\|\mathbf{f}\|^2 = \sum_k |\phi_k|^2 \qquad \text{(Parseval/Pythagoras)} \tag{a.31}$$

Equation (a.29) corresponds to the inverse transform in Fourier analysis. Equation (a.31) follows directly from (a.30), since $\|\mathbf{w}_k\| = 1, \forall k$. The equation shows that in the case of an orthonormal basis Bessel's inequality transforms to an equality.

A.4 LINEAR OPERATORS

Given two normed linear spaces R_1 and R_2, a mapping A of a subset of R_1 into R_2 is called an *operator* from R_1 to R_2. The subset of R_1 (possibly R_1 itself) for which the operator A is defined is called the *domain* D_A of A. The *range* R_A is the set $\{\mathbf{g}|\mathbf{g} = A\mathbf{f}, \mathbf{f} \in D_A\}$. In the sequel we will assume that D_A is a linear (sub)space. An operator is *linear* if for all vectors $\mathbf{f},\mathbf{g} \in D_A$ and for all α and β:

[3] In fact, $\mathbb{C}[-\infty,\infty]$ must satisfy some conditions in order to assure the existence of the Fourier expansion.

$$A(\alpha\mathbf{f} + \beta\mathbf{g}) = \alpha A\mathbf{f} + \beta A\mathbf{g} \tag{a.32}$$

Examples
Any orthogonal transform is a linear operation. In $\mathbb{R}^N$ and $\mathbb{C}^N$ any matrix–vector multiplication is a linear operation. In $\mathbb{R}^\infty$ and $\mathbb{C}^\infty$ left-sided shifts, right-sided shifts and any linear combination of them (i.e. discrete convolution) are linear operations. In $\mathbb{R}[a,b]$ convolution integrals and differential operators are linear operators.

Some special linear operators are:

- The *null operator 0* assigns the null vector to each vector: $0\mathbf{f} = \mathbf{0}$.
- The *identity operator I* carries each vector into itself: $I\mathbf{f} = \mathbf{f}$.

An operator A is *invertible* if for each $\mathbf{g} \in R_A$ the equation $\mathbf{g} = A\mathbf{f}$ has a unique solution $\mathbf{f} \in D_A$. The operator A^{-1} that uniquely assigns this solution $\mathbf{f}$ to $\mathbf{g}$ is called the *inverse operator* of A:

$$\mathbf{g} = A\mathbf{f} \Leftrightarrow \mathbf{f} = A^{-1}\mathbf{g} \tag{a.33}$$

The following properties are shown easily:

- $A^{-1}A = I$
- $AA^{-1} = I$
- The inverse of a linear operator – if it exists – is linear.

Suppose that in a linear space R two orthonormal bases $S_a = \{\,\mathbf{a}_0 \quad \mathbf{a}_1 \quad \cdots\}$ and $S_b = \{\,\mathbf{b}_0 \quad \mathbf{b}_1 \quad \cdots\}$ are defined. According to (a.29) each vector $\mathbf{f} \in R$ has two representations:

$$\mathbf{f} = \sum_k \alpha_k \mathbf{a}_k \quad \text{with:}\ \alpha_k = (\mathbf{f}, \mathbf{a}_k)$$
$$\mathbf{f} = \sum_k \beta_k \mathbf{b}_k \quad \text{with:}\ \beta_k = (\mathbf{f}, \mathbf{b}_k)$$

Since both Fourier series represent the same vector we conclude that:

$$\mathbf{f} = \sum_k \alpha_k \mathbf{a}_k = \sum_k \beta_k \mathbf{b}_k$$

The relationship between the Fourier coefficients α_k and β_k can be made explicit by the calculation of the inner product:

$$(\mathbf{f}, \mathbf{b}_n) = \sum_k \alpha_k(\mathbf{a}_k, \mathbf{b}_n) = \sum_k \beta_k(\mathbf{b}_k, \mathbf{b}_n) = \beta_n \tag{a.34}$$

The Fourier coefficients α_k and β_k can be arranged as vectors $\boldsymbol{\alpha} = (\alpha_0, \alpha_1, \cdots)$ and $\boldsymbol{\beta} = (\beta_0, \beta_1, \cdots)$ in $\mathbb{R}^N$ or $\mathbb{C}^N$ (if the dimension of R is finite), or in $\mathbb{R}^\infty$ and $\mathbb{C}^\infty$ (if the dimension of R is infinite). In one of these spaces equation (a.34) defines a linear operator U:

$$\boldsymbol{\beta} = U\boldsymbol{\alpha} \tag{a.35}$$

The inner product in (a.34) could equally well be accomplished with respect to a vector $\mathbf{a}_n$. This reveals that an operator U^* exists for which:

$$\boldsymbol{\alpha} = U^*\boldsymbol{\beta} \tag{a.36}$$

Clearly, from (a.33):

$$U^* = U^{-1} \tag{a.37}$$

Suppose we have two vectors $\mathbf{f}_1$ and $\mathbf{f}_2$ represented in S_a by $\boldsymbol{\alpha}_1$ and $\boldsymbol{\alpha}_2$, and in S_b by $\boldsymbol{\beta}_1$, and $\boldsymbol{\beta}_2$. Since the inner product $(\mathbf{f}_1,\mathbf{f}_2)$ must be independent of the representation, we conclude that $(\mathbf{f}_1,\mathbf{f}_2) = (\boldsymbol{\alpha}_1,\boldsymbol{\alpha}_2) = (\boldsymbol{\beta}_1,\boldsymbol{\beta}_2)$. Therefore:

$$(\boldsymbol{\alpha}_1, U^{-1}\boldsymbol{\beta}_2) = (U\boldsymbol{\alpha}_1,\boldsymbol{\beta}_2) \tag{a.38}$$

Each operator that satisfies (a.38) is called a *unitary* operator. A corollary of (a.38) is that any unitary operator preserves the Euclidean norm.

The *adjoint* A^* of an operator A is an operator that satisfies:

$$(A\mathbf{f}, \mathbf{g}) = (\mathbf{f}, A^*\mathbf{g}) \tag{a.39}$$

From this definition, and from (a.38), it follows that an operator U for which its adjoint U^* equals its inverse U^{-1} is a unitary operator. This is in accordance with the notation used in (a.37). An operator A is called *self-adjoint*, if $A^* = A$.

Suppose that A is a linear operator in a space R. A vector $\mathbf{e}_k$ that satisfies:

$$A\mathbf{e}_k = \lambda_k \mathbf{e}_k \quad \mathbf{e}_k \neq 0 \tag{a.40}$$

with λ_k a real or complex number is called an *eigenvector* of A. The number λ_k is the *eigenvalue*. The eigenvectors and eigenvalues of an operator are found by solving the equation $(A - \lambda_k I)\mathbf{e}_k = 0$. Note that if $\mathbf{e}_k$ is a solution of this equation, then so is $\alpha\mathbf{e}_k$ with α any real or complex number. If a unique solution is required, we should constrain the length of the eigenvector to unit, i.e. $\mathbf{v}_k = \mathbf{e}_k/\|\mathbf{e}_k\|$, yielding the so-called *normalized eigenvector*. However, since $+\mathbf{e}_k/\|\mathbf{e}_k\|$ and $-\mathbf{e}_k/\|\mathbf{e}_k\|$ are both valid eigenvectors, we still have to select one out of the two possible solutions. From now on, the phrase 'the normalized eigenvector' will denote both solutions.

Operators that are self-adjoint have – under mild conditions – some nice properties related to their eigenvectors and eigenvalues. The properties relevant in our case are:

1. All eigenvalues are real.
2. With each eigenvalue at least one normalized eigenvector is associated. However, an eigenvalue can also have multiple normalized eigenvectors. These eigenvectors span a linear subspace.
3. There is an orthonormal basis $V = \{\mathbf{v}_0 \quad \mathbf{v}_1 \quad \cdots\}$ formed by the normalized eigenvectors. Due to possible multiplicities of normalized eigenvalues (see above) this basis may not be unique.

A corollary of the properties is that any vector $\mathbf{f} \in R$ can be represented by a Fourier series with respect to V, and that in this representation the operation becomes simply a linear combination, that is:

$$\mathbf{f} = \sum_k \phi_k \mathbf{v}_k \quad \text{with:} \quad \phi_k = (\mathbf{f}, \mathbf{v}_k) \tag{a.41}$$

$$A\mathbf{f} = \sum_k \lambda_k \phi_k \mathbf{v}_k \tag{a.42}$$

The connotation of this decomposition of the operation is depicted in Figure A.2. The set of eigenvalues is called the *spectrum* of the operator.

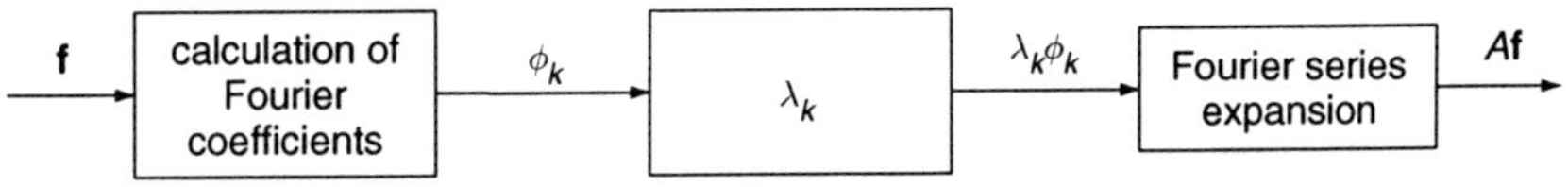

Figure A.2 Eigenvalue decomposition of a self-adjoint operator

A.5 REFERENCES

Kolmogorov, A.N. and Fomin, S.V., *Introductory Real Analysis*, Dover Publications, New York, 1970.

Pugachev, V.S. and Sinitsyn, I.N., *Lectures on Functional Analysis and Applications*, World Scientific, 1999.

Appendix B

Topics Selected from Linear Algebra and Matrix Theory

Whereas Appendix A deals with general linear spaces and linear operators, the current appendix restricts the attention to linear spaces with finite dimension, i.e. $\mathbb{R}^N$ and $\mathbb{C}^N$. With that, all that has been said in Appendix A also holds true for the topics of this appendix.

B.1 VECTORS AND MATRICES

Vectors in $\mathbb{R}^N$ and $\mathbb{C}^N$ are denoted by bold-faced letters, e.g. **f**, **g**. The elements in a vector are arranged either vertically (a column vector) or horizontally (a row vector). For example:

$$\mathbf{f} = \begin{bmatrix} f_0 \\ f_1 \\ \vdots \\ f_{N-1} \end{bmatrix} \quad \text{or:} \quad \mathbf{f}^T = [f_0 \quad f_1 \quad \cdots \quad f_{N-1}] \tag{b.1}$$

The superscript ^{T} is used to convert column vectors to row vectors. Vector addition and scalar multiplication are defined as in Section A.1.

A *matrix* $\mathbf{H}$ with dimension $N \times M$ is an arrangement of NM numbers $h_{n,m}$ (the elements) on an orthogonal grid of N rows and M columns:

Classification, Parameter Estimation and State Estimation: An Engineering Approach using MATLAB
F. van der Heijden, R.P.W. Duin, D. de Ridder and D.M.J. Tax
© 2004 John Wiley & Sons, Ltd ISBN: 0-470-09013-8

$$
\mathbf{H} = \begin{bmatrix} h_{0,0} & h_{0,1} & \cdots & h_{0,M-1} \\ h_{1,0} & h_{1,1} & \cdots & h_{1,M-1} \\ h_{2,0} & \vdots & & \vdots \\ \vdots & \vdots & & \vdots \\ h_{N-1,0} & \cdots & \cdots & h_{N-1,M-1} \end{bmatrix} \tag{b.2}
$$

The elements are real or complex. Vectors can be regarded as $N \times 1$ matrices (column vectors) or $1 \times M$ matrices (row vectors). A matrix can be regarded as an horizontal arrangement of M column vectors with dimension N, for example:

$$
\mathbf{H} = [\,\mathbf{h}_0 \quad \mathbf{h}_1 \quad \cdots \quad \mathbf{h}_{M-1}\,] \tag{b.3}
$$

Of course, a matrix can also be regarded as a vertical arrangement of N row vectors.

The *scalar–matrix multiplication* $\alpha\mathbf{H}$ replaces each element in $\mathbf{H}$ with $\alpha h_{n,m}$. The *matrix–addition* $\mathbf{H} = \mathbf{A} + \mathbf{B}$ is only defined if the two matrices $\mathbf{A}$ and $\mathbf{B}$ have equal size $N \times M$. The result $\mathbf{H}$ is an $N \times M$ matrix with elements $h_{n,m} = a_{n,m} + b_{n,m}$. These two operations satisfy the axioms of a linear space (Section A.1). Therefore, the set of all $N \times M$ matrices is another example of a linear space.

The *matrix–matrix product* $\mathbf{H} = \mathbf{AB}$ is defined only when the number of columns of $\mathbf{A}$ equals the number of rows of $\mathbf{B}$. Suppose that $\mathbf{A}$ is an $N \times P$ matrix, and that $\mathbf{B}$ is a $P \times M$ matrix, then the product $\mathbf{H} = \mathbf{AB}$ is an $N \times M$ matrix with elements:

$$
h_{n,m} = \sum_{p=0}^{P-1} a_{n,p} b_{p,m} \tag{b.4}
$$

Since a vector can be regarded as an $N \times 1$ matrix, this also defines the *matrix–vector product* $\mathbf{g} = \mathbf{Hf}$ with $\mathbf{f}$ an M-dimensional column vector, $\mathbf{H}$ an $N \times M$ matrix and $\mathbf{g}$ an N-dimensional column vector. In accordance with these definitions, the inner product between two real N-dimensional vectors introduced in Section A.1.2 can be written as:

$$
(\mathbf{f}, \mathbf{g}) = \sum_{n=0}^{N-1} f_n g_n = \mathbf{f}^T \mathbf{g} \tag{b.5}
$$

It is easy to show that a matrix–vector product $\mathbf{g} = \mathbf{Hf}$ defines a linear operator from $\mathbb{R}^M$ into $\mathbb{R}^N$ and $\mathbb{C}^M$ into $\mathbb{C}^N$. Therefore, all definitions and properties related to linear operators (Section A.4) also apply to matrices.

Some special matrices are:

- The *null matrix* $\mathbf{O}$. This is a matrix fully filled with zero. It corresponds to the null operator: $\mathbf{Of} = \mathbf{0}$.
- The *unit matrix* $\mathbf{I}$. This matrix is square ($N = M$), fully filled with zero, except for the diagonal elements which are unit:

$$\mathbf{I} = \begin{bmatrix} 1 & & 0 \\ & \ddots & \\ 0 & & 1 \end{bmatrix}$$

  This matrix corresponds to the unit operator: $\mathbf{If} = \mathbf{f}$.
- A *diagonal matrix* Λ is a square matrix, fully filled with zero, except for its diagonal elements $\lambda_{n,n}$:

$$\Lambda = \begin{bmatrix} \lambda_{0,0} & & 0 \\ & \ddots & \\ 0 & & \lambda_{N-1,N-1} \end{bmatrix}$$

  Often, diagonal matrices are denoted by upper case Greek symbols.
- The *transposed matrix* $\mathbf{H}^T$ of an $N \times M$ matrix $\mathbf{H}$ is an $M \times N$ matrix, its elements are given by $h^T_{m,n} = h_{n,m}$.
- A *symmetric matrix* is a square matrix for which $\mathbf{H}^T = \mathbf{H}$.
- The *conjugated* of a matrix $\mathbf{H}$ is a matrix $\overline{\mathbf{H}}$ the elements of which are the complex conjugated of the one of $\mathbf{H}$.
- The *adjoint* of a matrix $\mathbf{H}$ is a matrix $\mathbf{H}^*$ which is the conjugated and the transposed of $\mathbf{H}$, that is: $\mathbf{H}^* = \overline{\mathbf{H}}^T$. A matrix $\mathbf{H}$ is *self-adjoint* or *Hermitian* if $\mathbf{H}^* = \mathbf{H}$. This is the case only if $\mathbf{H}$ is square and $h_{n,m} = \overline{h}_{m,n}$.
- The *inverse* of a square matrix $\mathbf{H}$ is the matrix $\mathbf{H}^{-1}$ that satisfies $\mathbf{H}^{-1}\mathbf{H} = \mathbf{I}$. If it exists, it is unique. In that case the matrix $\mathbf{H}$ is called *regular*. If $\mathbf{H}^{-1}$ doesn't exist, $\mathbf{H}$ is called *singular*.
- A *unitary matrix* $\mathbf{U}$ is a square matrix that satisfies $\mathbf{U}^{-1} = \mathbf{U}^*$. A real unitary matrix is called *orthonormal*. These matrices satisfy $\mathbf{U}^{-1} = \mathbf{U}^T$.

- A square matrix **H** is *Toeplitz* if its elements satisfy $h_{n,m} = g_{(n-m)}$ in which g_n is a sequence of $2N - 1$ numbers.
- A square matrix **H** is *circulant* if its elements satisfy $h_{n,m} = g_{(n-m)\%N}$. Here, $(n - m)\%N$ is the remainder of $(n - m)/N$.
- A matrix **H** is *separable* if it can be written as the product of two vectors: $\mathbf{H} = \mathbf{f}\mathbf{g}^T$.

Some properties with respect to the matrices mentioned above:

$$(\mathbf{H}^*)^* = \mathbf{H} \tag{b.6}$$

$$(\mathbf{AB})^* = \mathbf{B}^*\mathbf{A}^* \tag{b.7}$$

$$(\mathbf{H}^{-1})^* = (\mathbf{H}^*)^{-1} \tag{b.8}$$

$$(\mathbf{AB})^{-1} = \mathbf{B}^{-1}\mathbf{A}^{-1} \tag{b.9}$$

$$(\mathbf{A}^{-1} + \mathbf{H}^T\mathbf{B}^{-1}\mathbf{H})^{-1} = \mathbf{A} - \mathbf{AH}^T\left(\mathbf{HAH}^T + \mathbf{B}\right)^{-1}\mathbf{HA} \tag{b.10}$$

The relations hold if the size of the matrices are compatible and the inverses exist. Property (b.10) is known as the *matrix inversion lemma*.

B.2 CONVOLUTION

Defined in a finite interval, the discrete convolution between a sequence f_k and g_k:

$$g_n = \sum_{k=0}^{N-1} h_{n-k} f_k \quad \text{with:} \quad n = 0, 1, \ldots, N-1 \tag{b.11}$$

can be written economically as a matrix–vector product $\mathbf{g} = \mathbf{Hf}$. The matrix **H** is a Toeplitz matrix:

$$\mathbf{H} = \begin{bmatrix} h_0 & h_{-1} & h_{-2} & \cdots & h_{1-N} \\ h_1 & h_0 & h_{-1} & & h_{2-N} \\ h_2 & h_1 & \ddots & \ddots & \vdots \\ \vdots & & \ddots & \ddots & h_{-1} \\ h_{N-1} & h_{N-2} & \cdots & h_1 & h_0 \end{bmatrix} \tag{b.12}$$

If this 'finite interval' convolution is replaced with a *circulant* (wrap-around) discrete convolution, then (b.11) becomes:

$$g_n = \sum_{k=0}^{N-1} h_{(n-k)\%N} f_k \quad \text{with:} \quad n = 0, 1, \ldots, N-1 \tag{b.13}$$

In that case, the matrix–vector relation $\mathbf{g} = \mathbf{Hf}$ still holds. However, the matrix $\mathbf{H}$ is now a circulant matrix:

$$\mathbf{H} = \begin{bmatrix} h_0 & h_{N-1} & h_{N-2} & \cdots & h_1 \\ h_1 & h_0 & h_{N-1} & & h_2 \\ h_2 & h_1 & \ddots & \ddots & \vdots \\ \vdots & & \ddots & \ddots & h_{N-1} \\ h_{N-1} & h_{N-2} & \cdots & h_1 & h_0 \end{bmatrix} \tag{b.14}$$

The *Fourier matrix* $\mathbf{W}$ is a unitary $N \times N$ matrix with elements given by:

$$w_{n,m} = \frac{1}{\sqrt{N}} \exp\left(\frac{-2\pi \mathrm{j} nm}{\mathrm{N}}\right) \quad \text{with:} \quad \mathrm{j} = \sqrt{-1} \tag{b.15}$$

The (row) vectors in this matrix are the complex conjugated of the basisvectors given in (a.24). Note that $\mathbf{W}^* = \mathbf{W}^{-1}$ because $\mathbf{W}$ is unitary. It can be shown that the circulant convolution in (b.13) can be transformed into an element-by-element multiplication provided that the vector $\mathbf{g}$ is represented by the orthonormal basis of (a.24). In this representation the circulant convolution $\mathbf{g} = \mathbf{Hf}$ becomes; see (a.36):

$$\mathbf{W}^*\mathbf{g} = \mathbf{W}^*\mathbf{Hf} \tag{b.16}$$

Writing $\mathbf{W} = [\,\mathbf{w}_0 \quad \mathbf{w}_1 \quad \ldots \quad \mathbf{w}_{N-1}\,]$ and carefully examining $\mathbf{Hw}_k$ reveals that the basisvectors $\mathbf{w}_k$ are the eigenvectors of the circulant matrix $\mathbf{H}$. Therefore, we may write $\mathbf{Hw}_k = \lambda_k \mathbf{w}_k$ with $k = 0, 1, \ldots, N-1$. The numbers λ_k are the eigenvalues of $\mathbf{H}$. If these eigenvalues are arranged in a diagonal matrix:

$$\Lambda = \begin{bmatrix} \lambda_0 & & 0 \\ & \ddots & \\ 0 & & \lambda_{N-1} \end{bmatrix} \tag{b.17}$$

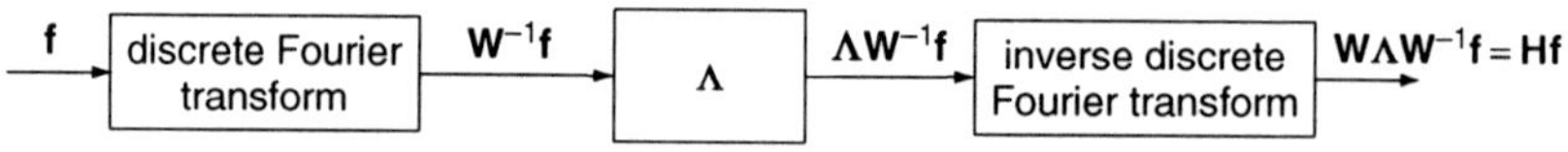

Figure B.1 Discrete circulant convolution accomplished in the Fourier domain

the N equations $\mathbf{Hw}_k = \lambda_k \mathbf{w}_k$ can be written economically as: $\mathbf{HW} = \mathbf{W\Lambda}$. Right-sided multiplication of this equation by $\mathbf{W}^{-1}$ yields: $\mathbf{H} = \mathbf{W\Lambda W}^{-1}$. Substitution in (b.16) gives:

$$\mathbf{W}^*\mathbf{g} = \mathbf{W}^*\mathbf{W\Lambda W}^{-1}\mathbf{f} = \mathbf{\Lambda W}^{-1}\mathbf{f} \tag{b.18}$$

Note that the multiplication Λ with the vector $\mathbf{W}^{-1}\mathbf{f}$ is an element-by-element multiplication because the matrix Λ is diagonal. The final result is obtained if we perform a left-sided multiplication in (b.18) by $\mathbf{W}$:

$$\mathbf{g} = \mathbf{W\Lambda W}^{-1}\mathbf{f} \tag{b.19}$$

The interpretation of this result is depicted in Figure B.1.

B.3 TRACE AND DETERMINANT

The trace $trace(\mathbf{H})$ of a square matrix $\mathbf{H}$ is the sum of its diagonal elements:

$$trace(\mathbf{H}) = \sum_{n=0}^{N-1} h_{n,n} \tag{b.20}$$

Properties related to the trace are ($\mathbf{A}$ and $\mathbf{B}$ are $N \times N$ matrices, $\mathbf{f}$ and $\mathbf{g}$ are N-dimensional vectors):

$$trace(\mathbf{AB}) = trace(\mathbf{BA}) \tag{b.21}$$
$$(\mathbf{f}, \mathbf{g}) = \mathbf{f}^*\mathbf{g} = trace(\mathbf{fg}^*) \tag{b.22}$$

The *determinant* $|\mathbf{H}|$ of a square matrix $\mathbf{H}$ is recursively defined with its co-matrices. The co-matrix $\mathbf{H}_{n,m}$ is an $(N-1) \times (N-1)$ matrix that is derived from $\mathbf{H}$ by exclusion of the n-th row and the m-th column. The following equations define the determinant:

$$\begin{aligned} &\text{If} \quad N = 1: \quad |\mathbf{H}| = h_{0,0} \\ &\text{If} \quad N > 1: \quad |\mathbf{H}| = \sum_{m=0}^{N-1} (-1)^m h_{0,m} |\mathbf{H}_{0,m}| \end{aligned} \tag{b.23}$$

Some properties related to the determinant:

$$|\mathbf{AB}| = |\mathbf{A}||\mathbf{B}| \tag{b.24}$$

$$|\mathbf{A}^{-1}| = \frac{1}{|\mathbf{A}|} \tag{b.25}$$

$$|\mathbf{A}^T| = |\mathbf{A}| \tag{b.26}$$

$$\mathbf{U} \text{ is unitary: } \Rightarrow \quad |\mathbf{U}| = \pm 1 \tag{b.27}$$

$$\Lambda \text{ is diagonal: } \Rightarrow \quad |\Lambda| = \prod_{n=0}^{N-1} \lambda_{n,n} \tag{b.28}$$

If λ_n are the eigenvalues of a square matrix $\mathbf{A}$, then:

$$trace(\mathbf{A}) = \sum_{n=0}^{N-1} \lambda_n \quad \text{and} \quad |\mathbf{A}| = \prod_{n=0}^{N-1} \lambda_n \tag{b.29}$$

The *rank* of a matrix is the maximum number of column vectors (or row vectors) that are linearly independent. The rank of a regular $N \times N$ matrix is always N. In that case, the determinant is always non-zero. The reverse holds true too. The rank of a singular $N \times N$ matrix is always less than N, and the determinant is zero.

B.4 DIFFERENTIATION OF VECTOR AND MATRIX FUNCTIONS

Suppose $f(\mathbf{x})$ is a real or complex function of the real N-dimensional vector $\mathbf{x}$. Then, the first derivative of $f(\mathbf{x})$ with respect to $\mathbf{x}$ is an N-dimensional vector function (the *gradient*):

$$\frac{\partial f(\mathbf{x})}{\partial \mathbf{x}} \quad \text{with elements:} \quad \frac{\partial f(\mathbf{x})}{\partial x_n} \tag{b.30}$$

If $f(\mathbf{x}) = \mathbf{a}^T\mathbf{x}$ (i.e. the inner product between $\mathbf{x}$ and a real vector $\mathbf{a}$), then:

$$\frac{\partial [\mathbf{a}^T\mathbf{x}]}{\partial \mathbf{x}} = \mathbf{a} \tag{b.31}$$

Likewise, if $f(\mathbf{x}) = \mathbf{x}^T\mathbf{H}\mathbf{x}$ (i.e. a quadratic form defined by the matrix $\mathbf{H}$), then:

$$\frac{\partial[\mathbf{x}^T\mathbf{H}\mathbf{x}]}{\partial\mathbf{x}} = 2\mathbf{H}\mathbf{x} \tag{b.32}$$

The second derivative of $f(\mathbf{x})$ with respect to $\mathbf{x}$ is an $N \times N$ matrix called the *Hessian matrix*:

$$\mathbf{H}(\mathbf{x}) = \frac{\partial^2 f(\mathbf{x})}{\partial\mathbf{x}^2} \quad \text{with elements:} \quad h_{n,m}(\mathbf{x}) = \frac{\partial^2 f(\mathbf{x})}{\partial x_n \partial x_m} \tag{b.33}$$

The determinant of this matrix is called the *Hessian*.

The *Jacobian matrix* of an N-dimensional vector function $\mathbf{f}()$: $\mathbb{R}^M \to \mathbb{R}^N$ is defined as an $N \times M$ matrix:

$$\mathbf{H}(\mathbf{x}) = \frac{\partial\mathbf{f}(\mathbf{x})}{\partial\mathbf{x}} \quad \text{with elements:} \quad h_{n,m}(\mathbf{x}) = \frac{\partial f_n(\mathbf{x})}{\partial x_m} \tag{b.34}$$

Its determinant (only defined if the Jacobian matrix is square) is called the *Jacobian*.

The differentiation of a function of a matrix, e.g. $f(\mathbf{H})$, with respect to this matrix is defined similar to the differentiation in (b.30). The result is a matrix:

$$\frac{\partial f(\mathbf{H})}{\partial\mathbf{H}} \quad \text{with elements:} \quad \frac{\partial f(\mathbf{H})}{\partial h_{n,m}} \tag{b.35}$$

Suppose that we have a square, invertible $R \times R$ matrix function $\mathbf{F}(\mathbf{H})$ of an $N \times M$ matrix $\mathbf{H}$, that is $\mathbf{F}()$: $\mathbb{R}^N \times \mathbb{R}^M \to \mathbb{R}^R \times \mathbb{R}^R$, then the derivative of $\mathbf{F}(\mathbf{H})$ with respect to one of the elements of $\mathbf{H}$ is:

$$\frac{\partial}{\partial h_{n,m}}\mathbf{F}(\mathbf{H}) = \begin{bmatrix} \frac{\partial}{\partial h_{n,m}} f_{0,0}(\mathbf{H}) & \cdots & \frac{\partial}{\partial h_{n,m}} f_{0,R-1}(\mathbf{H}) \\ \vdots & & \vdots \\ \frac{\partial}{\partial h_{n,m}} f_{R-1,0}(\mathbf{H}) & \cdots & \frac{\partial}{\partial h_{n,m}} f_{R-1,R-1}(\mathbf{H}) \end{bmatrix} \tag{b.36}$$

From this the following rules can be derived:

$$\frac{\partial}{\partial h_{n,m}}[\mathbf{F}(\mathbf{H}) + \mathbf{G}(\mathbf{H})] = \frac{\partial}{\partial h_{n,m}}\mathbf{F}(\mathbf{H}) + \frac{\partial}{\partial h_{n,m}}\mathbf{G}(\mathbf{H}) \tag{b.37}$$

$$\frac{\partial}{\partial h_{n,m}}[\mathbf{F}(\mathbf{H})\mathbf{G}(\mathbf{H})] = \left[\frac{\partial}{\partial h_{n,m}}\mathbf{F}(\mathbf{H})\right]\mathbf{G}(\mathbf{H}) + \left[\frac{\partial}{\partial h_{n,m}}\mathbf{G}(\mathbf{H})\right]\mathbf{F}(\mathbf{H}) \tag{b.38}$$

$$\frac{\partial}{\partial h_{n,m}}\mathbf{F}^{-1}(\mathbf{H}) = -\mathbf{F}^{-1}(\mathbf{H})\left[\frac{\partial}{\partial h_{n,m}}\mathbf{F}(\mathbf{H})\right]\mathbf{F}^{-1}(\mathbf{H}) \tag{b.39}$$

Suppose that $\mathbf{A}$, $\mathbf{B}$ and $\mathbf{C}$ are square matrices of equal size. Then, some properties related to the derivatives of the trace and the determinant are:

$$\frac{\partial}{\partial \mathbf{A}}\mathit{trace}(\mathbf{A}) = \mathbf{I} \tag{b.40}$$

$$\frac{\partial}{\partial \mathbf{A}}\mathit{trace}(\mathbf{BAC}) = \mathbf{B}^T\mathbf{C}^T \tag{b.41}$$

$$\frac{\partial}{\partial \mathbf{A}}\mathit{trace}(\mathbf{ABA}^T) = \mathbf{A}(\mathbf{B} + \mathbf{B}^T) \tag{b.42}$$

$$\frac{\partial}{\partial \mathbf{A}}|\mathbf{BAC}| = |\mathbf{BAC}|(\mathbf{A}^{-1})^T \tag{b.43}$$

$$\frac{\partial}{\partial a_{n,m}}|\mathbf{A}| = |\mathbf{A}|[\mathbf{A}^{-1}]_{m,n} \tag{b.44}$$

In (b.44), $[\mathbf{A}^{-1}]_{m,n}$ is the m, n-th element of $\mathbf{A}^{-1}$.

B.5 DIAGONALIZATION OF SELF-ADJOINT MATRICES

Recall from Section B.1 that a $N \times N$ matrix $\mathbf{H}$ is called self-adjoint or Hermitian if $\mathbf{H}^* = \mathbf{H}$. From the discussion on self-adjoint operators in Section A.4 it is clear that associated with $\mathbf{H}$, there exists an orthonormal basis $V = \{\mathbf{v}_0 \ \ \mathbf{v}_1 \ \ \cdots \ \ \mathbf{v}_{N-1}\}$ which we now arrange in a unitary matrix $\mathbf{V} = [\mathbf{v}_0 \ \ \mathbf{v}_1 \ \ \cdots \ \ \mathbf{v}_{N-1}]$. Each vector $\mathbf{v}_k$ is an eigenvector with corresponding (real) eigenvalue λ_k. These eigenvalues are now arranged as the diagonal elements in a diagonal matrix Λ.

The operation **Hf** can be written as; see (a.42) and (b.5):

$$\mathbf{Hf} = \sum_{n=0}^{N-1} \lambda_n (\mathbf{v}_n, \mathbf{f}) \mathbf{v}_n = \sum_{n=0}^{N-1} \lambda_n \mathbf{v}_n \mathbf{f}^* \mathbf{v}_n = \sum_{n=0}^{N-1} \lambda_n \mathbf{v}_n \mathbf{v}_n^* \mathbf{f} \tag{b.45}$$

Suppose that the rank of **H** equals R. Then there are exactly R non-zero eigenvalues. Consequently, the number of terms in (b.45) can be replaced with R. From this, it follows that **H** is a composition of its eigenvectors according to:

$$\mathbf{H} = \sum_{k=0}^{R-1} \lambda_k \mathbf{v}_k \mathbf{v}_k^* \tag{b.46}$$

The summation on the right-hand side can be written more economically as: $\mathbf{V\Lambda V}^*$. Therefore:

$$\mathbf{H} = \mathbf{V\Lambda V}^* \tag{b.47}$$

The unitary matrix $\mathbf{V}^*$ transforms the domain of **H** such that **H** becomes the diagonal matrix Λ in this new domain. The matrix **V** accomplishes the inverse transform. In fact, (b.47) is the matrix version of the decomposition shown in Figure A.2.

If the rank R equals N, there are exactly N non-zero eigenvalues. In that case, the matrix Λ is invertible, and so is **H**:

$$\mathbf{H}^{-1} = \mathbf{V\Lambda}^{-1}\mathbf{V}^* = \sum_{n=0}^{N-1} \frac{\mathbf{v}_n \mathbf{v}_n^*}{\lambda_n} \tag{b.48}$$

It can be seen that the inverse $\mathbf{H}^{-1}$ is also self-adjoint.

A self-adjoint matrix **H** is *positive definite* if its eigenvalues are all positive. In that case the expression $\rho(\mathbf{f},\mathbf{g}) = \sqrt{(\mathbf{f}-\mathbf{g})^*\mathbf{H}(\mathbf{f}-\mathbf{g})}$ satisfies the conditions of a distance measure (Section A.2). To show this it suffices to prove that $\sqrt{\mathbf{f}^*\mathbf{Hf}}$ satisfies the conditions of a norm; see Section A.2. These conditions are given in Section A.1. We use the diagonal form of **H**; see (b.47):

$$\mathbf{fHf}^* = \mathbf{f}^*\mathbf{V\Lambda V}^*\mathbf{f} \tag{b.49}$$

Since **V** is a unitary matrix, the vector $\mathbf{V}^*\mathbf{f}$ can be regarded as the representation of **f** with respect to the orthonormal basis defined by the vectors in V. Let this representation be denoted by: $\boldsymbol{\phi} = \mathbf{V}^*\mathbf{f}$. The expression $\mathbf{f}^*\mathbf{H}\mathbf{f}$ equals:

$$\mathbf{f}^*\mathbf{H}\mathbf{f} = \boldsymbol{\phi}^*\Lambda\boldsymbol{\phi} = \sum_{n=0}^{N-1} \lambda_n |\phi_n|^2 \tag{b.50}$$

Written in this form, it is easy to show that all conditions of a norm are met.

With the norm $\sqrt{\mathbf{f}^*\mathbf{H}\mathbf{f}}$ the sets of points equidistant to the origin, i.e. the vectors that satisfy $\mathbf{f}^*\mathbf{H}\mathbf{f} = \text{constant}$, are ellipsoids. See Figure B.2. This follows from (b.50):

$$\mathbf{f}^*\mathbf{H}\mathbf{f} = \text{constant} \quad \Leftrightarrow \quad \sum_{n=0}^{N-1} \lambda_n |\phi_n|^2 = \text{constant}$$

Hence, if we introduce a new vector $\mathbf{u}$ with elements defined as: $u_n = \phi_n/\sqrt{\lambda_n}$, we must require that:

$$\sum_{n=0}^{N-1} |u_n|^2 = \text{constant}$$

We conclude that in the $\mathbf{u}$ domain the ordinary Euclidean norm applies. Therefore, the solution space in this domain is a sphere (Figure B.2(a)). The operation $u_n = \phi_n/\sqrt{\lambda_n}$ is merely a scaling of the axes by factors $\sqrt{\lambda_n}$. This transforms the sphere into an ellipsoid. The principal axes of this ellipsoid line up with the basisvectors $\mathbf{u}$ (and $\boldsymbol{\phi}$), see Figure B.2(b). Finally, the unitary transform $\mathbf{f} = \mathbf{V}\boldsymbol{\phi}$ rotates the principal axes, but without affecting the shape of the ellipsoid (Figure B.2(c)). Hence, the orientation of these axes in the $\mathbf{f}$ domain is along the directions of the eigenvectors $\mathbf{v}_n$ of $\mathbf{H}$.

The metric expressed by $\sqrt{(\mathbf{f}-\mathbf{g})^*\mathbf{H}(\mathbf{f}-\mathbf{g})}$ is called is a *quadratic distance*. In pattern classification theory it is usually called the *Mahalanobis distance*.

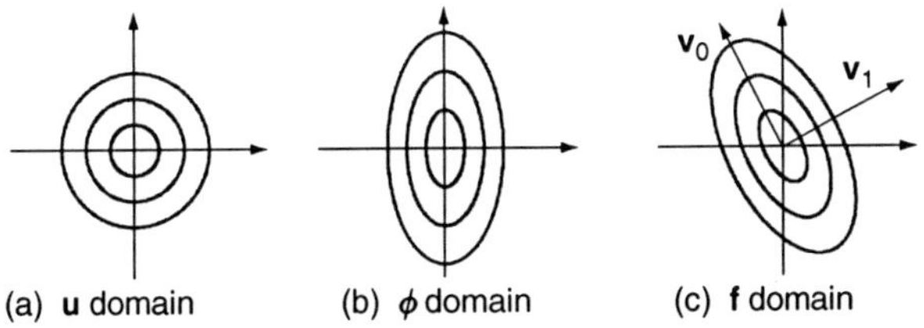

Figure B.2 'Circles' in $\mathbb{R}^2$ equipped with the Mahalanobis distance

B.6 SINGULAR VALUE DECOMPOSITION (SVD)

The *singular value decomposition* theorem states that an arbitrary $N \times M$ matrix $\mathbf{H}$ can be decomposed into the product:

$$\mathbf{H} = \mathbf{U}\Sigma\mathbf{V}^T \tag{b.51}$$

where $\mathbf{U}$ is an orthonormal $N \times R$ matrix, Σ is a diagonal $R \times R$ matrix, and $\mathbf{V}$ is an orthonormal $M \times R$ matrix. R is the rank of the matrix $\mathbf{H}$. Therefore, $R \leq \min(N,M)$.

The proof of the theorem is beyond the scope of this book. However, its connotation can be clarified as follows. Suppose for a moment, that $R = N = M$. We consider all pairs of vectors of the form $\mathbf{y} = \mathbf{Hx}$ where $\mathbf{x}$ is on the surface of the unit sphere in $\mathbb{R}^M$, i.e. $\|\mathbf{x}\| = 1$. The corresponding $\mathbf{y}$ must be on the surface in $\mathbb{R}^N$ defined by the equation $\mathbf{y}^T\mathbf{y} = \mathbf{x}^T\mathbf{H}^T\mathbf{Hx}$. The matrix $\mathbf{H}^T\mathbf{H}$ is a symmetric $M \times M$ matrix. By virtue of (b.47) there must be an orthonormal matrix $\mathbf{V}$ and a diagonal matrix $\mathbf{S}$ such that:[1]

$$\mathbf{H}^T\mathbf{H} = \mathbf{V}\mathbf{S}\mathbf{V}^T \tag{b.52}$$

The matrix $\mathbf{V} = [\,\mathbf{v}_0 \quad \cdots \quad \mathbf{v}_{M-1}\,]$ contains the (unit) eigenvectors $\mathbf{v}_i$ of $\mathbf{H}^T\mathbf{H}$. The corresponding eigenvalues are all on the diagonal of the matrix $\mathbf{S}$. Without loss of generality we may assume that they are sorted in descending order. Thus, $S_{i,i} \geq S_{i+1,i+1}$.

With $\boldsymbol{\xi} \overset{def}{=} \mathbf{V}^T\mathbf{x}$, the solutions of the equation $\mathbf{y}^T\mathbf{y} = \mathbf{x}^T\mathbf{H}^T\mathbf{Hx}$ with $\|\mathbf{x}\| = 1$ is the same set as the solutions of $\mathbf{y}^T\mathbf{y} = \boldsymbol{\xi}^T\mathbf{S}\boldsymbol{\xi}$. Clearly, if $\mathbf{x}$ is a unit vector, then so is $\boldsymbol{\xi}$ because $\mathbf{V}$ is orthonormal. Therefore, the solutions of $\mathbf{y}^T\mathbf{y} = \boldsymbol{\xi}^T\mathbf{S}\boldsymbol{\xi}$ is the set:

$$\{\mathbf{y}\} \quad \text{with} \quad \mathbf{y} = \Sigma\boldsymbol{\xi} \quad \text{and} \quad \|\boldsymbol{\xi}\| = 1$$

where:

$$\Sigma = \mathbf{S}^{\frac{1}{2}} \tag{b.53}$$

Σ is the matrix that is obtained by taking the square roots of all (diagonal) elements in $\mathbf{S}$. The diagonal elements of Σ are usually denoted

---

[1] We silently assume that $\mathbf{H}$ is a real matrix so that $\mathbf{V}$ is also real, and $\mathbf{V}^* = \mathbf{V}^T$.

by the *singular values* $\sigma_i = \Sigma_{i,i} = +\sqrt{S_{i,i}}$. With that, the solutions become a set:

$$\{\mathbf{y}\} \quad \text{with} \quad \mathbf{y} = \sum_{i=0}^{N-1} \sigma_i \xi_i \mathbf{v}_i \quad \text{and} \quad \sum_{i=0}^{N-1} \xi_i^2 = 1 \tag{b.54}$$

Equation (b.54) reveals that the solutions of $\mathbf{y}^T\mathbf{y} = \mathbf{x}^T\mathbf{H}^T\mathbf{H}\mathbf{x}$ with $\|\mathbf{x}\| = 1$ are formed by a scaled version of the sphere $\|\mathbf{x}\| = 1$. Scaling takes place with a factor σ_i in the direction of $\mathbf{v}_i$. In other words, the solutions form an ellipsoid.

Each eigenvector $\mathbf{v}_i$ gives rise to a principal axis of the ellipsoid. The direction of such an axis is the one of the vector $\mathbf{H}\mathbf{v}_i$. We form the matrix $\mathbf{HV} = [\,\mathbf{H}\mathbf{v}_0 \quad \ldots \quad \mathbf{H}\mathbf{v}_{M-1}\,]$ which contains the vectors that point in the directions of all principal axes. Using (b.52):

$$\mathbf{H}^T\mathbf{H}\mathbf{V} = \mathbf{V}\mathbf{S} \quad \Rightarrow \quad \mathbf{H}\mathbf{H}^T\mathbf{H}\mathbf{V} = \mathbf{H}\mathbf{V}\mathbf{S}$$

Consequently, the column vectors in the matrix $\mathbf{U}$ that fulfils

$$\mathbf{H}\mathbf{H}^T\mathbf{U} = \mathbf{U}\mathbf{S} \tag{b.55}$$

successively point in the directions of the principal axes. Since S is a diagonal matrix, we conclude from (b.47) that $\mathbf{U}$ must contain the eigenvectors $\mathbf{u}_i$ of the matrix $\mathbf{H}\mathbf{H}^T$, i.e. $\mathbf{U} = [\,\mathbf{u}_0 \quad \cdots \quad \mathbf{u}_{M-1}\,]$. The diagonal elements of S are the corresponding eigenvalues. U is an orthonormal matrix.

The operator $\mathbf{H}$ maps the vector $\mathbf{v}_i$ to a vector $\mathbf{H}\mathbf{v}_i$ whose direction is given by the unit vector $\mathbf{u}_i$, and whose length is given by σ_i. Therefore, $\mathbf{H}\mathbf{v}_i = \sigma_i\mathbf{u}_i$. Representing an arbitrary vector $\mathbf{x}$ as a linear combination of $\mathbf{v}_i$, i.e. $\boldsymbol{\xi} = \mathbf{V}^T\mathbf{x}$ gives us finally the SVD theorem stated in (b.51).

In the general case, N can differ from M and $R \leq \min(N, M)$. However, (b.51), (b.52), (b.54) and (b.55) are still valid, except that the dimensions of the matrices must be adapted. If $R < M$, then the singular values from σ_R up to σ_{M-1} are zero. These singular values and the associated eigenvectors are discarded then. The ellipsoid mentioned above becomes an R-dimensional ellipsoid that is embedded in $\mathbb{R}^N$.

The process is depicted graphically in Figure B.3. The operator $\boldsymbol{\xi} = \mathbf{V}^T\mathbf{x}$ aligns the operand to the suitable basis $\{\mathbf{v}_i\}$ by means of a rotation. This operator also discards all components $\mathbf{v}_R$ up to $\mathbf{v}_{M-1}$ if $R < M$. The operator $\boldsymbol{\mu} = \Sigma\boldsymbol{\xi}$ stretches the axes formed by the $\{\mathbf{v}_i\}$ basis.

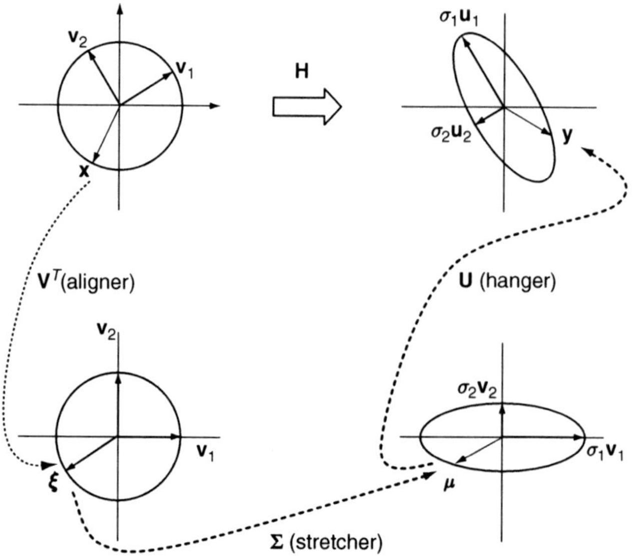

Figure B.3 Singular value decomposition of a matrix **H**

Finally, the operator $\mathbf{y} = \mathbf{U}\boldsymbol{\mu}$ hangs out each $\mathbf{v}_i$ component in $\mathbb{R}^M$ as a $\mathbf{u}_i$ component in $\mathbb{R}^N$.

An important application of the SVD is *matrix approximation*. For that purpose, (b.51) is written in an equivalent form:

$$\mathbf{H} = \sum_{i=0}^{R-1} \sigma_i \mathbf{u}_i \mathbf{v}_i^T \tag{b.56}$$

We recall that the sequence $\sigma_0, \sigma_1, \cdots$ is in descending order. Also, $\sigma_i > 0$. Therefore, if the matrix must be approximated by less then R terms, then the best strategy is to nullify the last terms. For instance, if the matrix $\mathbf{H}$ must be approximated by a vector product, i.e. $\mathbf{H} = \mathbf{f}\mathbf{g}^T$, then the best approximation is obtained by nullifying all singular values except σ_0. That is $\mathbf{H} \approx \sigma_0 \mathbf{u}_0 \mathbf{v}_0^T$.

The SVD theorem is useful to find the pseudo inverse of a matrix:

$$\mathbf{H}^+ = \mathbf{V}\Sigma^{-1}\mathbf{U}^T = \sum_{i=0}^{R-1} \frac{1}{\sigma_i} \mathbf{v}_i \mathbf{u}_i^T \tag{b.57}$$

The pseudo inverse gives the least squares solution to the equation $\mathbf{y} = \mathbf{H}\mathbf{x}$. In other words, the solution $\hat{\mathbf{x}} = \mathbf{H}^+\mathbf{y}$ is the one that minimizes

$||\mathbf{Hx} - \mathbf{y}||_2^2$. If the system is underdetermined, $R < M$, it provides a minimum norm solution. That is, $\hat{\mathbf{x}}$ is the smallest solution of $\mathbf{y} = \mathbf{Hx}$.

The SVD is also a good departure point for a stability analysis of the pseudo inverse, for instance by studying the ratio between the largest and smallest singular values. This ratio is a measure of the sensitivity of the inverse to noise in the data. If this ratio is too high, we may consider regularizing the pseudo inverse by discarding all terms in (b.57) whose singular values are too small.

Besides matrix approximation and pseudo inverse calculation the SVD finds application in image restoration, image compression and image modelling.

B.7 REFERENCES

Bellman, R.E., *Introduction to Matrix Analysis*, McGraw-Hill, New York, 1970.

Strang, G., *Linear Algebra and its Applications*, Harcourt Brace Jovanovich, San Diego, CA, 3rd edition, 1989.

Appendix C
Probability Theory

This appendix summarizes concepts from probability theory. This summary only concerns those concepts that are part of the mathematical background required for understanding this book. Mathematical peculiarities which are not relevant here are omitted. At the end of the appendix references to detailed treatments are given.

C.1 PROBABILITY THEORY AND RANDOM VARIABLES

The axiomatic development of *probability* involves the definitions of three concepts. Taken together these concepts are called an *experiment*. The three concepts are:

(a) A set Ω consisting of *outcomes* ω_i. A *trial* is the act of randomly drawing a single outcome. Hence, each trial produces one $\omega \in \Omega$.
(b) A is a set of certain[1] subsets of Ω.
Each subset $\alpha \in A$ is called an *event*. The event $\{\omega_i\}$, which consists of a single outcome, is called an *elementary event*. The

[1] This set of subsets must comply with some rules. In fact, A must be a so-called *Borel set*. But this topic is beyond the scope of this book.

Classification, Parameter Estimation and State Estimation: An Engineering Approach using MATLAB
F. van der Heijden, R.P.W. Duin, D. de Ridder and D.M.J. Tax
 ISBN: 0-470-09013-8

set Ω is called the *certain event*. The empty subset $\emptyset$ is called the *impossible event*. We say that an event α occurred if the outcome ω of a trial is contained in α, i.e. if $\omega \in \alpha$.

(c) A real function $P(\alpha)$ is defined on A. This function, called *probability*, satisfies the following axioms:
   I: $P(\alpha) \geq 0$
   II: $P(\Omega) = 1$
   III: If $\alpha, \beta \in A$ and $\alpha \cap \beta = \emptyset$ then $P(\alpha \cup \beta) = P(\alpha) + P(\beta)$

Example

The space of outcomes corresponding to the colours of a traffic-light is: $\Omega = \{\text{red, green, yellow}\}$. The set A may consist of subsets like: $\emptyset$, red, green, yellow, red $\cup$ green, red $\cap$ green, red $\cup$ green $\cup$ yellow, ... With that, P(green) is the probability that the light will be green. P(green $\cup$ yellow) is the probability that the light will be green or yellow or both. P(green $\cap$ yellow) is the probability that at the same time the light is green and yellow.

A *random variable* $\underline{x}(\omega)$ is a mapping of Ω onto a set of numbers, for instance: integer numbers, real numbers, complex numbers, etc. The *distribution function* $F_{\underline{x}}(x)$ is the probability of the event that corresponds to $\underline{x} \leq x$:

$$F_{\underline{x}}(x) = P(\underline{x} \leq x) \tag{c.1}$$

A note on the notation

In the notation $F_{\underline{x}}(x)$ the variable $\underline{x}$ is the random variable of interest. The variable x is the independent variable. It can be replaced by other independent variables or constants. So, $F_{\underline{x}}(s)$, $F_{\underline{x}}(x^2)$ and $F_{\underline{x}}(0)$ are all valid notations. However, to avoid lengthy notations the abbreviation $F(x)$ will often be used if it is clear from the context that $F_{\underline{x}}(x)$ is meant. Also, the underscore notation for a random variable will be omitted frequently.

The random variable $\underline{x}$ is said to be *discrete* if a finite number (or infinite countable number) of events $x_1, x_2, \ldots$ exists for which:

$$P(\underline{x} = x_i) > 0 \quad \text{and} \quad \sum_{\text{all } i} P(\underline{x} = x_i) = 1 \tag{c.2}$$

Notation
If the context is clear, the notation $P(x_i)$ will be used to denote $P(\underline{x} = x_i)$.

The random variable $\underline{x}$ is *continuous* if a function $p_{\underline{x}}(x)$ exists for which:

$$F_{\underline{x}}(x) = \int_{\xi=-\infty}^{x} p_{\underline{x}}(\xi)d\xi \qquad \text{(c.3)}$$

The function $p_{\underline{x}}(x)$ is called the *probability density* of $\underline{x}$. The discrete case can be included in the continuous case by permitting $p_{\underline{x}}(x)$ to contain Dirac functions of the type $P_i\delta(x - x_i)$.
Notation
If the context is clear, the notation $p(x)$ will be used to denote $p_{\underline{x}}(x)$.

Examples
We consider the experiment consisting of tossing a (fair) coin. The possible outcomes are {head, tail}. The random variable $\underline{x}$ is defined according to:

$$\text{head} \rightarrow \underline{x} = 0$$
$$\text{tail} \rightarrow \underline{x} = 1$$

This random variable is discrete. Its distribution function and probabilities are depicted in Figure C.1(a).

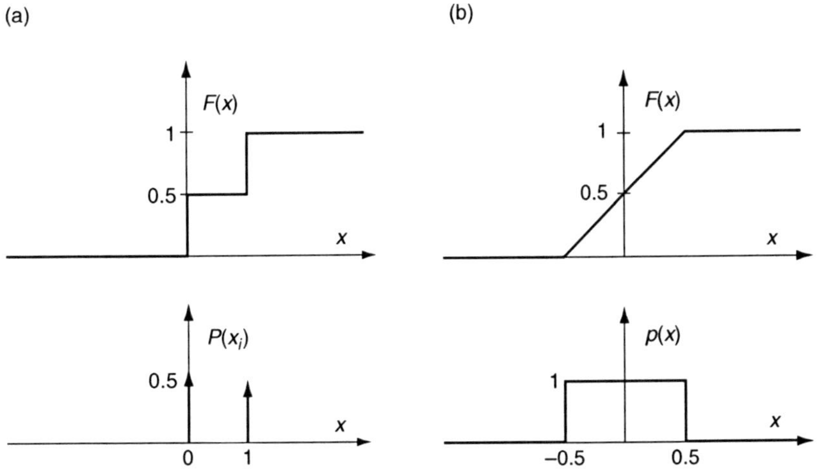

Figure C.1 (a) Discrete distribution function with probabilities. (b) Continuous distribution function with a uniform probability density

An example of a continuous random variable is the round-off error which occurs when a random real number is replaced with its nearest integer number. This error is uniformly distributed between −0.5 and 0.5. The distribution function and associated probability density are shown in Figure C.1(b).

Since $F(x)$ is a non-decreasing function of x, and $F(\infty) = 1$, the density must be a non-negative function with $\int p(x)dx = 1$. Here, the integral extends over the real numbers from $-\infty$ to ∞.

C.1.1 Moments

The *moment of order n* of a random variable is defined as:

$$\mathrm{E}[\mathrm{x}^{\mathrm{n}}] = \int_{x=-\infty}^{\infty} x^n p(x)dx \qquad \text{(c.4)}$$

Formally, the notation should be $\mathrm{E}[\underline{x}^n]$, but as said before we omit the underscore if there is no fear of ambiguity. Another notation of $\mathrm{E}[x^n]$ is $\overline{x^n}$.

The first order moment is called the *expectation*. This quantity is often denoted by μ_x or (if confusion is not to be expected) briefly μ. The *central moments of order n* are defined as:

$$\mathrm{E}[(x-\mu)^n] = \int_{x=-\infty}^{\infty} (x-\mu)^n p(x)dx \qquad \text{(c.5)}$$

The first central moment is always zero. The second central moment is called *variance*:

$$\mathrm{Var}[x] = \mathrm{E}[(x-\mu)^2] = \mathrm{E}[x^2] - \mu^2 \qquad \text{(c.6)}$$

The (*standard*) *deviation* of x denoted by σ_x, or briefly σ, is the square root of the variance:

$$\sigma_x = \sqrt{\mathrm{Var}[x]} \qquad \text{(c.7)}$$

C.1.2 Poisson distribution

The act of counting the number of times that a certain random event takes place during a given time interval often involves a *Poisson distribution*. A discrete random variable $\underline{n}$ which obeys the Poisson distribution has the probability function:

$$P(\underline{n} = n) = \frac{\lambda^n \exp(-\lambda)}{n!} \tag{c.8}$$

λ is a parameter of the distribution. It is the expected number of events. Thus, $\mathrm{E}[\underline{n}] = \lambda$. A special feature of the Poisson distribution is that the variance and the expectation are equal: $\mathrm{Var}[\underline{n}] = \mathrm{E}[\underline{n}] = \lambda$. Examples of this distribution are shown in Figure C.2.

Example

Radiant energy is carried by a discrete number of photons. In daylight situations, the average number of photons per unit area is in the order of 10^{12} $(1/(\mathrm{s} \cdot \mathrm{mm}^2))$. However, in fact the real number is Poisson distributed. Therefore, the relative deviation $\sigma_n/\mathrm{E}[\underline{n}]$ is $1/\sqrt{\lambda}$. An image sensor with an integrating area of 100 ($\mu\mathrm{m}^2$), an integration time of 25 (ms) and an illuminance of 250 (lx) receives about $\lambda = 10^6$ photons. Hence, the relative deviation is about 0.1%. In most imaging sensors this deviation is almost negligible compared to other noise sources.

C.1.3 Binomial distribution

Suppose that we have an experiment with only two outcomes, ω_1 and ω_2. The probability of the elementary event $\{\omega_1\}$ is denoted by P. Consequently, the probability of $\{\omega_2\}$ is $1 - P$. We repeat the experiment

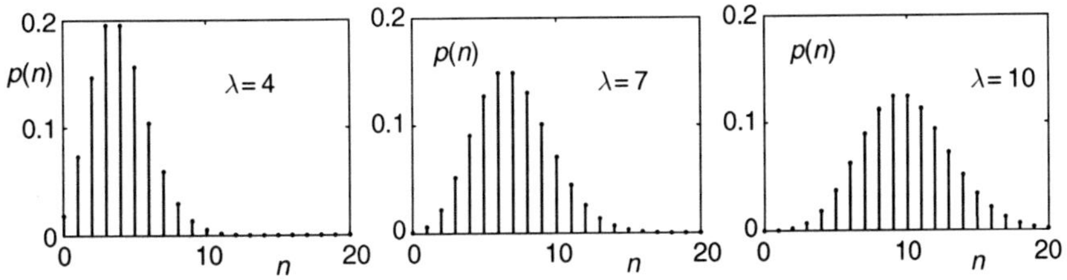

Figure C.2 Poisson distributions

N times, and form a random variable $\underline{n}$ by counting the number of times that $\{\omega_1\}$ occurred. This random variable is said to have a *binomial distribution* with parameters N and P.

The probability function of $\underline{n}$ is:

$$P(\underline{n} = n) = \frac{N!}{n!(N-n)!} P^n (1-P)^{N-n} \tag{c.9}$$

The expectation of $\underline{n}$ appears to be $\mathrm{E}[\underline{n}] = NP$ and its variance is $\mathrm{Var}[\underline{n}] = NP(1-P)$.

Example

The error rate of a classifier is E. The classifier is applied to N objects. The number of misclassified objects, n_{error}, has a binomial distribution with parameters N and E.

C.1.4 Normal distribution

A well-known example of a continuous random variable is the one with a *Gaussian* (or *normal*) probability density:

$$p(x) = \frac{1}{\sigma\sqrt{2\pi}} \exp\left(\frac{-(x-\mu)^2}{2\sigma^2}\right) \tag{c.10}$$

The parameters μ and σ^2 are the expectation and the variance, respectively. Two examples of the probability density with different μ and σ^2 are shown in Figure C.3.

Gaussian random variables occur whenever the underlying process is caused by the outcomes of many independent experiments and the associated random variables add up linearly (the *central limit theorem*). An example is thermal noise in an electrical current. The current is proportional to the sum of the velocities of the individual electrons. Another example is the Poisson distributed random variable mentioned above. The envelope of the Poisson distribution approximates the Gaussian distribution as λ tends to infinity. As illustrated in Figure C.2, the approximation looks quite reasonable already when $\lambda > 10$.

Also, the binomial distributed random variable is the result of an addition of many independent outcomes. Therefore, the envelope of

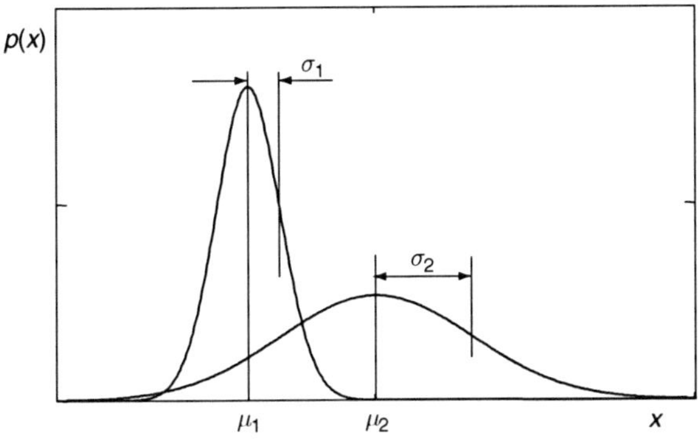

Figure C.3 Gaussian probability densities

the binomial distribution also approximates the normal distribution as $NP(1-P)$ tends to infinity. In practice, the approximation is already reasonably good if $NP(1-P) > 5$.

C.1.5 The Chi-square distribution

Another example of a continuous distribution is the χ_n^2 distribution. A random variable y is said to be χ_n^2 distributed (Chi-square distributed with n degrees of freedom) if its density function equals:

$$p(y) = \begin{cases} \dfrac{1}{2^{\frac{n}{2}}\Gamma\left(\dfrac{n}{2}\right)} y^{\frac{n}{2}-1} e^{-\frac{y}{2}} & y \geq 0 \\ 0 & \text{elsewhere} \end{cases} \tag{c.11}$$

with $\Gamma()$ the so-called *gamma function*. The expectation and variance of a Chi-square distributed random variable appear to be:

$$\begin{aligned} \mathrm{E}[y] &= n \\ \mathrm{Var}[y] &= 2n \end{aligned} \tag{c.12}$$

Figure C.4 shows some examples of the density and distribution.

The χ_n^2 distribution arises if we construct the sum of the square of normally distributed random variables. Suppose that $\underline{x}_j$ with $j = 1, \ldots, n$ are n Gaussian random variables with $\mathrm{E}[\underline{x}_j] = 0$ and $\mathrm{Var}[\underline{x}_j] = 1$.

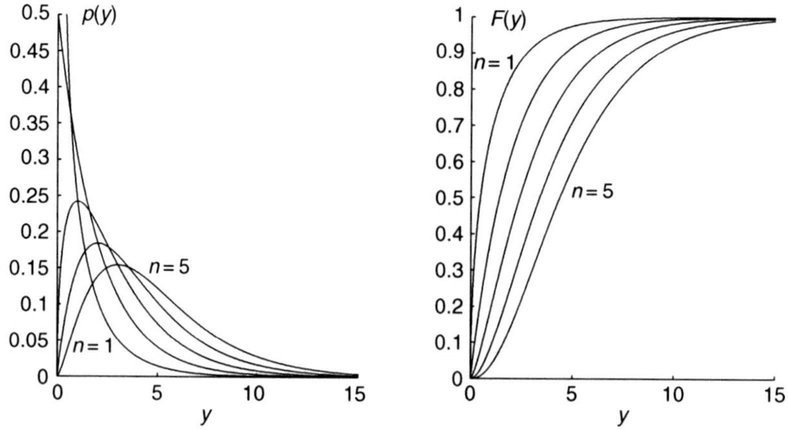

Figure C.4 Chi-square densities and (cumulative) distributions. The degrees of freedom, n, varies from 1 up to 5

In addition, assume that these random variables are mutually independent,[2] then the random variable:

$$\underline{y} = \sum_{j=1}^{n} \underline{x}_j^2 \qquad \text{(c.13)}$$

is χ_n^2 distributed.

C.2 BIVARIATE RANDOM VARIABLES

In this section we consider an experiment in which two random variables $\underline{x}$ and $\underline{y}$ are associated. The *joint distribution function* is the probability that $\underline{x} \leq x$ and $\underline{y} \leq y$, i.e.

$$F(x, y) = P(\underline{x} \leq x, \underline{y} \leq y) \qquad \text{(c.14)}$$

The function $p(x,y)$ for which:

$$F(x, y) = \int_{\xi=-\infty}^{x} \int_{\eta=-\infty}^{y} p(\xi, \eta) d\eta d\xi \qquad \text{(c.15)}$$

[2] For the definition of 'independent': see Section C.2.

is called the *joint probability density*. Strictly speaking, definition (c.15) holds true only when $F(x,y)$ is continuous. However, by permitting $p(x,y)$ to contain Dirac functions, the definition also applies to the discrete case.

From definitions (c.14) and (c.15) it is clear that the *marginal distribution* $F(x)$ and the *marginal density* $p(x)$ are given by:

$$F(x) = F(x, \infty) \tag{c.16}$$

$$p(x) = \int_{y=-\infty}^{\infty} p(x, y)dy \tag{c.17}$$

Two random variables $\underline{x}$ and $\underline{y}$ are *independent* if:

$$F_{\underline{x},\underline{y}}(x, y) = F_{\underline{x}}(x)F_{\underline{y}}(y) \tag{c.18}$$

This is equivalent to:

$$p_{\underline{x},\underline{y}}(x, y) = p_{\underline{x}}(x)p_{\underline{y}}(y) \tag{c.19}$$

Suppose that $h(\cdot, \cdot)$ is a function $\mathbb{R} \times \mathbb{R} \to \mathbb{R}$. Then $h(\underline{x},\underline{y})$ is a random variable. The expectation of $h(\underline{x},\underline{y})$ equals:

$$\mathrm{E}[h(\underline{x}, \underline{y})] = \int_{x=-\infty}^{\infty} \int_{y=-\infty}^{\infty} h(x, y)p(x, y)dydx \tag{c.20}$$

The *joint moments* m_{ij} of two random variables $\underline{x}$ and $\underline{y}$ are defined as the expectations of the functions $\underline{x}^i\underline{y}^j$:

$$m_{ij} = \mathrm{E}[x^i y^j] \tag{c.21}$$

The quantity $i + j$ is called the *order* of m_{ij}. It can easily be verified that: $m_{00} = 1$, $m_{10} = \mathrm{E}[x] = \mu_x$ and $m_{01} = \mathrm{E}[y] = \mu_y$.

The *joint central moments* μ_{ij} of order $i + j$ are defined as:

$$\mu_{ij} = \mathrm{E}\left[(x - \mu_x)^i (y - \mu_y)^j\right] \tag{c.22}$$

Clearly, $\mu_{20} = \text{Var}[x]$ and $\mu_{02} = \text{Var}[y]$. Furthermore, the parameter μ_{11} is called the *covariance* (sometimes denoted by Cov[x,y]). This parameter can be written as: $\mu_{11} = m_{11} - m_{10}m_{01}$. Two random variables are called *uncorrelated* if their covariance is zero. Two independent random variables are always uncorrelated. The reverse is not necessarily true.

Two random variables $\underline{x}$ and $\underline{y}$ are Gaussian if their joint probability density is:

$$p(x,y) = \frac{1}{2\pi\sigma_x\sigma_y\sqrt{1-r^2}} \exp\left(\frac{-1}{2(1-r^2)}\left(\frac{(x-\mu_x)^2}{\sigma_x^2} - \frac{2r(x-\mu_x)(y-\mu_y)}{\sigma_x\sigma_y} + \frac{(y-\mu_y)^2}{\sigma_y^2}\right)\right) \tag{c.23}$$

The parameters μ_x and μ_y are the expectations of $\underline{x}$ and $\underline{y}$, respectively. The parameters σ_x and σ_y are the standard deviations. The parameter r is called the *correlation coefficient*, defined as:

$$r = \frac{\mu_{11}}{\sqrt{\mu_{20}\mu_{02}}} = \frac{\text{Cov}[x,y]}{\sigma_x\sigma_y} \tag{c.24}$$

Figure C.5 shows a scatter diagram with 121 realizations of two Gaussian random variables. In this figure, a geometrical interpretation of (c.23) is also given. The set of points (x,y) which satisfy:

$$p(x,y) = p(\mu_x,\mu_y)\exp\left(-\frac{1}{2}\right)$$

(i.e. the 1σ-level contour) turns out to form an ellipse. The centre of this ellipse coincides with the expectation. The eccentricity, size and orientation of the 1σ contour describe the scattering of the samples around this centre. $\sqrt{\lambda_0}$ and $\sqrt{\lambda_1}$ are the standard deviations of the samples projected on the principal axes of the ellipse. The angle between the principal axis associated with λ_0 and the x axis is θ. With these conventions, the variances of $\underline{x}$ and $\underline{y}$, and the correlation coefficient r, are:

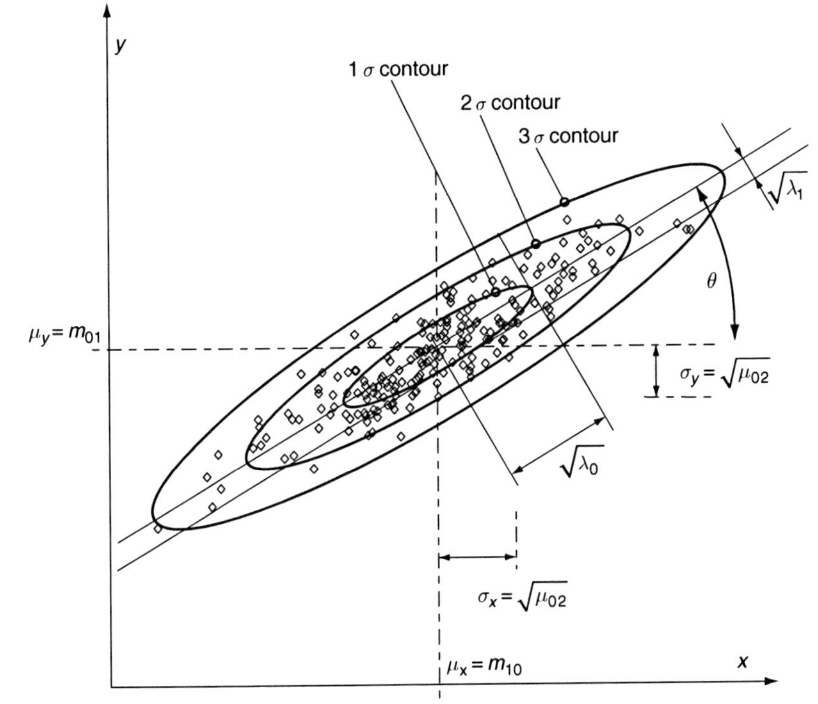

Figure C.5 Scatter diagram of two Gaussian random variables

$$
\begin{aligned}
\sigma_x^2 &= \lambda_0 \cos^2\theta + \lambda_1 \sin^2\theta \\
\sigma_y^2 &= \lambda_1 \cos^2\theta + \lambda_0 \sin^2\theta \\
r &= \frac{(\lambda_0 - \lambda_1)\sin\theta\cos\theta}{\sigma_x\sigma_y}
\end{aligned}
\qquad \text{(c.25)}
$$

Consequently, $r = 0$ whenever $\lambda_0 = \lambda_1$ (the ellipse degenerates into a circle), or whenever θ is a multiple of $\pi/2$ (the ellipse lines up with the axes). In both cases the random variables are independent. The conclusion is that two Gaussian random variables which are *uncorrelated* are also *independent*.

The situation $r = 1$ or $r = -1$ occurs only if $\lambda_0 = 0$ or $\lambda_1 = 0$. The ellipse degenerates into a straight line. Therefore, if two Gaussian random variables are completely *correlated*, then this implies that two constants a and b can be found for which $\underline{y} = a\underline{x} + b$.

The expectation and variance of the random variable defined by $\underline{z} = a\underline{x} + b\underline{y}$ are:

$$\mathrm{E}[z] = a\mathrm{E}[x] + b\mathrm{E}[y] \tag{c.26}$$

$$\sigma_z^2 = a^2\sigma_x^2 + b^2\sigma_y^2 + 2ab\ \mathrm{Cov}[x, y] \tag{c.27}$$

The expectation appears to be a linear operation, regardless of the joint distribution function. The variance is not a linear operation. However, (c.27) shows that if two random variables are uncorrelated, then $\sigma_{x+y}^2 = \sigma_x^2 + \sigma_y^2$.

Another important aspect of two random variables is the concept of conditional probabilities and moments. The *conditional distribution* $F_{\underline{x}|\underline{y}}(x|y)$ (or in shorthand notation $F(x|y)$) is the probability that $\underline{x} \le x$ given that $\underline{y} \le y$. Written symbolically:

$$F_{\underline{x}|\underline{y}}(x|y) = F(x|y) = P(\underline{x} \le x|\underline{y} \le y) \tag{c.28}$$

The *conditional probability density* associated with $F_{\underline{x}|\underline{y}}(x|y)$ is denoted by $p_{\underline{x}|\underline{y}}(x|y)$ or $p(x|y)$. Its definition is similar to (c.3). An important property of conditional probability densities is *Bayes' theorem for conditional probabilities*:

$$p_{\underline{x}|\underline{y}}(x|y)p_{\underline{y}}(y) = p_{\underline{x},\underline{y}}(x, y) = p_{\underline{y}|\underline{x}}(y|x)p_{\underline{x}}(x) \tag{c.29}$$

or in shorthand notation:

$$p(x|y)p(y) = p(x, y) = p(y|x)p(x)$$

Bayes' theorem is very important for the development of a classifier or an estimator; see Chapter 2 and 3.

The *conditional moments* are defined as:

$$\mathrm{E}[\underline{x}^n|\underline{y} = y] = \int_{x=-\infty}^{\infty} x^n p(x|y)dx \tag{c.30}$$

The shorthand notation is $\mathrm{E}[x^n|y]$. The *conditional expectation* and *conditional variance* are sometimes denoted by $\mu_{x|y}$ and $\sigma_{x|y}^2$, respectively.

C.3 RANDOM VECTORS

In this section, we discuss a finite sequence of N random variables: $\underline{x}_0, \underline{x}_1, \ldots, \underline{x}_{N-1}$. We assume that these variables are arranged in an N-dimensional random vector $\mathbf{x}$. The *joint distribution function* $F(\mathbf{x})$ is defined as:

$$F(\mathbf{x}) = P(\underline{x}_0 \leq x_0, \underline{x}_1 \leq x_1, \ldots, \underline{x}_{N-1} \leq x_{N-1}) \tag{c.31}$$

The *probability density* $p(\mathbf{x})$ of the vector $\mathbf{x}$ is the function that satisfies:

$$F(\mathbf{x}) = \int_{\xi=-\infty}^{\mathbf{x}} p(\xi)d\xi \tag{c.32}$$

with:

$$\int_{\xi=-\infty}^{\mathbf{x}} p(\xi)d\xi = \int_{\xi_0=-\infty}^{x_0} \int_{\xi_1=-\infty}^{x_1} \cdots \int_{\xi_{N-1}=-\infty}^{x_{N-1}} p(\xi)d\xi_{N-1} \cdots d\xi_1 d\xi_0$$

The *expectation* of a function $g(\mathbf{x})$: $\mathbb{R}^N \to \mathbb{R}$ is:

$$\mathrm{E}[g(\mathbf{x})] = \int_{\underline{\mathbf{x}}=-\infty}^{\infty} g(\mathbf{x})p(\mathbf{x})d\mathbf{x} \tag{c.33}$$

Similar definitions apply to vector-to-vector mappings ($\mathbb{R}^N \to \mathbb{R}^N$) and vector-to-matrix mappings ($\mathbb{R}^N \to \mathbb{R}^N \times \mathbb{R}^N$). Particularly, the *expectation vector* $\boldsymbol{\mu}_\mathbf{x} = \mathrm{E}[\mathbf{x}]$ and the *covariance matrix* $\mathrm{C}_\mathbf{x} = \mathrm{E}[(\mathbf{x} - \boldsymbol{\mu}_\mathbf{x})(\mathbf{x} - \boldsymbol{\mu}_\mathbf{x})^T]$ are frequently used.

A Gaussian random vector has a probability density given by:

$$p(\mathbf{x}) = \frac{1}{\sqrt{(2\pi)^N |\mathrm{C}_\mathbf{x}|}} \exp\left(\frac{-(\mathbf{x} - \boldsymbol{\mu}_\mathbf{x})^T \mathrm{C}_\mathbf{x}^{-1} (\mathbf{x} - \boldsymbol{\mu}_\mathbf{x})}{2}\right) \tag{c.34}$$

The parameters $\boldsymbol{\mu}_\mathbf{x}$ (expectation vector) and $\mathrm{C}_\mathbf{x}$ (covariance matrix) fully define the probability density.

A random vector is called *uncorrelated* if its covariance matrix is a diagonal matrix. If the elements of a random vector $\mathbf{x}$ are *independent*, the probability density of $\mathbf{x}$ is the product of the probability densities of the elements:

$$p(\mathbf{x}) = \prod_{n=0}^{N-1} p(x_n) \tag{c.35}$$

Such a random vector is uncorrelated. The reverse holds true in some specific cases, e.g. for all Gaussian random vectors.

The *conditional probability density* $p(\mathbf{x}|\mathbf{y})$ of two random vectors $\mathbf{x}$ and $\mathbf{y}$ is the probability density of $\mathbf{x}$ if $\mathbf{y}$ is known. Bayes' theorem for conditional probability densities becomes:

$$p(\mathbf{x}|\mathbf{y})p(\mathbf{y}) = p(\mathbf{x}, \mathbf{y}) = p(\mathbf{y}|\mathbf{x})p(\mathbf{x}) \tag{c.36}$$

The definitions of the conditional expectation vector and the conditional covariance matrix are similar.

C.3.1 Linear operations on Gaussian random vectors

Suppose the random vector $\mathbf{y}$ results from a linear (matrix) operation $\mathbf{y} = \mathbf{A}\mathbf{x}$. The input vector of this operator $\mathbf{x}$ has expectation vector $\boldsymbol{\mu}_\mathbf{x}$ and covariance matrix $\mathbf{C}_\mathbf{x}$, respectively. Then, the expectation of the output vector and its covariance matrix are:

$$\begin{aligned} \boldsymbol{\mu}_\mathbf{y} &= \mathbf{A}\boldsymbol{\mu}_\mathbf{x} \\ \mathbf{C}_\mathbf{y} &= \mathbf{A}\mathbf{C}_\mathbf{x}\mathbf{A}^T \end{aligned} \tag{c.37}$$

These relations hold true regardless of the type of distribution functions. In general, the distribution function of $\mathbf{y}$ may be of a type different from the one of $\mathbf{x}$. For instance, if the elements from $\mathbf{x}$ are uniformly distributed, then the elements of $\mathbf{y}$ will not be uniform except for trivial cases (e.g. when $\mathbf{A} = \mathbf{I}$). However, if $\mathbf{x}$ has a Gaussian distribution, then so has $\mathbf{y}$.

Example

Figure C.6 shows the scatter diagram of a Gaussian random vector $\mathbf{x}$, the covariance matrix of which is the identity matrix $\mathbf{I}$. Such a vector is called *white*. Multiplication of $\mathbf{x}$ by a diagonal matrix $\Lambda^{1/2}$ yields a vector $\mathbf{y}$ with covariance matrix Λ. This vector is still uncorrelated. Application of a unitary matrix $\mathbf{V}$ to $\mathbf{z}$ yields the random vector $\mathbf{z}$, which is correlated.

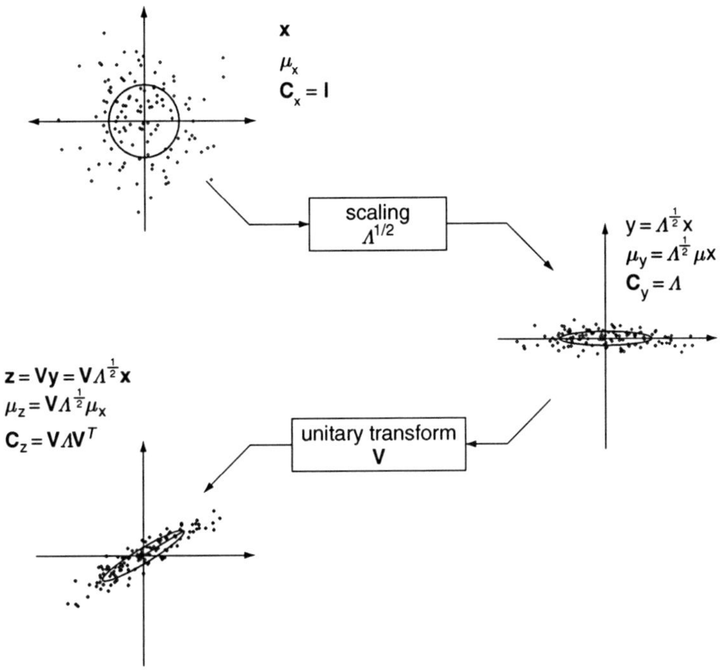

Figure C.6 Scatter diagrams of Gaussian random vectors applied to two linear operators

C.3.2 Decorrelation

Suppose a random vector $\mathbf{z}$ with covariance matrix $\mathbf{C_z}$ is given. *Decorrelation* is a linear operation $\mathbf{A}$ which, when applied to $\mathbf{z}$, will give a white random vector $\mathbf{x}$. (The random vector $\mathbf{x}$ is called white if its covariance matrix $\mathbf{C_x} = T$.) The operation $\mathbf{A}$ can be found by diagonalization of the matrix $\mathbf{C_z}$. To see this, it suffices to recognize that the matrix $\mathbf{C_z}$ is self-adjoint, i.e. $\mathbf{C_z} = \mathbf{C_z}^T$. According to Section B.5 a unitary matrix $\mathbf{V}$ and a (real) diagonal matrix Λ must exist such that $\mathbf{C_z} = \mathbf{V}\Lambda\mathbf{V}^T$. The matrix $\mathbf{V}$ consists of the normalized eigenvectors $\mathbf{v}_n$ of $\mathbf{C_z}$, i.e. $\mathbf{V} = [\mathbf{v}_0 \ \cdots \ \mathbf{v}_{N-1}]$. The matrix Λ contains the eigenvalues λ_n at the diagonal. Therefore, application of the unitary transform $\mathbf{V}^T$ yields a random vector $\mathbf{y} = \mathbf{V}^T\mathbf{z}$ the covariance matrix of which is Λ. Furthermore, the operation $\Lambda^{-1/2}$ applied to $\mathbf{y}$ gives the white vector $\mathbf{x} = \Lambda^{-1/2}\mathbf{y}$. Hence, the decorrelation/whitening operation $\mathbf{A}$ equals $\Lambda^{-1/2}\mathbf{V}^T$. Note that the operation $\Lambda^{-1/2}\mathbf{V}^T$ is the inverse of the operations shown in Figure C.6.

We define the 1σ-level contour of a Gaussian distribution as the solution of:

$$(\mathbf{z} - \boldsymbol{\mu}_\mathbf{z})^T \mathbf{C}_\mathbf{z}^{-1} (\mathbf{z} - \boldsymbol{\mu}_\mathbf{z}) = 1 \qquad \text{(c.38)}$$

The level contour is an ellipse ($N = 2$), an ellipsoid ($N = 3$) or a hyperellipsoid ($N > 3$). A corollary of the above is that the principal axes of these ellipsoids point in the direction of the eigenvectors of $\mathbf{C}_\mathbf{z}$, and that the ellipsoids intersect these axes at a distance from $\boldsymbol{\mu}_\mathbf{z}$ equal to the square root of the associated eigenvalue. See Figure C.6.

C.4 REFERENCE

Papoulis, A., *Probability, Random Variables and Stochastic Processes*, McGraw-Hill, New York, 1965 (third edition: 1991).

Appendix D

Discrete-time Dynamic Systems

This brief review is meant as a refresher for readers who are familiar with the topic. It summarizes those concepts that are used within the textbook. It also introduces the notations adopted in this book.

D.1 DISCRETE-TIME DYNAMIC SYSTEMS

A dynamic system is a system whose variables proceed in time. A concise representation of such a system is the *state space model*. The model consists of a *state vector* $\mathbf{x}(i)$ where i is an integer variable representing the discrete time. The dimension of $\mathbf{x}(i)$, called the *order of the system*, is M. We assume that the state vector is real-valued. The finite-state case is introduced in Chapter 4.

The process can be influenced by a *control vector* (*input vector*) $\mathbf{u}(i)$ with dimension L. The output of the system is given by the *measurement vector* (*observation vector*) $\mathbf{z}(i)$ with dimension N. The output is modelled as a memoryless vector that depends on the current values of the state vector and the control vector.

By definition, the state vector holds the minimum number of variables which completely summarize the past of the system. Therefore, the state

Classification, Parameter Estimation and State Estimation: An Engineering Approach using MATLAB
F. van der Heijden, R.P.W. Duin, D. de Ridder and D.M.J. Tax
 ISBN: 0-470-09013-8

vector at time $i+1$ is derived from the state vector and the control vector, both valid at time i:

$$\begin{aligned}\mathbf{x}(i+1) &= \mathbf{f}(\mathbf{x}(i), \mathbf{u}(i), i)\\ \mathbf{z}(i) &= \mathbf{h}(\mathbf{x}(i), \mathbf{u}(i), i)\end{aligned} \tag{d.1}$$

$\mathbf{f}(.)$ is a possible nonlinear vector function, the *system function*, that may depend explicitly on time. $\mathbf{h}(.)$ is the *measurement function.*

Note that if the time series starts at $i = i_0$, and the *initial condition* $\mathbf{x}(i_0)$ is given, along with the input sequence $\mathbf{u}(i_0)$, $\mathbf{u}(i_0+1), \ldots$, then (d.1) can be used to solve $\mathbf{x}(i)$ for every $i > i_0$. Such a solution is called a *trajectory.*

D.2 LINEAR SYSTEMS

If the system is *linear*, then the equations evolve into matrix–vector equations according to:

$$\begin{aligned}\mathbf{x}(i+1) &= \mathbf{F}(i)\mathbf{x}(i) + \mathbf{L}(i)\mathbf{u}(i)\\ \mathbf{z}(i) &= \mathbf{H}(i)\mathbf{x}(i) + \mathbf{D}(i)\mathbf{u}(i)\end{aligned} \tag{d.2}$$

The matrices $\mathbf{F}(i)$, $\mathbf{L}(i)$, $\mathbf{H}(i)$ and $\mathbf{D}(i)$ must have the appropriate dimensions. They have the following names:

$\mathbf{F}(i)$ *system matrix*
$\mathbf{L}(i)$ *distribution matrix (gain matrix)*
$\mathbf{H}(i)$ *measurement matrix*
$\mathbf{D}(i)$ *feedforward gain*

Often, the feedforward gain is zero.

The solution of the linear system is found by introduction of the *transition matrix* $\Phi(i, i_0)$, recursively defined by:

$$\begin{aligned}\Phi(i_0, i_0) &= \mathbf{I}\\ \Phi(i+1, i_0) &= \mathbf{F}(i)\Phi(i, i_0) \quad \text{for} \quad i = i_0, i_0+1, \ldots\end{aligned} \tag{d.3}$$

Given the initial condition $\mathbf{x}(i_0)$ at $i = i_0$, and assuming that the feedforward gain is zero, the solution is found as:

$$\mathbf{x}(i) = \Phi(i, i_0)\mathbf{x}(i_0) + \sum_{j=i_0}^{i-1} \Phi(i, j+1)\mathbf{L}(j)\mathbf{u}(j) \tag{d.4}$$

D.3 LINEAR TIME INVARIANT SYSTEMS

In the linear time invariant case, the matrices become constants:

$$\begin{aligned} \mathbf{x}(i+1) &= \mathbf{F}\mathbf{x}(i) + \mathbf{L}\mathbf{u}(i) \\ \mathbf{z}(i) &= \mathbf{H}\mathbf{x}(i) + \mathbf{D}\mathbf{u}(i) \end{aligned} \tag{d.5}$$

and the transition matrix simplifies to:

$$\Phi(i, i_0) = \Phi(i - i_0) = \mathbf{F}^{i-i_0} \tag{d.6}$$

With that, the input/output relation of the system is:

$$\mathbf{z}(i) = \mathbf{H}\mathbf{F}^{i-i_0}\mathbf{x}(i_0) + \sum_{j=i_0}^{i-1} \mathbf{H}\mathbf{F}^{i-j-1}\mathbf{L}\mathbf{u}(j) + \mathbf{D}\mathbf{u}(i) \tag{d.7}$$

The first term at the r.h.s. is the *free response*; the second term is the *particular response*; the third term is the *feedforward response.*

D.3.1 Diagonalization of a system

We assume that the system matrix F has M distinct eigenvectors $\mathbf{v}_k$ with corresponding (distinct) eigenvalues λ_k. Thus, by definition $\mathbf{F}\mathbf{v}_k = \lambda_k \mathbf{v}_k$. We define the eigenvalue matrix Λ as the $M \times M$ diagonal matrix containing the eigenvalues at the diagonal, i.e. $\Lambda_{k,k} = \lambda_k$, and the eigenvector matrix V as the matrix containing the eigenvectors as its column vectors, i.e. $\mathbf{V} = [\,\mathbf{v}_0 \;\cdots\; \mathbf{v}_{M-1}\,]$. Consequently:

$$\mathbf{F}\mathbf{V} = \mathbf{V}\Lambda \quad \Rightarrow \quad \mathbf{F} = \mathbf{V}\Lambda\mathbf{V}^{-1} \tag{d.8}$$

A corollary of (d.8) is that the power of F is calculated as follows:

$$\mathbf{F}^k = (\mathbf{V}\Lambda\mathbf{V}^{-1})^k = \mathbf{V}\Lambda^k\,\mathbf{V}^{-1} \tag{d.9}$$

Equation (d.9) is useful to decouple the system into a number of (scalar) first order systems, i.e. to diagonalize the system. For that purpose, define the vector

$$\mathbf{y}(i) = \mathbf{V}^{-1}\mathbf{x}(i) \tag{d.10}$$

Then, the state equation transforms into:

$$\begin{aligned}\mathbf{y}(i+1) &= \mathbf{V}^{-1}\mathbf{x}(i+1) \\ &= \mathbf{V}^{-1}\mathbf{F}\mathbf{x}(i) + \mathbf{V}^{-1}\mathbf{L}\mathbf{u}(i) \\ &= \mathbf{V}^{-1}\mathbf{F}\mathbf{V}\mathbf{y}(i) + \mathbf{V}^{-1}\mathbf{L}\mathbf{u}(i) \\ &= \Lambda\mathbf{y}(i) + \mathbf{V}^{-1}\mathbf{L}\mathbf{u}(i)\end{aligned} \tag{d.11}$$

and the solution is found as:

$$\mathbf{y}(i) = \Lambda^{i-i_0}\mathbf{y}(i_0) + \sum_{j=i_0}^{i-1} \Lambda^{i-j-1}\mathbf{V}^{-1}\mathbf{L}\mathbf{u}(i) \tag{d.12}$$

The k-th component of the vector $\mathbf{y}(i)$ is:

$$y_k(i) = \lambda_k^{i-i_0} y_k(i_0) + \sum_{j=i_0}^{i-1} \lambda_k^{i-j-1}(\mathbf{V}^{-1}\mathbf{L}\mathbf{u}(i))_k \tag{d.13}$$

The measurement vector is obtained from $\mathbf{y}(i)$ by substitution of $\mathbf{x}(i) = \mathbf{V}\mathbf{y}(i)$ in equation (d.5).

If the system matrix does not have distinct eigenvalues, the situation is a little more involved. In that case, the matrix Λ gets the Jordan form with eigenvalues at the diagonal, but with ones at the superdiagonal. The free response becomes combinations of terms $p_k(i - i_0)\lambda_k^{i-i_0}$ where $p_k(i)$ are polynomials in i with order one less than the multiplicity of λ_k.

D.3.2 Stability

In dynamic systems there are various definitions of the term stability. We return to the general case (d.1) first, and then check to see how the definition applies to the linear time invariant case. Let $\mathbf{x}_a(i)$ and $\mathbf{x}_b(i)$ be the solutions of (d.1) with a given input sequence and with initial conditions $\mathbf{x}_a(i_0)$ and $\mathbf{x}_b(i_0)$, respectively.

The solution $\mathbf{x}_a(i)$ is *stable* if for every $\varepsilon > 0$ we can find a $\delta > 0$ such that for every $\mathbf{x}_b(i_0)$ with $\|\mathbf{x}_b(i_0) - \mathbf{x}_a(i_0)\| < \delta$ is such that $\|\mathbf{x}_b(i) - \mathbf{x}_a(i)\| < \varepsilon$ for all $i \geq i_0$. Loosely speaking, the system is stable if small changes in the initial condition of stable solution cannot lead to very large changes of the trajectory.

The solution is $\mathbf{x}_a(i)$ is *asymptotically stable*, if it is stable, and if $\|\mathbf{x}_b(i) - \mathbf{x}_a(i)\| \to 0$ as $i \to \infty$ provided that $\|\mathbf{x}_b(i_0) - \mathbf{x}_a(i_0)\|$ is not too large. Loosely speaking, the additional requirement for a system to be asymptotically stable is that the initial condition does not influence the solution in the long range.

For linear systems, the stability of one solution assures the stability of all other solutions. Thus, stability is a property of the system then; not just of one of its solutions. From (d.13) it can be seen that a linear time invariant system is asymptotically stable if and only if the magnitude of the eigenvalues are less than one. That is, the eigenvalues must all be within the (complex) unit circle.

A linear time invariant system has *BIBO stability* (bounded input, bounded output) if a bounded input sequence always gives rise to a bounded output sequence. Note that asymptotical stability implies BIBO stability, but the reverse is not true. A system can be BIBO stable, while it is not asymptotical stable.

D.4 REFERENCES

Åström, K.J. and Wittenmark, B., *Computer-Controlled Systems – Theory and Design*, Second Edition, Prentice Hall International, Englewood Cliffs, NJ, 1990.

Luenberger, D.G., *Introduction to Dynamic Systems, Theory, Models and Applications*, Wiley, Toronto, 1979.

Appendix E
Introduction to PRTools

E.1 MOTIVATION

In statistical pattern recognition we study techniques for the generalization of decision rules to be used for the recognition of patterns in experimental data sets. This area of research has a strong computational character, demanding a flexible use of numerical programs for data analysis as well as for the evaluation of the procedures. As new methods keep being proposed in the literature, a programming platform is needed that enables a fast and flexible implementation of such algorithms. Because of its widespread availability, its simple syntax and general nature, MATLAB is a good choice for such a platform.

The pattern recognition routines and support functions offered by PRTools represent a basic set covering largely the area of statistical pattern recognition. Many methods and proposals, however, are not yet implemented. Neural networks are only implemented partially, as MATLAB already includes a very good toolbox in that area. PRTools has a few limitations. Due to the heavy memory demands of MATLAB, very large problems with learning sets of tens of thousands of objects cannot be handled on moderate machines. Moreover, some algorithms are slow as it can be difficult to avoid nested loops. In the present version, the handling of missing data has been prepared, but no routines are implemented yet.

Classification, Parameter Estimation and State Estimation: An Engineering Approach using MATLAB
F. van der Heijden, R.P.W. Duin, D. de Ridder and D.M.J. Tax
 ISBN: 0-470-09013-8

Table E.1 Notation differences between this book and the PRTools documentation

Mathematical notation	Notation in pseudo-code	PRTools notation	Meaning
T	`x, z`	`a, b`	*data set*
n	`n`	`m`	*number of objects*
N, D	`N, D`	`k, n`	*number of features, dimensions*
K	`K`	`C`	*number of classes*

The use of fuzzy or symbolic data is not supported, except for soft (and thereby also fuzzy) labels which are used by just a few routines. Multi-dimensional target fields are allowed, but at this moment no procedure makes use of this possibility.

The notation used in the PRTools documentation and code differs slightly from that used in the code throughout this book. In this appendix we try to follow the notation in the book. In Table E.1 notation differences between this book and the PRTools documentation are given.

E.2 ESSENTIAL CONCEPTS IN PRTOOLS

For recognizing the classes of objects they are first scanned by sensors, then represented, e.g. in a feature space, and after some possible feature reduction steps they are finally mapped by a classifier to the set of class labels. Between the initial representation in the feature space and this final mapping to the set of class labels the representation may be changed several times: simplified feature spaces (feature selection), normalization of features (e.g. by scaling), linear or nonlinear mappings (feature extraction), classification by a possible set of classifiers, combining classifiers and the final labelling. In each of these steps the data is transformed by some mapping.

Based on this observation. PRTools defines the following two basic concepts:

- *Data sets*: matrices in which the rows represent the objects and the columns the features, class memberships or other fixed sets of properties (e.g. distances to a fixed set of other objects).
- *Mappings*: transformations operating on data sets.

As pattern recognition has two stages, *training* and *execution*, mappings also have two main types:

- An *untrained mapping* refers to just the concept of a method, e.g. forward feature selection, PCA or Fisher's linear discriminant. It may have some parameters that are needed for training, e.g. the desired number of features or some regularization parameters. If an untrained mapping is applied to a data set it will be trained by it (*training*).
- A *trained mapping* is specific for its training set. This data set thereby determines the input dimensionality (e.g. the number of input features) as well as the output dimensionality (e.g. the number of output features or the number of classes). If a trained mapping is applied to a data set, it will transform the data set according to its definition (*execution*).

In addition, *fixed mappings* are used. They are almost identical to trained mappings, except that they do not result from a training step, but are directly defined by the user: e.g. the transformation of distances by a sigmoid function to the $[0, 1]$ interval.

PRTools deals with sets of *labelled* or *unlabelled objects* and offers routines for the generalization of such sets into functions for *mapping* and *classification*. A *classifier* is a special case of a *mapping*, as it maps objects on class labels or on $[0, 1]$ intervals that may be interpreted as *class memberships*, *soft labels* or *posterior probabilities*. An *object* is a N-dimensional vector of *feature values, distances, (dis)similarities* or *class memberships*. Within PRTools they are usually just called features. It is assumed that for all objects in a problem all values of the same set of features are given. The space defined by the actual set of features is called the *feature space*. Objects are represented as points or vectors in this space. New objects in a feature space are usually gradually converted to labels by a series of *mappings* followed by a final *classifier*.

E.3 IMPLEMENTATION

PRTools uses the object-oriented features of the MATLAB programming language. Two object classes (not to be confused with the objects and classes in pattern recognition) have been defined: `dataset` and `mapping`. A large number of operators (like *, [] etc.) and MATLAB commands have been overloaded and have a special meaning when applied to a `dataset` and/or a `mapping`.

The central data structure of PRTools is the `dataset`. It primarily consists of a set of objects represented by a matrix of feature vectors. Attached to this matrix is a set of labels, one for each object and a set of feature names, also called feature labels. Labels can be integer numbers or character strings. Moreover, a set of prior probabilities, one for each class, is stored. In most help files of PRTools, a `dataset` is denoted by `a`. We will use `z` in this text to stay consistent with the rest of the book. In almost any routine this is one of the inputs. Almost all routines can handle multi-class object sets. It is possible that for some objects no label is specified (a `NaN` is used, or an empty string). Such objects are, unless otherwise mentioned, skipped during training.

Data structures of the object class `mapping` store data transformations ('mappings'), classifiers, feature extraction results, data scaling definitions, nonlinear projections, etc. They are usually denoted by `w`. The easiest way of applying a mapping `w` to a data set `z` is by `z*w`. The matrix multiplication symbol `*` is overloaded for this purpose. This operation may also be written as `map(z, w)`. Like everywhere else in MATLAB, longer series of operations are possible, e.g. `z*w1*w2*w3`, and are executed from left to right.

Listing E.1

```
z=gendath([50 50]);           % generate data, 50 objects/class
[y,x]=gendat(z, [20 20]);     % split into training and test set
w1=ldc(y);                    % linear classifier
w2=qdc(y);                    % quadratic
w3=parzenc(y);                % Parzen density-based
w4=bpxnc(y,3);                % neural net with 3 hidden units
disp([testc(x*w1),testc(x*w2),testc(x*w3),testc(x*w4)]);
                   % compute and display classification errors
scatterd(z);                  % scatter plot of data
plotc({w1,w2,w3,w4});         % plot decision boundaries
```

A typical example is given in Listing E.1. This command file first generates two sets of labelled objects using `gendath`, both containing 50 two-dimensional object vectors, and stores them, their labels and prior probabilities in the `dataset z`. The distribution follows the so-called 'Higleyman classes'. The next call to `gendat` takes this `dataset` and splits it randomly into a `dataset y`, further on used for training, and a `dataset x`, used for testing. This training set `y` contains 20 objects from each class. The remaining 2×30 objects are collected in `x`.

In the next lines four classification functions (discriminants) are computed, called `w1, w2, w3` and `w4`. The first three are in fact density estimators based on various assumptions (class priors stored in `y` are taken into account). Programmatically, they are just mappings, as `xx = x*w1` computes the class densities for the objects stored in `d`.

xx has as many columns as there are classes in the training set for w1 (here two). The test routine testc assigns objects (represented by the rows in xx) to the class corresponding to the highest density (times prior probability), the mappings w1, ... , w4 can be used as classifiers. The linear classifier w1 (*ldc*) and quadratic classifier w2 (*qdc*) are both based on the assumption of normally distributed classes. The first assumes equal class covariance matrices. The Parzen classifier estimates the class densities by the Parzen density estimation and has a built-in optimization of the kernel width. The fourth classifier uses a feed forward neural network with three hidden units. It is trained by the back propagation rule using a varying step size.

The results are then displayed and plotted. The test data set x is used in a routine testc (test classifier) on each of the four discriminants. They are combined in a cell array, but individual calls are possible as well. The estimated probabilities of error are displayed in the MATLAB command window and may look like (note that they may differ due to different random seeds used in the generation of the data):

```
0.1500  0.0333  0.1333  0.0833
```

Finally the classes are plotted in a scatter diagram (scatterd) together with the discriminants (plotc); see Figure E.1

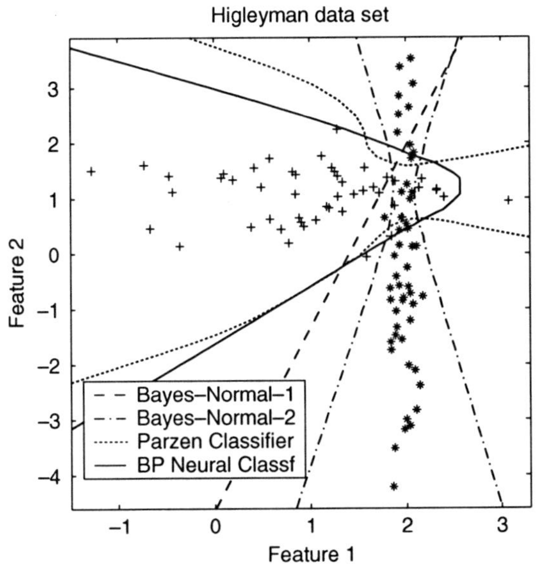

Figure E.1 Example output of Listing E.1

For more advanced examples, see the Examples section in `help prtools`.

E.4 SOME DETAILS

The command help files and the examples should give sufficient information to use the toolbox with a few exceptions. These are discussed in the following sections. They deal with the ways classifiers and mappings are represented. As these are the constituting elements of a pattern recognition analysis, it is important that the user understands these issues.

E.4.1 Data sets

A `dataset` consists of a set of `n` objects, each given by `D` features. In PRTools such a data set is represented by an `n` by `D` matrix: `n` rows, each containing an object vector of `D` features. Usually a data set is labelled. An example of a definition is:

```
>> z=dataset([1 2 3; 2 3 4; 3 4 5; 4 5 6],[3; 3; 5; 5])
4 by 3 dataset with 2 classes: [2 2]
```

The 4×3 data matrix (four objects given by three features) is accompanied by a label list of four labels, connecting each of the objects to one of the two classes, labelled 3 and 5. Class labels can be numbers or strings and should always be given as rows in the label list. If the label list is not given all objects are given the default label 1. In addition it is possible to assign labels to the columns (features) of a data set:

```
>> z=dataset(rand(100,3),genlab([50 50],[3; 5]));
>> z=setfeatlab(z,['r1';'r2';'r3'])
100 by 3 dataset with 2 classes: [50 50]
```

The routine `genlab` generates 50 labels with value 3, followed by 50 labels with value 5. Using `setfeatlab` the labels ('`r1`', '`r2`', '`r3`') for the three features are set. Various other fields can be set as well. One of the ways to see these fields is by converting the data set to a structure, using `struct(x)`. Fields can be inspected individually by the .- extension, also defined for data sets:

```
≫ x.lablist
ans =
    3
    5
```

The possibility to set prior probabilities for each of the classes by `setprior(x, prob, lablist)` is important. The prior values in `prob` should sum to one. If `prob` is empty or if it is not supplied the prior probabilities are computed from the data set label frequencies. If `prob` equals zero then equal class probabilities are assumed.

Various items stored in a data set can be retrieved by commands like `getdata`, `getlablist` and `getnlab`. The last one retrieves the numeric labels for the objects (1, 2, ...) referring to the true labels stored in the rows of lablist. The size of the data set can be found by:

```
[n,D] = size(x); or [n,D,K] = getsize(xa);
```

in which `n` is the number of objects, `D` the number of features and `K` the number of classes (equal to `max(nlab)`). Data sets can be combined by `[x; y]` if `x` and `y` have equal numbers of features and by `[x y]` if they have equal numbers of objects. Creating subsets of data sets can be done by `z(I,J)` in which `I` is a set of indices defining the desired objects and `J` is a set of indices defining the desired features.

The original data matrix can be retrieved by `double(z)` or by `+z`. The labels in the objects of `x` can be retrieved by `labels = getlabels(z)`, which is equivalent to

```
[nlab,lablist] = get (z,'nlab','lablist');
labels = lablist(nlab,:);
```

Be aware that the order of classes returned by `getprob` and `getlablist` is the standard order used in PRTools and may differ from the one used in the definition of `z`. For more information, type `help datasets`.

E.4.2 Classifiers and mappings

There are many commands to train and use mappings between spaces of different (or equal) dimensionalities. In general, the following applies:

if	z	is an n by D data set (n objects in a D-dimensional space)
and	w	is a D by K mapping (map from D to K dimensions)
then	z*w	is an n by K data set (n objects in an K-dimensional space)

Mappings can be linear or affine (e.g. a rotation and a shift) as well as nonlinear (e.g. a neural network). Typically they can be used as classifiers. In that case a D by K mapping maps a D-feature data vector on the output space of an K-class classifier (exception: two-class classifiers like discriminant functions may be implemented by a mapping to a one-dimensional space like the distance to the discriminant, K = 1).

Mappings are of the data type 'mapping' (class(w) is 'mapping'), have a size of [D, K] if they map from D to K dimensions. Mappings can be instructed to assign labels to the output columns, e.g. the class names. These labels can be retrieved by

```
labels=getlabels(w); %before the mapping, or
labels=getlabels (z*w); %after the data set z is mapped by w.
```

Mappings can be learned from examples, (labelled) objects stored in a data set z, for instance by training a classifier:

```
w1=ldc(z);          %the normal densities based linear classifier
w2=knnc(z,3);       %the 3-nearest neighbour rule
w3=svc(z,'p',2);    %the support vector classifier based on a 2nd
                       order polynomial kernel
```

Untrained or empty mappings are supported. They may be very useful. In this case the data set is replaced by an empty set or entirely skipped:

```
v1=ldc; v2=knnc([],3); v3=svc([],'p',2);
```

Such mappings can be trained later by

```
w1=z*v1; w2=z*v2; w3=z*v3;
```

(which is equivalent to the statements a few lines above) or by using cell arrays

```
v={ldc, knnc ([], 3), svc([],'p',2)}; w=z*v;
```

The mapping of a test set y by y*w1 is now equivalent to y*(z*v1). Note that expressions are evaluated from left to right, so y*z*v1 will result in an error as the multiplication of the two data sets (y*x) is executed first.

Some trainable mappings do not depend on class labels and can be interpreted as finding a feature space that approximates as good as possible the original data set given some conditions and measures. Examples are the Karhunen–Loève mapping (klm), principal component analysis (pca) and kernel mapping (kernelm) by which nonlinear, kernel PCA mappings can be computed.

In addition to trainable mappings, there are fixed mappings, the parameters of which cannot be trained. A number of them can be set by cmapm; others are sigm and invsigm.

The result x of mapping a test set on a trained classifier, x = y*w1, is again a data set, storing for each object in y the output values of the classifier. For discriminants they are sigmoids of distances, mapped on the [0, 1] interval, for neural networks their unnormalized outputs and for density based classifiers the densities. For all of them the following holds: the larger, the more similar with the corresponding class. The values in a single row (object) don't necessarily sum to one. This can be achieved by the fixed mapping classc:

```
x=y*w1*classc
```

The values in x can be interpreted as posterior probability estimates or classification confidences. Such a classification data set has column labels (feature labels) for the classes and row labels for the objects. The class labels of the maximum values in each object row can be retrieved by

```
labels=x*labeld; or labels=labeld(x);
```

A global classification error follows from

```
e=x*testc; or e=testc(x);
```

Mappings can be inspected by entering w by itself, or using display(w). This lists the size and type of a mapping or classifier as well as the routine used for computing a mapping z*w. The data stored in the mapping might be retieved using +w. Note that the type of data stored in each mapping depends on the type of mapping.

Affine mappings (e.g. constructed by `klm`) may be transposed. This is useful for back projection of data into the original space. For instance:

```
w=klm(x,3);     % compute 3-dimensional KL transform
y=x*w;          % map x using w, resulting in y
z=y*w';         % back-projection to the original space
```

A mapping may be given an output selection by `w = w(: , J)`, in which `J` is a set of indices pointing to the desired classes; `y = z*w(: , J);` is equivalent to `y = z*w;  y = y(: , J);` . Input selection is not possible for a mapping.

For more information, type `help mappings`.

E.5 HOW TO WRITE YOUR OWN MAPPING

Users can add new mappings or classifiers by a single routine that should support the following type of calls:

- `w = newmapm([], par1, ...);`
  Defines the untrained, empty mapping.
- `w = newmapm(z, par1, ...);`
  Defines the map based on the training data set `z`.
- `y = newmapm(z, w);`

Defines the mapping of data set `z` using `w`, resulting in a data set `y`. For an example, list the routine `subsc` (using `typesubsc`). This classifier approximates each class by a linear subspace and assigns new objects to the class of the closest subspace found in the training set. The dimensionalities of the subspaces can be directly set by `w = subsc(z, n)`, in which the integer `n` determines the dimensionality of all class subspaces, or by `w = subsc(z, alf)`, in which `alf` is the desired fraction of variance to retain, e.g. `alf = 0.95`. In both cases the class subspaces `v` (see the listing) are determined by a principal component analysis of the single-class data sets.

The three possible types of calls, listed above, are handled in the three main parts of the routine. If no input parameters are given (`nargin<1`) or no input data set is found (`z` is empty) an untrained classifier is returned. This is useful for calls like `w = subsc([],n)`, defining an untrained classifier to be used in routines like `cleval(z,w,...)` that operate on arbitrary untrained classifiers, but also to facilitate training by constructions as `w = z*subsc` or `w = z*subsc([],n)`.

The training section of the routine is accessed if z is not empty and n is either not supplied or set by the user as a double (i.e. the subspace dimensionality or the fraction of the retained variance). PRTools takes care that calls like w = z*subsc([],n) are executed as w = subsc(z,n). The first parameter in the mapping definitions w = mapping(mfilename, ... is substituted by MATLAB as 'subsc' (mfilename is a MATLAB function that returns the name of the calling file). This string is stored by PRTools in the mapping_file field of the mapping w and used to call subsc whenever it has to be applied to a data set. For some special mappings, like ldc, another file might be used (in the case of ldc it is normal_map).

The trained mapping w can be applied to a test data set y by x = y*w or by z = map(y,w). Such a call is converted by PRTools to x = subsc(y,w). Consequently, the second parameter of subsc(), n, is now substituted by the mapping w. This is executed in the final part of the routine. Here, the data stored in the data field of w during training is retrieved (class mean, rotation matrix and mean square distances of the training objects) and used to find normalized distances of the test objects to the various subspaces. Finally they are converted to a density, assuming a normal distribution of distances. These values are returned in a data set using the setdata routine. This data set is thereby similar to the input data set: same object labels, object identifiers, etc. Just the data matrix itself is changed and the columns now refer to classes instead of features. The definition of the mappings and data sets form the core of the PRTools toolbox. There are numerous supporting algorithms for inspecting and visualizing classifiers and classfication results. Also the possibilities for combining classifiers is not discussed. In the file Contents.m (which can be inspected in MATLAB by help prtools) the most important commands are listed. A final possibility is to inspect all *.m. files and use help.

Appendix F

MATLAB Toolboxes Used

Apart from the PRTools toolbox,[1] this book has used the following standard MATLAB toolboxes:

- *Control System Toolbox*
  This toolbox contains many functions and data structures for the modelling and design of control systems. However, the linear time invariant cases are emphasized. As such, the toolbox contains a number of functions that are of interest for state estimation. But, unfortunately, the scope of these functions is limited to the steady state case.
- *Signal Processing Toolbox*
  This toolbox is a collection of processing operations. Much attention is paid to the realization aspect of filters, i.e. how to build a filter with given properties, e.g. a given cut-off frequencies, constrained phase characteristics and so on. This aspect is of less importance for the design of state estimators. Nevertheless, the toolbox contains a number of functions that are of interest with respect to the scope of this book:
  - correlation and covariance estimation
  - spectral analysis, e.g. the periodogram

[1] And with PRTools, implicitly the Neural Network Toolbox.

Classification, Parameter Estimation and State Estimation: An Engineering Approach using MATLAB
F. van der Heijden, R.P.W. Duin, D. de Ridder and D.M.J. Tax
 ISBN: 0-470-09013-8

- estimation of parameters of random processes, e.g. the AR coefficients based on, for instance, the Yule–Walker equations
- some various tools, such as waveform generation and resampling functions.

- *Optimization Toolbox*
  This toolbox contains many functions for optimization problems including constrained or unconstrained, linear or nonlinear minimization. The toolbox also contains functions for curve fitting.
- *Statistics Toolbox*
  This toolbox includes functions for the following topics (among others):
  - many distribution and density functions of random variables and corresponding random number generators
  - parameter estimation for some standard distributions
  - linear and nonlinear regression methods
  - methods for principal components analysis
  - functions for the analysis of hidden Markov models.
- *System Identification Toolbox*
  This toolbox contains various methods for the identification of dynamic systems. Many models are available, but nevertheless the emphasis is on linear time invariant systems.

MATLAB offers many more interesting toolboxes, e.g. for the analysis of financial time series (Financial Time Series Toolbox; Garch Toolbox), and for calibration and curve fitting (Curve Fitting Toolbox, Model-based Calibration Toolbox), but these toolboxes are not used in this book.

Index

Classification, Parameter Estimation and State Estimation: An Engineering Approach using MATLAB
F. van der Heijden, R.P.W. Duin, D. de Ridder and D.M.J. Tax

 ISBN: 0-470-09013-8

Printed in the United States
117035LV00002B/73/P

9 780470 090138